AF581635

Recent Developments in Non-conventional Welding of Materials

Recent Developments in Non-conventional Welding of Materials

Editors

Rui Manuel Leal
Ivan Galvão

MDPI • Basel • Beijing • Wuhan • Barcelona • Belgrade • Manchester • Tokyo • Cluj • Tianjin

Editors

Rui Manuel Leal
ESAD.CR/Polytechnic Institute of Leiria
Portugal
CEMMPRE/University of Coimbra
Portugal

Ivan Galvão
ISEL/Polytechnic Institute of Lisbon
Portugal
CEMMPRE/University of Coimbra
Portugal

Editorial Office
MDPI
St. Alban-Anlage 66
4052 Basel, Switzerland

This is a reprint of articles from the Special Issue published online in the open access journal *Materials* (ISSN 1996-1944) (available at: https://www.mdpi.com/journal/materials/special_issues/Recent_Materials).

For citation purposes, cite each article independently as indicated on the article page online and as indicated below:

LastName, A.A.; LastName, B.B.; LastName, C.C. Article Title. *Journal Name* **Year**, *Volume Number*, Page Range.

ISBN 978-3-0365-3873-0 (Hbk)
ISBN 978-3-0365-3874-7 (PDF)

Cover image courtesy of Prof. Dr. Ivan Galvão and Prof. Dr. Rui Manuel Leal

Contents

About the Editors

Rui Manuel Leal has a PhD in Mechanical Engineering (Production Technology field), MSc in Mechanical Engineering (Technology and Materials field) and BSc in Mechanical Engineering (Production field) from the Faculty of Sciences and Technology at the University of Coimbra (Portugal). He is currently an Assistant Professor at ESAD.CR—Polytechnic Institute of Leiria, where he teaches curricular units of Technology and Materials. He is also an integrated researcher at the CEMMPRE—University of Coimbra and a collaborative researcher at the Research Laboratory in Design and Arts (LIDA). His main research areas are focused on production technologies, having collaborated on several research projects. As a result of this research, he has participated in national and international events, such as conferences and congresses, and has published several works in scientific journals.

Ivan Galvão is an Assistant Professor at the Department of Mechanical Engineering of ISEL —Polytechnic Institute of Lisbon (Portugal) and Integrated Researcher at CEMMPRE—University of Coimbra (Portugal). He has a PhD in Mechanical Engineering (Production Technology field) from the University of Coimbra. His research field is the joining and processing of materials, specifically, welding technology and metallurgy, the mechanical behavior of welded joints, the solid-state welding of similar and dissimilar materials, and the solid-state processing of materials. He has published many scientific papers in high-impact journals and conference proceedings, and he has participated in several international conferences. Other activities developed by him include his involvement in research projects, his collaboration with international journals and scientific/engineering associations, and his supervision of research works and theses.

Editorial

Recent Developments in Non-Conventional Welding of Materials

Rui M. Leal [1,2] and Ivan Galvão [2,3,*]

1 LIDA-ESAD.CR, Instituto Politécnico de Leiria, Rua Isidoro Inácio Alves de Carvalho, 2500-321 Caldas da Rainha, Portugal; rui.leal@ipleiria.pt
2 CEMMPRE, Departamento de Engenharia Mecânica, University Coimbra, Rua Luís Reis Santos, 3030-788 Coimbra, Portugal
3 ISEL, Departamento de Engenharia Mecânica, Instituto Politécnico de Lisboa, Rua Conselheiro Emídio Navarro 1, 1959-007 Lisbon, Portugal
* Correspondence: ivan.galvao@dem.uc.pt

1. Introduction and Scope

Welding is one of the technological fields with the greatest impact in many industries, such as automotive, aerospace, energy production, electronics, the health sector, etc. Welding technologies are currently used to join the most diverse materials, from metallic alloys to polymers, composites, or even biological tissues. Despite the relevance and wide application of traditional welding technologies, these processes do not meet the demanding requirements of some industries. This has driven strong research efforts in non-conventional welding processes, such as laser welding, ultrasonic welding, impact welding, friction stir welding (FSW), diffusion welding, and many other welding technologies. Important studies have been recently developed all over the world on the application of these processes to the joining of cutting-edge materials and material combinations, enabling the production of joints with improved properties. Thus, this Special Issue presents a sample of the most recent developments in the non-conventional welding of materials, which will drive the design of future industrial solutions with increased efficiency and sustainability.

Citation: Leal, R.M.; Galvão, I. Recent Developments in Non-Conventional Welding of Materials. *Materials* **2022**, *15*, 171. https://doi.org/10.3390/ma15010171

Received: 6 November 2021
Accepted: 24 December 2021
Published: 27 December 2021

2. Contributions

The present Special Issue encompasses the publication of eight research papers and three review papers. Literature analysis, experimental and experimental/numerical research approaches were followed in the published papers, which were focused on several welding technologies (FSW; explosive welding; laser welding, etc.) and on varied materials and material combinations (steels; aluminium, copper, and high entropy alloys; aluminium/titanium/magnesium and zirconium/titanium/steel combinations, etc.). Profound multi-scale characterisations, focused on a large range of engineering topics, such as weld macro- and microstructures, mechanical behaviour, chemical and phase composition, and supported literature-based discussions are presented in these works.

FSW was the most addressed welding process in the works published in this Special Issue. Khan et al. [1] studied the effect of the strain rate and the heat generation on the traverse force registered during dissimilar FSW of the 2219 and 7475 aluminium alloys. These authors estimated the strain rate by grain size and welding temperature analysis, and they found that maximum values of traverse force are associated with higher strain rates. Additionally, the authors found that the strain rate mostly depends on the tool traverse speed than on the tool rotation speed. However, according to them, the effect of the rotation speed on the flow stress, which manifests by the frictional heating, is dominant when compared to the strain rate strengthening. In the same year, Tamadon et al. [2] conducted a work focused on characterising the microstructure of the flow patterns in bobbin FSW of the 6082 aluminium alloy. The authors reported that one of the main difficulties in bobbin FSW is the visualisation of the weld microstructure, especially when materials with a fine grain structure are welded. Thus, they were able to characterise the microstructure of the flow lines observed in the weld macrostructure by combining optical microscopy and electron backscatter diffraction (EBSD). These structures were found to be composed of fine layers of materials of about 10 μm thickness, which were formed by dynamic recrystallisation.

Unlike the previous works, in which butt joining was addressed, Manuel et al. [3] studied the influence of the base material properties and the traverse speed on the morphology and mechanical behaviour of tri-dissimilar AA2017/AA6082/AA5083 T-joints produced by FSW. These authors reported that the traverse speed and the base material properties and relative position in the joint have a strong influence on nugget formation. According to them, improved static and fatigue behaviour of the welds can be achieved by increasing the traverse speed. Moreover, they found that higher peak temperatures and better fatigue properties are achieved when the 2017 aluminium alloy is located at the advancing side of the weld.

The application of the FSW process to other materials beyond the aluminium alloys and to dissimilar material combinations was also addressed in the present Special Issue. Machniewicz et al. [4] studied the effect of the traverse speed on the mechanical properties of copper FS welds. For the traverse speed range tested by the authors, the effect of this parameter on the static and especially on the fatigue behaviour of the welds was found to be stronger than its effect on the weld macro and microstructure. Considering the base material and the tested rotation speed and tool design, the authors reported that intermediate values of the traverse speed range provide welds with better mechanical properties. They still stressed that, among all the inspections performed in this work, fatigue testing was revealed as the most effective procedure for assessing the weld quality. Regarding the FSW of dissimilar materials, Chitturi et al. [5] studied the influence of the tilt angle and the pin depth on the structural and mechanical properties of AA5052/304 stainless steel joints. Following a design of experiments based on Taguchi's orthogonal array, a very strong influence of the tilt angle on the weld morphology was noticed. The authors reported that the increase in the tilt angle prevented the formation of defects and promoted the generation of intermetallic compounds (IMCs). According to them, the formation of IMCs was also influenced by the pin depth, increasing with increments in this parameter. The authors also observed that the maximum tensile strength was achieved for higher values of pin depth and tilt angle.

Besides the original research papers, one review work on FSW was also published in this Special Issue. This work, which was conducted by Laska and Szkodo [6], was focused on presenting an overview of the influence of the welding parameters on the microstructural and mechanical properties of similar and dissimilar butt welds in a large range of materials, i.e., aluminium, magnesium, steel and ferrous alloys, titanium, copper, polymers, and composites. The authors highlighted the influence of the tool traverse and rotational speeds, tool geometry and material, tilt angle, tool offset, plunge depth, and axial force on the welding conditions. For dissimilar welding, they stressed the recent trends of using interlayers or water environments to minimise the formation of detrimental IMCs. The authors also recommended more understanding of the material flow and further work on the characterisation of the weld electrochemical properties.

FSW was not the only solid-state process addressed in the present Special Issue. Two of the published works were on explosive welding. One of these works, which was developed by Wachowski et al. [7], was aimed at studying the effect of the post-welding hot-rolling on the structural and mechanical properties of magnesium/aluminium/titanium composites. The authors reported that the composites were affected by the strain and temperature experienced during hot-rolling. However, they referred that important differences in structure and mechanical behaviour existed for both weld interfaces. While a clear trend in the evolution of the interfacial strength and the rolling temperature was registered for the magnesium–aluminium interface, a less clear trend was registered for the other interface. In terms of microstructure, the hot-rolling promoted the formation of continuous IMC layers at the magnesium/aluminium interface and the precipitation of IMCs in the molten regions of the aluminium/titanium interface. In the other work on explosive welding,

Karolczuk et al. [8] studied the influence of the impact velocity on the residual stresses, strength, and structural properties of zirconium/titanium/steel composites. According to the authors, the tested differences in impact velocity did not promote strong changes in the microhardness distribution and in the tensile yield force. Regarding the residual stresses, they found that the compressive stresses initially present in the as-received zirconium alloy decreased in the welding process, resulting in tensile stresses. Additionally, the authors observed that a higher impact velocity led to higher tensile stresses in the zirconium, and therefore, a lower impact velocity is recommended to protect the composite from stress-based corrosion cracking.

A paper on laser welding was published by Danielewski et al. [9], who studied dissimilar lap welding of the S355J2 and 316L steels under the two alternative positions of the base material plates. The authors, following an experimental/numerical approach, concluded that the relative position of the base materials has a strong effect on the welding results. According to them, better weld properties are obtained by placing the stainless steel (316L) on the top of the joint. The authors found that this position of the plates promoted a maximum principal stress value, lower differences in the distribution of the chemical elements over the weld, and a more uniform weld structure and fusion rate.

Review works addressing the welding issues of two specific material classes were also published in the present Special Issue. The study conducted by Parker and Siefert [10] was developed based on the Electric Power Research Institute's works on the manufacture and performance of welds in creep strength-enhanced ferritic steels. The authors refer that these works have enabled identifying and quantifying the factors affecting the high-temperature performance of these steels. According to them, the resulting knowledge has been used to support recommendations for improving the production and control of components in creep strength-enhanced ferritic steels. In turn, Filho and Monteiro [11] addressed the recent trends in the welding of high entropy alloys. The authors found that both fusion welding and solid-state welding techniques could be used to join these materials. They also stressed that the works already developed in this field have focused on varying aspects, such as grain refinement, dynamic recrystallization, hardness enhancement, secondary phase precipitation, ductility decrease, and strengthening of the material. Furthermore, the authors reported that, for welding techniques with high power density, the loss of elements with low melting and evaporation temperatures was a challenge for controlling the chemical composition of the welded alloys.

3. Conclusions and Outlook

A sample of the cutting-edge research currently conducted in non-conventional welding of materials was presented in this Special Issue. The demanding methodological approaches followed in the different works, coupled with the supported discussions, allowed solid conclusions to be achieved, which strongly contributed to enriching the knowledge in this field. Even so, future research in this area keeps pertinent and relevant so that engineering solutions meeting the increasingly demanding industrial criteria can be developed.

As the Editors of the present Special Issue, we consider that it was a very successful project. This would be impossible without the precious contribution of all the authors and the reviewers. The quality, rigour, scientific relevance, and actuality of the papers submitted by the authors, which were ensured and potentiated by supported and rigorous reviews, are the main strength of this project. We are profoundly grateful to all the authors and the reviewers. Finally, we acknowledge all support provided by the Materials editorial team, especially by Elsa Qiu, during the development of this Special Issue.

Funding: This research is sponsored by FEDER funds through the program COMPETE—Programa Operacional Factores de Competitividade—and by national funds through FCT—Fundação para a Ciência e a Tecnologia—under the project UIDB/00285/2020.

Conflicts of Interest: The authors declare no conflict of interest.

References

1. Khan, N.Z.; Bajaj, D.; Siddiquee, A.N.; Khan, Z.A.; Abidi, M.H.; Umer, U.; Alkhalefah, H. Investigation on Effect of Strain Rate and Heat Generation on Traverse Force in FSW of Dissimilar Aerospace Grade Aluminium Alloys. *Materials* **2019**, *12*, 1641. [CrossRef] [PubMed]
2. Tamadon, A.; Pons, D.J.; Clucas, D.; Sued, K. Texture Evolution in AA6082-T6 BFSW Welds: Optical Microscopy and EBSD Characterisation. *Materials* **2019**, *12*, 3215. [CrossRef] [PubMed]
3. Manuel, N.; Galvão, I.; Leal, R.M.; Costa, J.D.; Loureiro, A. Nugget Formation and Mechanical Behaviour of Friction Stir Welds of Three Dissimilar Aluminum Alloys. *Materials* **2020**, *13*, 2664. [CrossRef] [PubMed]
4. Machniewicz, T.; Nosal, P.; Korbel, A.; Hebda, M. Effect of FSW Traverse Speed on Mechanical Properties of Copper Plate Joints. *Materials* **2020**, *13*, 1937. [CrossRef] [PubMed]
5. Chitturi, V.; Pedapati, S.R.; Awang, M. Effect of Tilt Angle and Pin Depth on Dissimilar Friction Stir Lap Welded Joints of Aluminum and Steel Alloys. *Materials* **2019**, *12*, 3901. [CrossRef] [PubMed]
6. Laska, A.; Szkodo, M. Manufacturing Parameters, Materials, and Welds Properties of Butt Friction Stir Welded Joints–Overview. *Materials* **2020**, *13*, 4940. [CrossRef] [PubMed]
7. Wachowski, M.; Kosturek, R.; Śnieżek, L.; Mróz, S.; Stefanik, A.; Szota, P. The Effect of Post-Weld Hot-Rolling on the Properties of Explosively Welded Mg/Al/Ti Multilayer Composite. *Materials* **2020**, *13*, 1930. [CrossRef] [PubMed]
8. Karolczuk, A.; Kluger, K.; Derda, S.; Prażmowski, M.; Paul, H. Influence of Impact Velocity on the Residual Stress, Tensile Strength, and Structural Properties of an Explosively Welded Composite Plate. *Materials* **2020**, *13*, 2686. [CrossRef] [PubMed]
9. Danielewski, H.; Skrzypczyk, A.; Hebda, M.; Tofil, S.; Witkowski, G.; Długosz, P.; Nigrovič, R. Numerical and Metallurgical Analysis of Laser Welded, Sealed Lap Joints of S355J2 and 316L Steels under Different Configurations. *Materials* **2020**, *13*, 5819. [CrossRef] [PubMed]
10. Parker, J.; Siefert, J. Manufacture and Performance of Welds in Creep Strength Enhanced Ferritic Steels. *Materials* **2019**, *12*, 2257. [CrossRef] [PubMed]
11. Garcia Filho, F.C.; Monteiro, S.N. Welding Joints in High Entropy Alloys: A Short-Review on Recent Trends. *Materials* **2020**, *13*, 1411. [CrossRef] [PubMed]

Article

Investigation on Effect of Strain Rate and Heat Generation on Traverse Force in FSW of Dissimilar Aerospace Grade Aluminium Alloys

Noor Zaman Khan [1], Dhruv Bajaj [2], Arshad Noor Siddiquee [2,*], Zahid A. Khan [2], Mustufa Haider Abidi [3,*], Usama Umer [3] and Hisham Alkhalefah [3]

1 Department of Mechanical Engineering, National Institute of Technology, Srinagar 190006, Jammu and Kashmir, India; noor_0315@yahoo.com
2 Department of Mechanical Engineering, Jamia Millia Islamia (A Central University), New Delhi 110025, India; maildhruv08@gmail.com (D.B.); zakhan@jmi.ac.in (Z.A.K.)
3 King Saud University, Advanced Manufacturing Institute, Riyadh 11421, Saudi Arabia; uumer@ksu.edu.sa (U.U.); halkhalefah@ksu.edu.sa (H.A.)
* Correspondence: arshadnsiddiqui@gmail.com (A.N.S.); mabidi@ksu.edu.sa (M.H.A.)

Received: 24 April 2019; Accepted: 17 May 2019; Published: 20 May 2019

Abstract: The emergence of the aerospace sector requires efficient joining of aerospace grade aluminium alloys. For large-scale industrial practices, achievement of optimum friction stir welding (FSW) parameters is chiefly aimed at obtaining maximum strain rate in deforming material with least application of traverse force on the tool pin. Exact computation of strain rate is not possible due to complex and unexposed material flow kinematics. Estimation using micro-structural evolution serves as one of the very few methods applicable to analyze the yet unmapped interdependence of strain rate and traverse force. Therefore, the present work assessed strain rate in the stir zone using Zener Holloman parameter. The maximum and minimum strain rates of 6.95 and 0.31 s^{-1} were obtained for highest and least traverse force, respectively.

Keywords: aluminium alloys; friction stir welding; strain rate; traverse force

1. Introduction

Friction stir welding (FSW) of aluminium alloys has opened new avenues of building lightweight and yet high strength structures. This process overcomes various problems associated with conventional fusion welding processes in joining of high strength aluminium alloys. Benefits of FSW over conventional fusion welding processes include the absence of brittle inter-dendritic phases in the weld microstructure [1], low distortion, improved mechanical properties of the joint and good dimensional stability [2]. FSW, invented in 1991 by W.M. Thomas at TWI [3], finds numerous applications [4–7] in a short time span compared to other welding processes. However, a coherent understanding of FSW process has not yet been achieved and there persists the quest for even better understanding of various aspects of the process, such as strain, strain rate, evolution of process forces, etc.

In addition, owing to various desirable properties such as good formability, high plasticity [8] and high strength to weight ratio [9], the usage of aluminium has increased significantly in various engineering applications [10]. Considering (1) growing applications of FSW, (2) increased usage of aluminium across various industrial sectors and (3) the need for thorough and more comprehensive understanding of the process, FSW of aluminium alloys was chosen for the subject of the present study.

FSW is a solid-state welding process [11], in which a non-consumable rotating tool with a pin is plunged in the abutting surfaces of the base materials (BM). After the plunge is complete, dwell time may be provided to the tool before the tool's traverse in order to preheat the BM [12]; and then the rotating tool is traversed along the joint line. Heat generated by friction and plastic deformation softens the BM. The softened material is stirred by the tool pin and then consolidated behind the tool, forming a solid phase joint. According to numerical simulations, about 2 to 20% of total heat is obtained from plastic deformation [13,14], and the rest of the heat is obtained from frictional heating due to rubbing of the tool with the BM.

Plastic deformation due to stirring induces severe plastic deformation (SPD) in the BM at a very high strain rate. During plastic deformation, shear stress plays an important role in determining the material flow characteristics. Shear stress induced in the BM during FSW is also a key topic which needs further understanding. However, shear stress in the BM also varies with strain rate [12]. Thus, studies based upon strain rate undergone by the material are very important for the coherent understanding of FSW process.

Due to plastic deformation of the material by traversing the tool, a force, termed as traverse force, is experienced by the tool in the direction opposite to tool movement. High process torque and traverse forces may lead to shear failure of the tool pin. Since tool pin failure is the most common cause of tool failure in FSW [15], understanding of traverse force holds significance for the development of tool materials, tool design and the application of FSW in manufacturing industries involving high strength materials. Su and Wu [16] estimated traverse force during FSW using radius of the recrystallized zone for different pin profiles using calculated strain rate. They found that tool pin profiles significantly affect the strain rate, plastic deformation and traverse force exerted on the base metal.

Experimental investigations and numerical analysis have been performed by researchers in order to estimate the temperature fields, strain and strain rate during FSW [17–22]. Frigaard et al. [23] used electron backscattered diffraction (EBSD) technique and a three-dimensional heat flow model to estimate the strain rate. Strain rates of 1–2 and 10–18 s^{-1} were reported for non-recrystallized and recrystallized region, respectively. Masaki et al. [22] simulated the recrystallized micro-structure of aluminium alloy using plain strain compression and estimated a strain rate of 1.8 s^{-1} during FSW. A combined three-dimensional heat transfer and visco-plastic flow model was used by Nandan et al. [24] and Arora et al. [25] to compute strain rate during FSW. Detailed variation of strain rate was computed at different depths along the weld thickness by Nandan et al. [24]. Kumar et al. [26] used particle image velocimetry to measure the strain rate around the tool pin during FSW. A transparent visco-plastic material consisting of micro-glass tracers was used in this study. Maximum strain rate of 20 s^{-1} was observed using this method and an attempt was made to predict strain rate variation with respect to change in tool traverse and rotational speeds. Chen and Cui [27] used a broken pin embedded into the work-piece in order to estimate the strain and strain rate ahead of the tool pin during FSW. A strain of 3.5 and strain rate of 85 s^{-1} was calculated by analyzing the deformed dendrites. Luo et al. [28] analyzed the welding characteristics of 2A14-T6 aluminum alloy using FSW. Tensile testing shows that the strength coefficient of the joint reaches 82.5%. Dong et al. [29] studied the micro-structures of the joints of dissimilar aluminium alloys (AA7003-T4 and 6060-T4) prepared using FSW. It was concluded that the weak area exists in the heat-affected zone (HAZ) of 6060 alloy, which was placed in the retreating side during FSW.

It is pertinent to mention that the material flow and joint consolidation closely relate to the flow stress, which also depends on the prevailing temperature. Further, the prevalent strain rate, in turn, also affects the flow stress. Under this situation the material flow and process forces become significantly dependent on the strain rate and welding temperature. Incidentally, high temperature is detrimental to the strength of age-hardened material, while low temperature makes the material flow difficult. Importantly, the specific relation between the traverse force and shear stress with regard to the FSW is not available and this becomes the motivation for conducting the present research work by correlating the strain rate and process force at different welding conditions. Though several researchers have experimentally measured and analyzed traverse force exerted on the tool [30–32], and induced strain rate [22–24] during FSW, the available literature on the inherent correlation between traverse force and strain rate is scarcely reported. In the present paper, the relationship between traverse force and strain rate has been studied with reference to two important FSW parameters, i.e., tool traverse speed and tool rotational speed.

2. Materials and Methods

Aluminium alloys AA2219-O and AA7475-T761 were selected as base materials for welding. The chemical composition of AA2219 and AA7475 alloys is given in Tables 1 and 2, respectively. Mechanical and thermal properties of the base materials are shown in Table 3.

Table 1. Chemical composition of AA2219-O.

Elements	Al	Cu	Sn	Mn	Fe	Si	Ti	Zn	Ni	Zr	V
AA2219 (wt.%)	91.97	6.8	0.02	0.315	0.16	0.06	0.04	0.06	0.023	0.203	0.165

Table 2. Chemical composition of AA7475-T761.

Elements	Al	Cu	Mg	Mn	Fe	Si	Ti	Zn	Ni	Zr	Cr
AA7475 (wt.%)	90.99	1.34	1.93	0.006	0.101	0.06	0.023	5.36	0.002	0.0076	0.155

Table 3. Mechanical and thermal properties of AA2219-O and AA7475-T761 alloys.

Aluminium Alloy	UTS (MPa)	Yield Strength (MPa)	Specific Heat Capacity (J/g°C)	Thermal Conductivity (W/mK)	Incipient Melting-Liquidus Temperature (°C)
AA7475-T761	468	430	0.88	163	538-635
AA2219-O	256	210	0.864	120	543–643

Welding experiments were performed on a robust vertical milling machine adapted for performing FSW. A cylindrical tool having threaded pin with 14 mm shoulder diameter and 4 mm pin diameter was used for performing FSW. High carbon high chromium steel was selected as the tool material. Tool tilt angle and tool insertion were kept constant at 2.5° and 2.25 mm, respectively. 2.5 mm thick plates of AA2219 and AA7475 alloys were welded in butt-joint configuration. AA2219 and AA7475 alloys were kept on the advancing side (AS) and the retreating side (RS), respectively. Traverse force was measured using load cells attached with a computer interfaced data acquisition system. Customized data acquisition system was devised for recording and plotting real-time values of traverse force with distance traversed by the tool. The traverse force during FSW is strongly dependent upon the contact area between the tool and the deforming material [15]. Moreover, for the same pin length, the optimum plunge depth during FSW is dependent upon the shoulder diameter of the tool [33]. Even minor variation in plunge depth significantly affects the temperature and traverse force during welding. Therefore, the variation in shoulder diameter could prove detrimental to the rationality of results in the present analysis. Hence, only the traverse speed and the rotational speed of the tool were varied. FSW experimental plan is shown in Table 4. The set of tool traverse and rotational speeds was carefully devised after thorough trial experimentation. For metals, the temperature dependence of flow stress is stronger as compared to the dependence on strain rate [34]. Thus, in order to enhance the effect of strain rate, a large difference of about 57% with respect to lower level was selected for both traverse and rotational speed. Simultaneously, a relatively smaller but continuous increase in temperature is also achieved at this set of traverse and rotational speeds. This is required to accomplish the cumulative effect of heat generation and strain rate. In conjunction, these parameters are also able to yield sound welds. Experimental setup for FSW and force measurement is shown in Figure 1.

Figure 1. Experimental setup used for performing friction stir welding.

Table 4. Experimental plan.

Experiment No.	Tool Rotational Speed (rpm)	Tool Traverse Speed (mm/min)
1	710	250
2	1120	250
3	710	160
4	1120	160

Wire electrical discharge machine (WEDM; manufacturer—Steer Corporation, City—Shanghai, Country—China) was used for cutting specimens for micro-structural examination. After the standard metallographic procedure, Keller's reagent was used for etching. Etched specimens were observed under an optical microscope and mean sub-grain size was estimated according to ASTM E 112 [35] using line intercept method through a MIAS software (Version 2.0) [36].

3. Results and Discussion

Traverse force is an indirect measure of flow-stress required for plastic deformation of the material during FSW. Flow-stress in the deforming material depends upon (1) heat input per unit weld length and (2) the strain rate experienced by the deforming material. Higher heat input softens the material and, hence, reduces flow-stress. On the other hand, higher strain rate increases flow-stress [20]. The SPD produced due to high strain rates during stirring produces ultrafine grains through dynamic recrystallization and consequently enhances the joint strength. Thus, greater strain rate yet lower traverse force on the tool is desirable for FSW. This is because higher load augments wear, deformation and degradation of the tool, causing possible weld contamination and increased tool replacement frequency [37].

3.1. Estimation of Local Strain Rate

Due to the complexities involved during stirring and material flow during FSW, exact calculation of actual strain rate is very difficult. However, local strain rate can be estimated using the following relationship described by McQueen and Jonas for aluminium alloys [38]:

$$d_s = \left[-0.6 + 0.08 \log_{10} Z_h\right]^{-1} \tag{1}$$

where d_s is the mean sub-grain diameter and Z_h is the Zener Holloman parameter given by:

$$Z_h = \dot{\varepsilon}\ \exp\left(\frac{Q}{RT_p}\right) \tag{2}$$

Here, $\dot{\varepsilon}$ is the strain rate and T_p is the peak temperature (in Kelvin) reached during FSW at the given location, Q is the apparent activation energy (taken as 156 kJ/mol [23]) and R is the universal gas constant (8.314 J/mol). The average grain size and peak temperature of the stir zone of four welded samples was measured using the line intercept method (Figure 2) and customized thermocouple, respectively. Due to SPD, only equi-axed grains are observed in the stir zone (SZ) unlike the bi-model sized grains of the thermo-mechanically affected zone (TMAZ) and heat affected zone (HAZ). Thus, nominal grain size was used for representing the results. Morphology of grains is also dependent upon the section of the weld under analysis. However, microstructural analysis of sections parallel to the traverse force can result in noise and needs extensive care. Notably, transverse cross-section of the welded joints was analyzed in this study.

Using the measured mean sub-grain diameter and peak temperature, local strain rate is estimated (at region A indicated in Figure 3) and presented in Table 5.

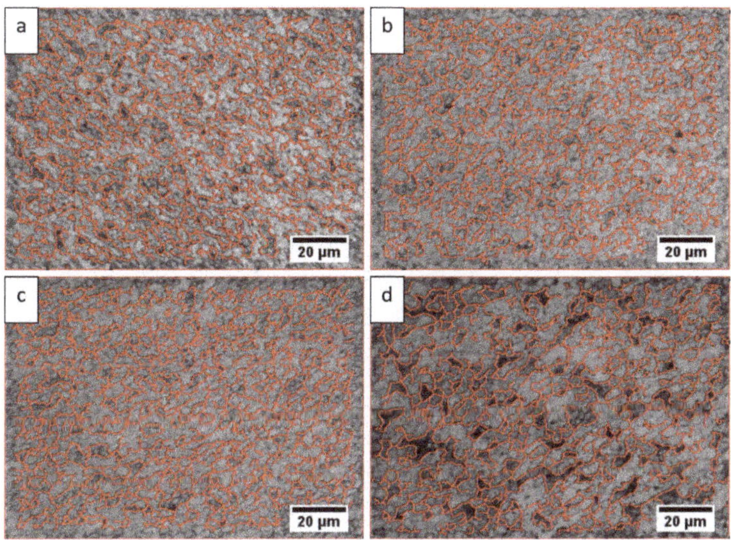

Figure 2. Optical Micro-graphs at region A for grain size measurement of (**a**) Experiment 1, (**b**) Experiment 2, (**c**) Experiment 3, (**d**) Experiment 4.

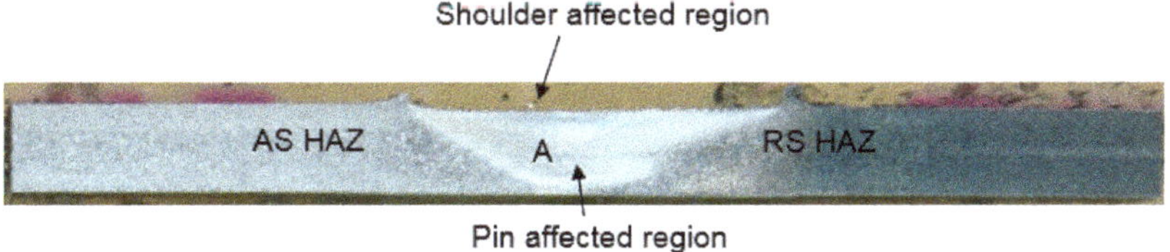

Figure 3. Schematic diagram showing the region A where optical micro-graphs are analyzed and local strain rate is computed.

Table 5. Estimated local strain rates and sub-grain diameter.

Experiment	d_s (μm)	T_p (°C)	Z_h	$\dot{\varepsilon}$ (s^{-1})
1	2.5	426	3.16×10^{12}	6.95
2	3.1	448	3.41×10^{11}	1.69
3	3.4	462	1.50×10^{11}	1.23
4	4.8	495	1.27×10^{10}	0.31

Table 6 shows the peak and average (for the specific distance of 60 mm (from 60 to 120 mm) values of traverse force exerted by the deforming material on the tool pin for different welding parameters. Figure 4 shows the variation in traverse force exerted on the tool with distance traversed for different welding parameters.

Table 6. Peak values of traverse force for different welding parameters.

Experiment Number	Tool Rotational Speed (rpm)	Traverse Speed (mm/min)	Peak Value of Traverse Force (N)	Average Values of Traverse Force (N)
1	710	250	1078	999
2	1120	250	830	737
3	710	160	764	729
4	1120	160	705	678

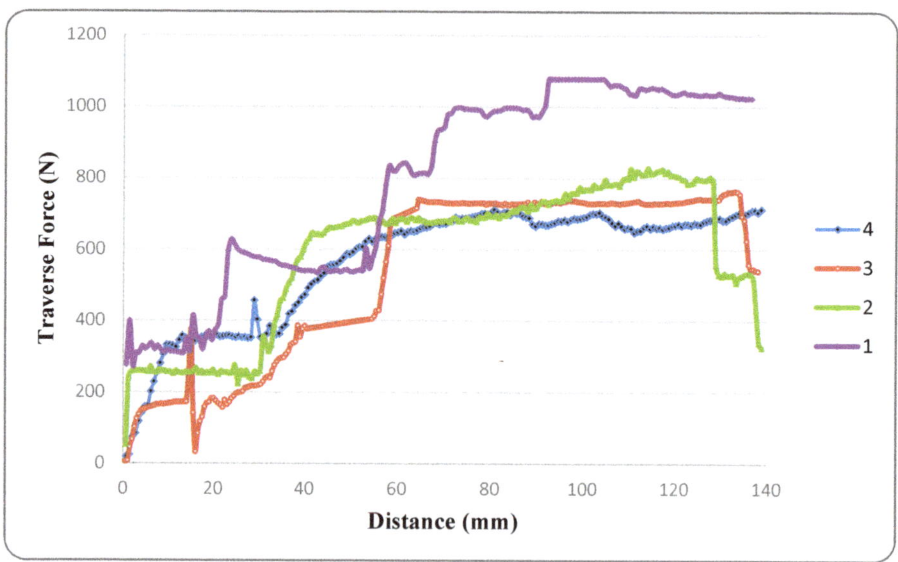

Figure 4. Traverse force exerted on the tool pin vs. distance travelled by the friction stir welding (FSW) tool for Experiments 1–4.

Experimental data was analyzed using standard statistical software Minitab-17 [39]. Lower the better (minimization) criterion was chosen for traverse force as its lower value results in less tool wear, low load on machine spindle and work fixture, etc. ANOVA was performed on the values of Table 7 to assess the significance of FSW parameters on traverse force and its results are given in Table 8. Results of ANOVA reveal that traverse speed is found to be the most dominating factor affecting traverse force with percentage contribution of 63.26%.

Table 7. Experimental values for ANOVA.

Experiment Number	Tool Rotational Speed (rpm)	Traverse Speed (mm/min)	Peak Traverse Force (N)
1	710	250	1078
2	1120	250	830
3	710	160	764
4	1120	160	705

Table 8. Analysis of variance.

Source	Sum of Squares	DF	Mean Square	F Value	% Contribution
A	2.2036	1	2.2036	3.56	28.69
B	4.8582	1	4.8582	7.86	63.26
Residual	0.6184	1	0.6184		8.05
Total	7.6802	3			

For a fixed traverse speed, higher rotational speed results in greater heat generation per unit weld length and increased material stirring action induced by the tool and thus, greater shear strain rate. However, enhanced strain rate strengthening induced in the material may not significantly contribute to the force acting along traverse direction. This is due to the fact that any change in force due to the strain rate strengthening on the tool normal to its axis is symmetrically distributed in plane normal to tool axis. Thus, change in net force in traversing direction may not significantly depend on strain rate strengthening.

Furthermore, high heat input associated with increased rotational speed induces material softening which dominates strain rate strengthening. Thus, increase in the tool rotational speed leads to decrease in flow-stress. By increasing rotational speed from 710 to 1120 rpm, the peak traverse force reduced from 1078 to 830 N (at 250 mm/min) and 764 to 705 N (at 160 mm/min), whereas average traverse force reduced from 999 to 737 N (at 250 mm/min) and 729 to 678 N (at 160 mm/min) which is evidence of reduction in flow stress.

Conversely, at fixed rotational speed, higher traverse speed results in lower heat input per unit weld length and higher strain rate. Both of these effects contribute to increase in flow-stress. Moreover, due to the increased traversing speed, the net traverse force acting on the tool witnesses a sharp increase.

Even though for Experiments 2 and 3 the change in traverse speeds is much lower than the change in rotational speeds, the traverse force in Experiment 2 was higher than that in Experiment 3. This is due to the dominant effect of traverse speed on traverse force; because apart from the heat input, drag force for the tool's traverse adds to the traverse force. However, it can be observed from Figure 4 that, for a small displacement due to tool traverse (60–85 mm), the traverse force for Experiment 3 exceeds that of Experiment 2. This is due to the fact that excess heat was generated due to higher rotational speed in Experiment 2 during the first half of the tool traverse. This excess heat overcomes the cooling effect provided by greater traverse speed for a small distance of tool traverse.

Maximum traverse force is obtained for the combination of higher traverse speed and lower rotational speed. This is because both higher traverse speed and lower rotational speed increase the flow-stress due to lower heat input per unit length and increased traverse speed. Due to similar reasons, the peak value of traverse force is lowest for the combination of lower traverse speed and higher rotational speed.

During experimentation, before the beginning of the tool traverse, a dwell time is given during which the rotating tool remains stationary at its position after plunging. Due to the plunging and the dwell time provided, the material ahead of the tool gets pre-heated. This leads to the softening of material in the vicinity of the rotating tool. As a result, the traverse force is less in the initial phase of the tool traverse and increases as the tool progresses towards the relatively colder material as shown in Figure 4.

3.2. Micro-Structure

The micro-structural analysis of welded joints having dynamically re-crystallized grains provides an important means to analyze the degree of severe plastic deformation for different welding parameters. Dynamic re-crystallization in stir zone is a consequence of plastic deformation and high temperature experienced by the material during FSW. Deformation, recovery and recrystallization of the base material, which occur during FSW, lead to dynamic recrystallization. The rate of transformation of sub-grains to grains decreases and grain growth rate increases as temperature rises above recrystallization point. In addition, the increase in degree of plastic deformation/strain causes greater grain refinement in the stir zone [40].

Figure 5 shows the micro-structures of the SZ, TMAZ and HAZ for all experiments. Table 9 represents the grain sizes of TMAZ and HAZ corresponding to these micro-graphs. Since grains of bi-model size exist in the TMAZ and HAZ, the grain size has been measured along the transverse/rolling direction for a comparative analysis. The strain experienced by the base metal decreases on moving from AS to RS [40]. Thus, the effect of strain rate decreases on moving from AS to RS. Thereby, estimation of strain rate was performed from the region A of the AS SZ, as shown in Figure 2.

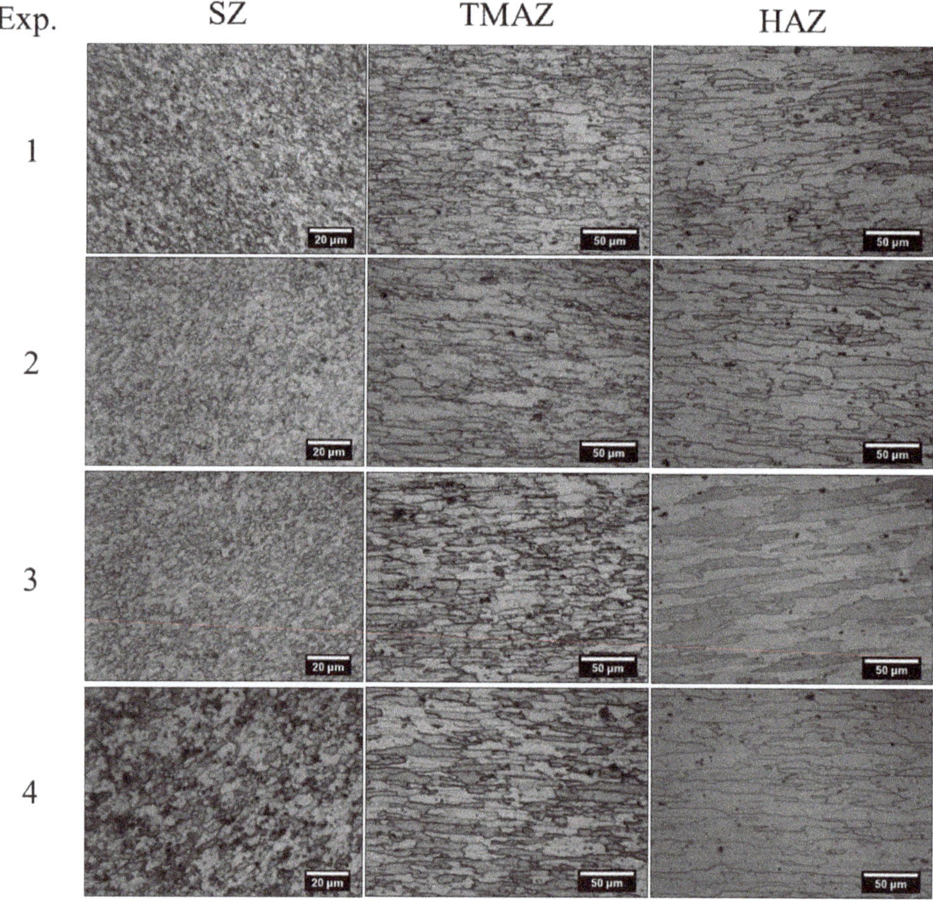

Figure 5. Micro-structural characterization for Experiments 1–4.

Table 9. Grain size of thermo-mechanically affected zone (TMAZ) and heat affected zone (HAZ) for Experiments 1–4.

Exp. No.	Grain Size (μm)	
	TMAZ	HAZ
1	31.6	43.8
2	32.5	48.4
3	32.8	55.8
4	36.6	61.0

In the SZ, maximum and minimum grain diameter of 4.8 and 2.7 μm were observed in Experiments 4 and 1, respectively. This might be due to (1) higher heat generation in Experiment 4 resulting in lower nucleation rate and higher growth rate; or (2) higher strain rate in Experiment 1 resulting in an increase in nucleation and transformation rate, resulting in smaller grain size. These results are in agreement with the work of Bird et al. [41]. Grain sizes ranging from 31.6 to 36.6 μm in the TMAZ, and 43.8 to 61.0 μm in the HAZ have been observed. Partial recrystallization in the TMAZ and induced strain from the SPD in SZ is the reason for a smaller grain size in SZ and TMAZ as compared to that of HAZ. Moreover, grain coarsening in the HAZ also contributes to the higher grain size in this zone, as shown in Table 9. The difference in the grain size of successive experiments is least for the SZ and maximum for the HAZ. This is due to the fact that prevalence of heat generation factor increasingly dominates the grain refinement due to strain as the distance from the weld center increases. However, similar order of grain size has been observed for TMAZ and HAZ for Experiments 1 to 4. The exact opposite order of magnitudes of strain rate and grain size as shown in Table 5 reflects the inherent inverse relation between the grain size and degree of plastic deformation.

4. Conclusions

This work has attempted to estimate strain rate through measured values of grain size and welding temperature and establish its correlation with flow stress. Optical microscopy and Zener Holloman parameter were used for the estimation of local strain rate. The following conclusions are drawn in light of the results of the conducted experimental investigations:

(1) Greater strain rate leads to the higher flow stress required for plastic deformation; however, the dependence of strain rate on tool traverse speed is more severe as compared to tool rotational speed.

(2) Effect of rotational speed on flow stress with regard to frictional heating is dominant compared to that with regard to strain rate strengthening.

(3) Maximum traverse force of 1078 N was observed at higher strain rate and vice versa.

(4) Strain rate during FSW can vary between 0.31 to 6.95 s^{-1}. The same order of traverse force and estimated local strain rate (at the stir zone) for all the experiments is found. This suggests that a permissible window of strain rate can be estimated to prevent the shear failure of the tool during FSW.

Future work will include the computational simulation of the FSW process, and the results of the finite element simulation will be validated with the experimental results.

Author Contributions: Conceptualization, N.Z.K. and A.N.S.; Methodology, A.N.S. and Z.A.K.; Formal Analysis, A.N.S., N.Z.K. and M.H.A.; Investigation, D.B., N.Z.K. and A.N.S.; Resources, A.N.S., U.U. and H.A.; Data Curation, D.B. and M.H.A.; Writing—Original Draft Preparation, A.N.S., N.Z.K. and Z.A.K.; Writing—Review and Editing, A.N.S., U.U. and M.H.A.; Supervision, A.N.S. and H.A.; Funding Acquisition, U.U., H.A. and M.H.A.

Funding: This research was funded by Deanship of Scientific Research, King Saud University through research group no. (RG-1440-026).

Acknowledgments: The authors extend their appreciation to the Deanship of Scientific Research at King Saud University for funding this work through research group no (RG-1440-026).

Conflicts of Interest: The authors declare no conflict of interest.

References

1. Rhodes, C.G.; Mahoney, M.W.; Bingel, W.H.; Spurling, R.A.; Bampton, C.C. Effects of friction stir welding on microstructure of 7075 aluminum. *Scr. Mater.* **1997**, *36*, 69–75. [CrossRef]
2. Mishra, R.S.; Ma, Z.Y. Friction stir welding and processing. *Mater. Sci. Eng. R Rep.* **2005**, *50*, 1–78. [CrossRef]
3. Thomas, W.; Nicholas, E.; Needham, J.; Murch, M.; Templesmith, P.; Dawes, C. Friction stir butt welding. USA Patent No. 9125978.8, 6 December 1991.
4. Johnsen, M.R. Friction stir welding takes off at boeing. *Weld. J.* **1999**, *78*, 35–39.
5. Nicholas, E.D.; Thomas, W.M. A review of friction processes for aerospace applications. *Int. J. Mater. Prod. Technol.* **1998**, *13*, 45–55.
6. Thomas, W.M.; Nicholas, E.D. Friction stir welding for the transportation industries. *Mater. Des.* **1997**, *18*, 269–273. [CrossRef]
7. Balokhonov, R.; Romanova, V.; Batukhtina, E.; Martynov, S.; Zinoviev, A.; Zinovieva, O. A mesomechanical analysis of the stress–strain localisation in friction stir welds of polycrystalline aluminium alloys. *Meccanica* **2016**, *51*, 319–328. [CrossRef]
8. Konstantinov, I.L.; Gubanov, I.Y.; Astrashabov, I.O.; Sidel'nikov, S.B.; Belan, N.A. Simulation of die forging of an ak6 aluminum alloy forged piece. *Russ. J. Non Ferr. Met.* **2015**, *56*, 177–180. [CrossRef]
9. Shabani, M.O.; Mazahery, A. Automotive copper and magnesium containing cast aluminium alloys: Report on the correlation between yttrium modified microstructure and mechanical properties. *Russ. J. Non Ferr. Met.* **2014**, *55*, 436–442. [CrossRef]
10. Saravanan, S.D.; Senthilkumar, M. Prediction of tribological behaviour of rice husk ash reinforced aluminum alloy matrix composites using artificial neural network. *Russ. J. Non Ferr. Met.* **2015**, *56*, 97–106. [CrossRef]
11. Yang, Y.-P. Developing friction stir welding process model for icme application. *J. Mater. Eng. Perform.* **2015**, *24*, 202–208. [CrossRef]
12. Mishra, R.S.; De, P.S.; Kumar, N. *Friction Stir Welding and Processing*; Springer: Cham, Switzerland, 2014; p. 338.
13. Russell, M.; Shercliff, H. Analytical modeling of microstructure development in friction stir welding. In Proceedings of the 1st International Symposium on Friction Stir Welding, Thousand Oaks, CA, USA, 14–16 June 1999.
14. Colegrove, P.; Painter, M.D.; Graham, D.; Miller, T. 3-dimensional flow and thermal modelling of the friction stir welding process. In Proceedings of the 2nd International Symposium on Friction Stir Welding, Gothenburg, Sweden, 27–29 June 2000.
15. Arora, A.; Mehta, M.; De, A.; DebRoy, T. Load bearing capacity of tool pin during friction stir welding. *Int. J. Adv. Manuf. Technol.* **2012**, *61*, 911–920. [CrossRef]
16. Su, H.; Wu, C. Determination of the traverse force in friction stir welding with different tool pin profiles. *Sci. Technol. Weld. Join.* **2018**, *24*, 209–217. [CrossRef]
17. Song, M.; Kovacevic, R. Thermal modeling of friction stir welding in a moving coordinate system and its validation. *Int. J. Mach. Tools Manuf.* **2003**, *43*, 605–615. [CrossRef]
18. Sharghi, E.; Farzadi, A. Simulation of strain rate, material flow, and nugget shape during dissimilar friction stir welding of AA6061 aluminum alloy and Al-Mg2Si composite. *J. Alloy. Compd.* **2018**, *748*, 953–960. [CrossRef]
19. Colegrove, P.A.; Shercliff, H.R. Experimental and numerical analysis of aluminium alloy 7075-T7351 friction stir welds. *Sci. Technol. Weld. Join.* **2003**, *8*, 360–368. [CrossRef]
20. Colegrove, P.A.; Shercliff, H.R. CFD modelling of friction stir welding of thick plate 7449 aluminium alloy. *Sci. Technol. Weld. Join.* **2006**, *11*, 429–441. [CrossRef]
21. Tang, J.; Shen, Y. Numerical simulation and experimental investigation of friction stir lap welding between aluminum alloys AA2024 and AA7075. *J. Alloy. Compd.* **2016**, *666*, 493–500. [CrossRef]
22. Masaki, K.; Sato, Y.S.; Maeda, M.; Kokawa, H. Experimental estimation of strain rate during fsw of al-alloy using plane-strain compression. *Mater. Sci. Forum* **2008**, *580–582*, 299–302. [CrossRef]
23. Frigaard, Ø.; Grong, Ø.; Midling, O.T. A process model for friction stir welding of age hardening aluminum alloys. *Metall. Mater. Trans. A* **2001**, *32*, 1189–1200. [CrossRef]
24. Nandan, R.; Roy, G.G.; Debroy, T. Numerical simulation of three-dimensional heat transfer and plastic flow during friction stir welding. *Metall. Mater. Trans. A* **2006**, *37*, 1247–1259. [CrossRef]

25. Arora, A.; Zhang, Z.; De, A.; DebRoy, T. Strains and strain rates during friction stir welding. *Scr. Mater.* **2009**, *61*, 863–866. [CrossRef]
26. Kumar, R.; Pancholi, V.; Bharti, R.P. Material flow visualization and determination of strain rate during friction stir welding. *J. Mater. Process. Technol.* **2018**, *255*, 470–476. [CrossRef]
27. Chen, Z.W.; Cui, S. Strain and strain rate during friction stir welding/processing of Al-7Si-0.3Mg alloy. *IOP Conf. Ser. Mater. Sci. Eng.* **2009**, *4*, 012026. [CrossRef]
28. Luo, H.; Wu, T.; Fu, J.; Wang, W.; Chen, N.; Wang, H. Welding characteristics analysis and application on spacecraft of friction stir welded 2A14-T6 aluminum alloy. *Materials* **2019**, *12*, 480. [CrossRef]
29. Dong, J.; Zhang, D.; Zhang, W.; Zhang, W.; Qiu, C. Microstructure evolution during dissimilar friction stir welding of AA7003-T4 and AA6060-T4. *Materials* **2018**, *11*, 342. [CrossRef]
30. Buchibabu, V.; Reddy, G.M.; De, A. Probing torque, traverse force and tool durability in friction stir welding of aluminum alloys. *J. Mater. Process. Technol.* **2017**, *241*, 86–92. [CrossRef]
31. Atharifar, H.; Lin, D.; Kovacevic, R. Numerical and experimental investigations on the loads carried by the tool during friction stir welding. *J. Mater. Eng. Perform.* **2009**, *18*, 339–350. [CrossRef]
32. Mehta, M.; Chatterjee, K.; De, A. Monitoring torque and traverse force in friction stir welding from input electrical signatures of driving motors. *Sci. Technol. Weld. Join.* **2013**, *18*, 191–197. [CrossRef]
33. Khan, N.Z.; Siddiquee, A.N.; Khan, Z.A. Proposing a new relation for selecting tool pin length in friction stir welding process. *Measurement* **2018**, *129*, 112–118. [CrossRef]
34. Nunes, A.C., Jr. Friction Stir Welding at MSFC: Kinematics. In Proceedings of the 4th Conference on Aerospace Materials, Processes, and Environmental Technology, Huntsville, AL, USA, 18–20 September 2000.
35. ASTM International. *Standard Test Methods for Determining Average Grain Size*; ASTM International Standard E 112–04; ASTM International: West Conshohocken, PA, USA, 2006.
36. Metallurgical Image Analyser by QS Metrology. Available online: https://www.qsmetrology.com/metallurgical-image-analyzer.html (accessed on 7 May 2019).
37. Rai, R.; De, A.; Bhadeshia, H.K.D.H.; DebRoy, T. Review: Friction stir welding tools. *Sci. Technol. Weld. Join.* **2011**, *16*, 325–342. [CrossRef]
38. McQueen, H.J.; Jonas, J.J. Plastic deformation of materials. In *Treatise on Materials Science & Technology*; Arsenault, R.J., Ed.; Academic Press: New York, NY, USA, 1975; pp. 393–493.
39. Minitab 17: Statistical Analysis Software. Available online: http://www.minitab.com/en-us/ (accessed on 5 May 2019).
40. Khan, N.Z.; Siddiquee, A.N.; Khan, Z.A.; Mukhopadhyay, A.K. Mechanical and microstructural behavior of friction stir welded similar and dissimilar sheets of AA2219 and AA7475 aluminium alloys. *J. Alloy. Compd.* **2017**, *695*, 2902–2908. [CrossRef]
41. Bird, J.E.; Mukherjee, A.K.; Dorn, J.E. *Quantitative Relation between Properties and Microstructure*; Israel Universities Press: Jerusalem, Israel, 1969.

Article

Texture Evolution in AA6082-T6 BFSW Welds: Optical Microscopy and EBSD Characterisation

Abbas Tamadon [1,*], Dirk J. Pons [1], Don Clucas [1] and Kamil Sued [2]

[1] Department of Mechanical Engineering, University of Canterbury, Christchurch 8140, New Zealand; dirk.pons@canterbury.ac.nz (D.J.P.); don.clucas@canterbury.ac.nz (D.C.)

[2] Fakulti Kejuruteraan Pembuatan, Universiti Teknikal Malaysia Melaka, Durian Tunggal 76100, Malaysia; kamil@utem.edu.my

* Correspondence: abbas.tamadon@pg.canterbury.ac.nz; Tel.: +64-021-028-12680

Received: 23 August 2019; Accepted: 28 September 2019; Published: 1 October 2019

Abstract: One of the difficulties with bobbin friction stir welding (BFSW) has been the visualisation of microstructure, particularly grain boundaries, and this is especially problematic for materials with fine grain structure, such as AA6082-T6 aluminium as here. Welds of this material were examined using optical microscopy (OM) and electron backscatter diffraction (EBSD). Results show that the grain structures that form depend on a complex set of factors. The motion of the pin and shoulder features transports material around the weld, which induces shear. The shear deformation around the pin is non-uniform with a thermal and strain gradient across the weld, and hence the dynamic recrystallisation (DRX) processes are also variable, giving a range of observed polycrystalline and grain boundary structures. Partial DRX was observed at both hourglass boundaries, and full DRX at mid-stirring zone. The grain boundary mapping showed the formation of low-angle grain boundaries (LAGBs) at regions of high shear as a consequence of thermomechanical nature of the process.

Keywords: AA6082-T6; bobbin friction stir welding; microstructure; optical microscopy; EBSD

---

1. Introduction

1.1. Context

Friction stir welding (FSW) [1,2] is a solid-phase joining technique whereby a bond is formed between two plates by a severe plastic deformation induced by mechanical friction and the heat generated by a rotating tool. Due to the deformation nature of the process, ductile materials such as aluminium [3] are suitable candidates to be processed by the FSW. One of the difficulties with FSW has been the visualisation of microstructure, particularly grain boundaries [4,5]. This is especially a problem for those materials that intrinsically have a fine grain structure. A case in point is aluminium AA6082-T6 [6,7]. This is a marine-grade aluminium alloy [8], with high-strength mechanical properties (Elastic Young's Modulus of 71 GPa, Fatigue Strength of 95 MPa, Shear Modulus of 26 GPa, Shear Strength of 220 MPa and Ultimate Tensile Strength of 330 MPa) [9], compared to other Al-series. Furthermore, the T6 tempering cycle achieves an artificially aged super-saturated solid solution to meet a high-strength structure compared to other 6xxx-series alloys [10,11]. The standard chemical composition for AA6082-T6 is shown in Table 1. Although this corrosion-resistance Al alloy is a suitable choice for machining, it suffers from poor weldability [8]. It has historically been difficult to demonstrate the microscopic features for this material, which has hindered the diagnosis of the causes of its poor weldability [12].

Recent novel developments have yielded an etchant that is capable of showing microstructures using optical microscopy (OM) [5]. Applications of the etchants have elucidated the grain size and morphology in different regions of the weld texture [7], however, the grain boundary network and thermomechanical features (e.g., dynamic recrystallisation evolution and grain refinement mechanisms) need more advanced and precision measurement such as electron microscopy [13,14]. There is also a need to compare and contrast the different features evident in the optical and electron methods, and to validate the etchant method. Furthermore, there is a need to better understand the linear features, or flow lines, evident in the cross section.

Table 1. Element composition of the AA6082-T6 aluminium alloy (wt %) [15].

Chemical Element	Present (wt %)
Silicon (Si)	(0.70–1.30)
Magnesium (Mg)	(0.60–1.20)
Manganese (Mn)	(0.40–1.00)
Iron (Fe)	(0.0–0.50)
Chromium (Cr)	(0.0–0.25)
Zinc (Zn)	(0.0–0.20)
Titanium (Ti)	(0.0–0.10)
Copper (Cu)	(0.0–0.10)
Other (Each)	(0.0–0.05)
Other (Total)	(0.0–0.15)
Aluminium (Al)	Balance

1.2. Background Literature

Due to the severe shear deformation during friction stir welding, it is not straightforward to evaluate the microstructure evolution of the FSW weld [16]. The grain map of the crystallographic texture can provide an accurate analysis to investigate the relationship between the microstructure and the thermomechanical characteristics of the FSW process [17–19]. However, the rotating nature of the tool during the FSW process makes it different to conventional deformation processes, such as rolling, extrusion, or compression. While the shear-bands in these processes are aligned with the deformation direction, the deformation orientation induced within the texture varies across the weld region, as a function of the position of the rotating tool during the FSW process [19,20].

Most of the published research is focused on the characterisation of the grain structure within the FSW weld structure [21]. Another important research strand has been to better understand the material flow [19,20,22] and this requires visualisation of the texture variations in the aluminium FSW welds [17–19].

Furthermore, due to differences in heat generation [23] and flow mechanism [24] between FSW and BFSW, the texture evaluation is expected to be different. This has been observed in the microstructural evolution of the AA6082-T6 BFSW weld structure [4,5]. The flow arms at the hourglass-borders of the BFSW weld [25] are different to the onion ring patterns in the basin-shaped FSW weld structure. Therefore, these two welding processes show different features, and there is a need to better understand the dynamic recrystallised grain structure and flow-based characteristics of the weld region.

1.3. Approach

The present paper compares the optical microscopy results (using etchant), against electron backscatter diffraction (EBSD) results to analyse the bobbin friction stir welding (BFSW) weld texture of aluminium alloy AA6082-T6 with a focus on the thermomechanical details of the microstructure.

EBSD analysis was used to determine the DRX details. This also allows the flow features to be identified at greater resolution compared to OM. Additionally, the grain boundary network and the content of the grain orientation within the weld texture can be investigated with EBSD.

It should be noted that although the transmission electron microscope (TEM) is capable of resolving the fine feature of the microstructure of FSW welds, this research focused on EBSD analysis because of the accessibility and ease of sample preparation.

2. Materials and Methods

The AA6082-T6 aluminium plate of 4 mm thickness was used as the workpiece material for the welding trial.

A fixed-gap bobbin-tool, manufactured from H13 tool steel was used for the BFSW welding (Figure 1). The tool was fully-featured (threaded tri-flat pin, and 360-degree spiral scrolled shoulders) to create a better stirring flow. The weld was arranged in the butt-joint position no gap between the plates, also no preheating. A 3-axis CNC machine (2000 Richmond VMC Model, 600 Group brand, Sydney, Australia) was used for the welding trial, while the plates were rigidly fixed by strap clamps at the outer faces during the process.

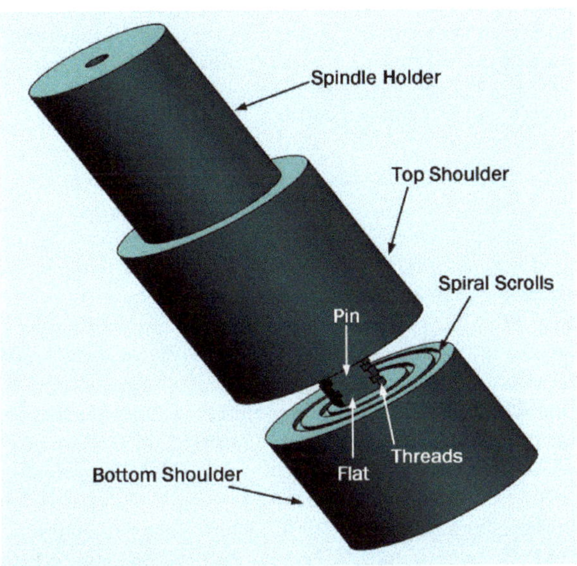

Figure 1. Schematic of the fully-featured Bobbin-Tool (tri-flat threaded pin and spiral scrolled shoulders).

Since the aim of the research was to evaluate the weld texture in a defect-free structure, a variety of welding speeds (rotational speed; ω and advancing speed; V) were used to validate the welding process. After running some tests in ω (350–650 rpm) and V (300–400 mm/min) [5,7,12,26], the optimum welding trial was performed in clockwise rotational speed (ω) of 600 rpm with an advancing speed (V) of 400 mm/min in the traveling direction. This set of speeds (ω, V) achieved a weld with no crack or void defect on surface, neither any material loss through the weld-seam. Lower speeds were unable to create a bonded weld between the aluminium plates, and higher speeds deteriorated the quality of the weld by material loss. The details of the welding test are listed in Table 2.

Table 2. The specification of the operation parameters for the AA6082-T6 BFSW weld sample.

Workpiece	Tool Material	Work Temp °C	$D_{Shoulder}$ (mm)	D_{Pin} (mm)	Plate Thickness (mm)	Feed ω (rpm)	Speed V (mm/min)	Thread Pitch (mm)	Number of Threads in the Gap
AA6082-T6	H13 Tool Steel	18	21	7	6	600	400	1.5	4

After conducting a 150 mm single-pass weld line, the sample was cut along the transverse direction of the plates for metallographic analysis.

The AA6082-T6 weld samples were prepared first for etching, and then repolished for EBSD. On both cases the polish method was per: standard mechanical polishing with different grades of SiC sand papers (600-grit, 800-grit and 1200-grit). To achieve a mirror surface, the micro-polishing step was conducted on a micro-cloth pad with a 3 μm diamond paste, and finally a 0.05 μm colloidal silica solution [5].

The samples then were etched in a reagent etchant solution with the composition of (0.5 g $(NH_4)_2MoO_4$ + 3.0 g NH_4Cl + 1 mL HF + 18 mL HNO_3 + 80 mL H_2O). The immersion etching was done in an ultrasonic bath for 90 s, at 70 °C. The microstructure of the etched cross-sections was examined by optical microscopy (OM).

Samples were repolished back to 600-grit between optical and EBSD examination. Based on sizes of polishing particles, this might correspond to about 100 μm surface removal. Our experience in repolishing for optical work shows that flow features are reasonably consistent after repolishing. Samples were positioned by geometric measurements from the edges of the weld, in such a way to view the same region of the weldment. Repositioning accuracy is estimated to be with 20 μm. For both these reasons the repolished surface features evident in the results may not correspond exactly to each other in the pairs of images.

For EBSD the following process applied: the mounted specimens were examined with a scanning electron microscope (SEM) (JEOL 6100, JEOL Inc., Peabody, MA, USA) with an HKL Nordlys III EBSD detector (Oxford Instruments plc, Abingdon, UK). The EBSD plots for different regions of the weld region were reconstructed with HKL Tango software (HKL Channel 5 Tango software version 5.12.60.0, Oxford Instruments plc, Abingdon, UK) [27].

For the mid-Stirring Zone (SZ) region with the ultrafine grain structure, a magnification step size of 0.75 μm with an overall acquisition area of (~400 × 300 μm) was used for EBSD mapping. Alternatively, for other regions of the weld with larger grains, a step size of 3 μm with an overall acquisition area of 2 mm^2 was used, while all other control parameters such as binning, probe current, accelerating voltage, and exposure time were held constant [7,27]. The average indexing rate of all collected samples was 97.5%, where the unindexed pixels were filled in with the software [28].

Using the average Taylor factor, the EBSD data was further analysed for the distribution of the grain boundaries within the microstructure [29,30]. The low-angle grain boundaries (LAGBs) with misorientation degree of 2°–10° were highlighted in blue and the high-angle grain boundaries (HAGBs) with misorientation degree larger than 10° were shown by red [7,31]. It should be noted that the twinning boundaries were not analysed but instead included in the LAGBs distribution map.

3. Results

3.1. Characterisation of the Sample Regions with OM

The samples were from various areas of the weld cross-section as described below and in Figure 2. The cross-section is perpendicular to the welding direction as the Advancing Side (AS) of the weld region is situated in the left and the Retreating Side (RS) is on the right at the cross-section.

1. Base metal (BM). This is the parent metal of the workpiece and situated outside of the weld region far away from the active region of the stirring, and thermally and mechanically unaffected by the welding process. In the AA6082-T6 workpiece it is expected that BM would conserve the columnar-shaped directional grain morphology of the rolled structure with the primary average grain size unchanged during the BFSW welding process, and this is what was observed.

2. AS/RS Hourglass Borders. The interfaces between the plastic deformation region or Stirring Zone (SZ) and the transition region adjacent to the weld region are distinguished as hourglass shaped borders at both the AS and RS of the weld. The bent shape of the border at the middle of the cross-section can be attributed to the interaction of the pin and shoulders with the substrate. The pin induces more shear compared to shoulders, therefore the borders are stretched towards the pin position. The grains

size and morphology in the texture of hourglass border is observed to be different than both BM and SZ. This is due to the different thermal and mechanical characteristics through the cross-section.

3. Flow-arms. These are characteristic microscopic feature of the BFSW weld structure, evident as elongated bands in direction of the stirring flow lines from the centre of the weld drawn towards the top and bottom shoulders. The formation of the flow-arms is generally attributed to a direct outcome of the shear banding in a continuous plastic deformation, where the layered mass flow is deposited by the advancing of the rotating tool.

4. Sub-shoulder region. A severe plastic deformation is experienced underneath the shoulders. The scrolled features of the shoulders may increase the frictional heat generated in this region of the weld. Study of the sub-shoulder texture in microscopic scale can potentially reveal thermomechanical details of the BFSW process as the thermal and mechanical stress/strain fields are in a maximum rate in this region.

5. Mid-SZ. This is the main region of the weld represented by the ultrafine equiaxed grain texture in comparison with the BM. As will be shown below, the Mid-SZ region experiences full dynamic recrystallisation (DRX) transformation including grain refinement and precipitation, more than any other region in the weldment. However, because of the specific ultrafine characteristics of the microstructure of the SZ, a precise micro-analysis of the Mid-SZ texture requires electron microscopy rather than the OM metallographic measurements.

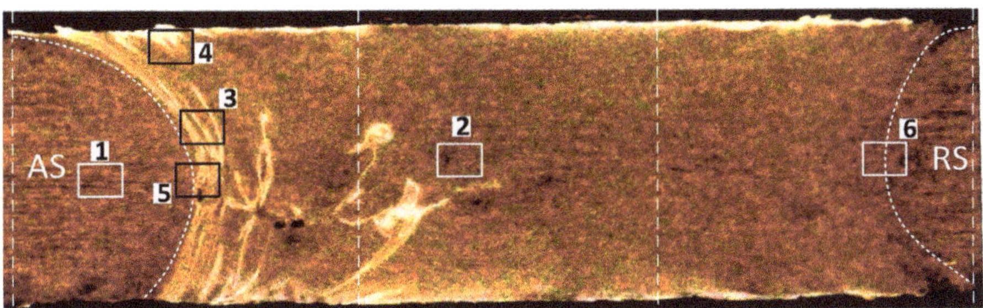

Figure 2. Weld cross-section of the AA6082-T6 plates, (1) Base Metal; (2) Middle Stirring Zone; (3) Flow-arms; (4) Sub-shoulder region; (5) Advancing Side (AS) Hourglass border and (6) Retreating Side (RS) Hourglass border. The two dashed-lines in the middle are representative of the position of the pin, the dashed-line in the corners of the cross-section are representative of the width of the shoulders.

3.2. Texture Evaluation with EBSD

The crystal orientation texture results from the EBSD mapping should be referenced compared to the RD-TD-ND perpendicular directions as a reference frame. In BFSW welded plates, the proprietary EBSD parameters need to be redefined in TD-ND-WD direction frame, as:

- Transverse direction (TD); perpendicular to the welding direction, parallel to the cross section of the weld, where the AS is in (−) and the RS is situated in the (+) of the TD axis.
- Normal direction (ND); perpendicular to the plate surface, representative of the distance between the top and bottom surface.
- Welding direction (WD); the direction of the advancement of the tool, parallel to the weld-line.

For the texture observation of the AA6082-T6 BFSW weld, the grain maps were indicated in the TD-ND plane for different regions of the weld cross-section. This plane is identical to the cross-section etched for the optical microscopy. The directions of the orientation used for the texture analysis of the EBSD plots are shown in Figure 3a.

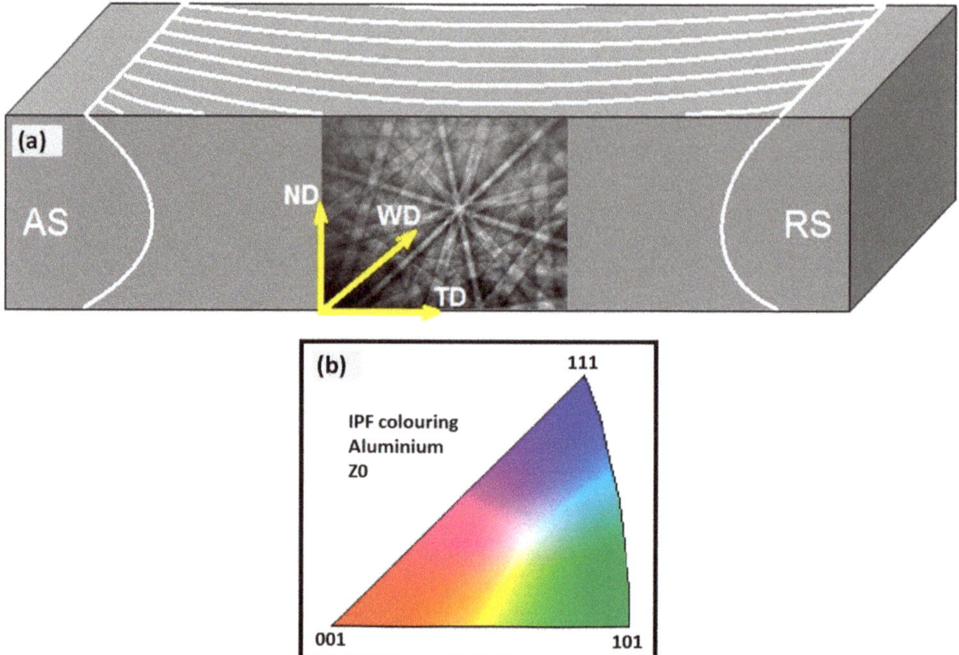

Figure 3. Crystallography directions in Aluminium alloys, (**a**) Schematic of ED, ND and TD reference directions, applied for EBSD analysis; (**b**) IPF colour triangle. (Source for (**b**) Ex EBSD machine).

The maps from the EBSD analysis present the following microscopic information: grain orientation map and grains boundary map. In the first phase, the colour mapping based on the crystallography directions of the grains indicates the crystal orientation map distribution for the analysed region. For the EBSD the standard inverse pole figures (IPF) map was used per Figure 3b. This shows the three main crystallographic directions within the grains distinguished by different colours; red for (001), green for (101) and blue for (111).

The colours in Figure 3b show the corresponding orientation of grains and crystals with respect to (001), (101) and (111) orientations. While the main crystallographic orientations are demonstrated by these three colours of red, green and blue, the other crystallographic directions between these three main crystallographic orientations are shown with the mixed colours. To obtain the grain boundary misorientations it would be necessary to specifically measure the individual misorientation.

In the second phase of the EBSD analysis, the grain boundary maps were derived in the post processing data procedure to measure the density of the high-angle grain boundaries (HAGBs) and low-angle grain boundaries (LAGBs) within the texture. Similar to the grain misorientation mapping, the HAGBs were defined by a misorientation angle greater than 10° and LAGBs were defined as the misorientation angles between 2° and 10°.

In some analyses 15° is used as the LAGBs/HAGBs transition demarcation rather than 10°. This is because at such higher angles the misorientation is more definitive to show the grain boundaries. Nonetheless the misorientation angle between 10° and 15° also can be counted as the HAGBs.

The following sections compare the results from the optical and EBSD approaches.

3.3. Base Metal

The optical and EBSD microscopy results (IPF map) are shown in Figure 4. This shows that the overall structure is represented similarly between the two methods: columnar grains of rolled base material are evident. The crystal direction (grain orientation) is similar. The colours in the EBSD results show that a variety of orientation occurs, and the rolling effect is observable by a columnar directional alignment in the orientation distribution, similar to the OM micrograph. The grain boundaries are evident in the optical results but not in the EBSD. The magnification scale was kept similar to the OM micrograph, and post-processing/zooming of the EBSD map shows the GBs.

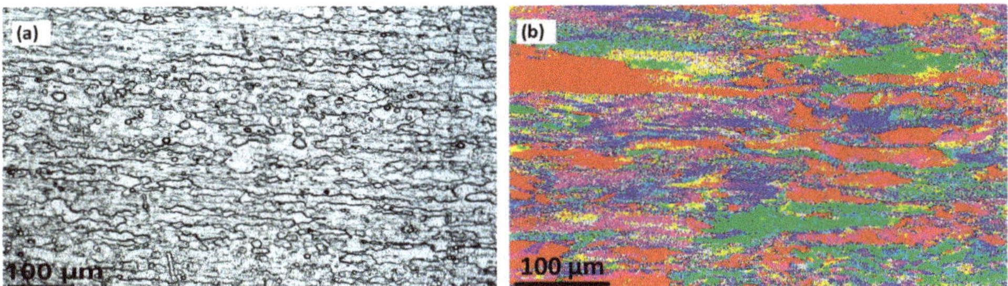

Figure 4. Base Metal microstructure, (**a**) Optical microscopy and (**b**) EBSD colour map.

To apply a post-processing clean-up of the EBSD maps, the Euler map is presented below, with a common orientation in each grain to better view the grain details. However, we followed the same EBSD IPF mapping for the weld region, to compare the similarities of the weld texture with OM micrographs.

3.4. Stirring Zone

The stirring zone microstructure for optical microscopy and EBSD is shown in Figure 5. EBSD verifies (which was not possible with OM) that the grain size average is below 10 microns. The randomly distributed grain with equiaxed morphology is evident by the EBSD map. The EBSD IPF colour map (Figure 5b) also demonstrates the microscopic distribution of the misorientation through the SZ texture, in which is not observable via the OM micrograph. The grain boundaries pattern is still bulged which may be related to the ultrafine grain size within the SZ.

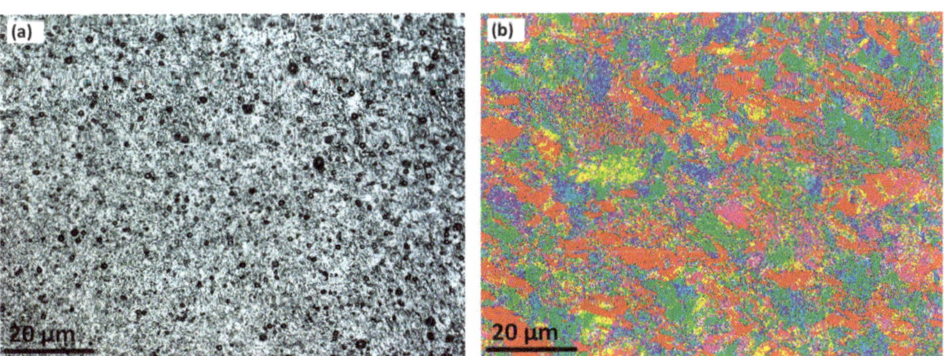

Figure 5. The microstructure of the Mid-Stirring Zone (SZ) region, (**a**) Optical microscopy and (**b**) EBSD colour map.

3.5. Flow Layers

The elongated micropattern of the flow-arms region was compared by optical microscopy and EBSD method (see Figure 6). The continuous flow direction in Figure 6b in red colour is situated at the position of each flow-arm, confirming that each flow-arm possess a specific crystallographic direction compared to the neighbouring region. These macro-bands and macro-regions were readily detected by the EBSD detector and is also present in the etched sample. There is some distance between each branch of flow-arms which has been filled by other crystals with different orientations (distinguished by corresponding colours). In this regard, our interpretation is that the macro-regions of shear bands are aligned along the (001) orientation, in a background of grains with a variation in crystallographic orientation.

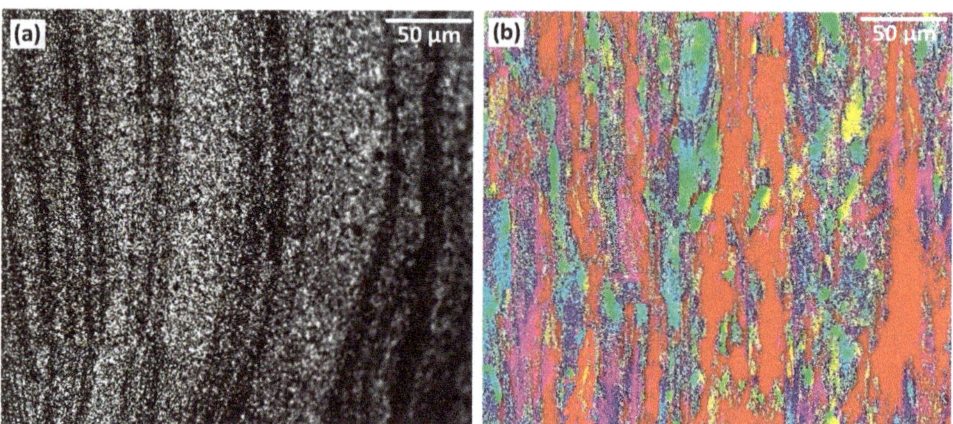

Figure 6. The microstructure of Flow layer patterns, (**a**) Optical microscopy and (**b**) EBSD colour map.

This is attributed to the discrete transport mechanism whereby packets of stirred material are deposited at the trailing edge of the tool. In turn this is attributed to the interaction between thread-flat features of the pin tool with the substrate material.

3.6. *Heat Flow (Sub-Shoulder Region)*

The microstructural analysis of the sub-shoulder region is shown in Figure 7, for OM and the related EBSD mapping. The optical microscopy result of the sub-shoulder weld region (Figure 7a) demonstrates the shoulder induced flow. These flow layers represent a transverse transportation of the mass, caused by the frictional behaviour of the shoulder action. Although the microstructure shows a severe plastic deformation underneath the shoulder, it does not contain any specific surface defect. This is attributed to the high compaction effect induced by the direct contact between the shoulder and the top surface of the weld region. The flow patterns are not observed in the deeper areas. This may be representative of the decrease of the compaction effect of the shoulder, and the decreasing effect of the surface shear strain. The flow of the mass at the deeper regions of the stirring zone is more affected by the pin action.

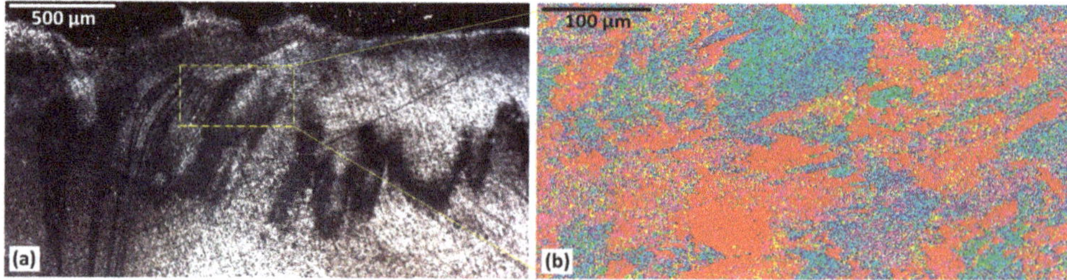

Figure 7. The microstructure of the Sub-shoulder region, (**a**) Optical microscopy and (**b**) EBSD colour map.

The selected area from the sub-shoulder region shown in Figure 7a has been mapped by the EBSD analysis in Figure 7b. The crystallographic texture at the sub-shoulder region is complex. This is attributed to the complex combination of thermal and strain effects. The approximate position of the flow patterns, delineated in Figure 7a, can be seen in a colour map in Figure 7b. The random dispersion of the grain orientation can be related to the complexity of the strain history in the location, as flow patterns are not evident. However, it should be noted that based on the EBSD standard IPF contouring (colour triangle, Figure 3b), the (001) crystallographic direction (red colour) is more pronounced compared to other orientations. This is the preferred crystallographic direction for the shear planes in the lattice of the Al atoms. Therefore, we interpret this as the shoulder-induced stirring action activating a network of the shear planes in the (001) direction.

3.7. Hourglass-Border (AS)

The highly deformed region at the hourglass-border of the weld as the triple junction is characterised in Figures 7 and 8 for the AS and RS of the weld region, respectively. As shown in Figure 8, the tapered-shape triple junction at the middle of the AS hourglass-border is characterised by a highly deformed elongated grain structure of the base material, stretched towards two opposite directions parting in the middle upwards and downwards. This deformed structure is identified with an upward-downward flow pattern (Figure 8a).

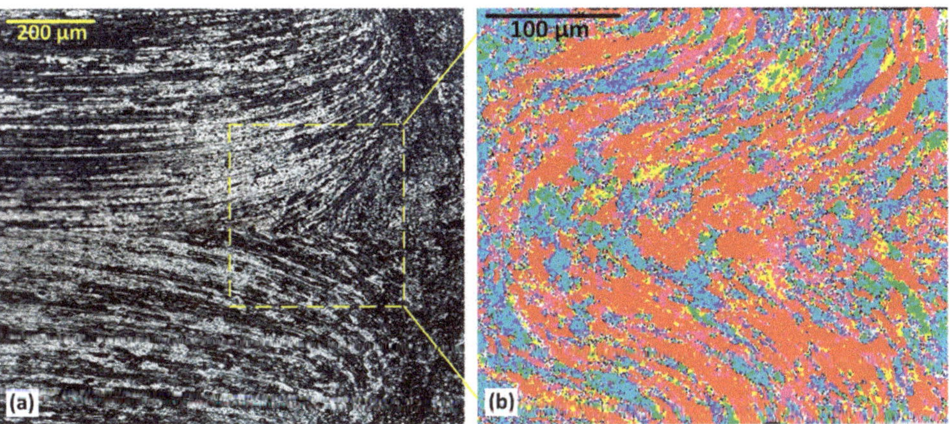

Figure 8. The microstructure of the AS Hourglass border, (**a**) Optical microscopy and (**b**) EBSD colour map.

The EBSD analysis of the tip of the tapered-shaped triple junction reveals more details regarding the deformed crystalline structure at the outer part of the hourglass border. The grain orientation of the texture shows the grain structure to be elongated towards the top and bottom shoulders. This micro-pattern is interpreted as DRX features of the FSW process whereby the stored strain induced by plastic deformation affects the grain structure of the TMAZ by grain misorientation. This can be detected as the microscopic change in etching response in the optical microscopy (Figure 8a), and the grain misorientation alteration in the EBSD colour mapping.

The EBSD pattern (Figure 8b) identifies a preferred crystallographic orientation for the deformation-induced grains structure by the red colour representative of the (001) direction from the EBSD standard triangle (Figure 3). Similar to the sub-shoulder region (Figure 7), we identify the red colour as the preferred crystallographic orientation of the Al-based texture for the formation of the shear bands undergone the plastic deformation.

3.8. Hourglass-Border (RS)

Similar to the AS hourglass-border in Figure 8, the RS of the weld also has a particular transition appearance at the hourglass-border, where an abrupt alteration in microstructure is visible at the TMAZ.

The delineated micropattern in Figure 8a reveals an ellipse-shape feature as a separating boundary between the TMAZ and the hourglass-border. This structure shows elongated grains at the edge of the TMAZ. This is not stirred material. The effect presumably results from the strain induced by the rotational movement of the tool and the frictional heating.

The micrograph showing in Figure 9b exhibits the EBSD colour map for the ellipse-shape region at the RS, corresponding to the OM microstructure in Figure 9a. It is noted that the EBSD colour map, which shows grain orientation, does not show the same grain morphology apparent in the OM microstructure. Evidently orientation and morphology are decoupled in this region, presumably because of partial recrystallisation. Partly as a result of this somewhat unexpected outcome, we explored the grain angle boundaries in further detail (see the next section).

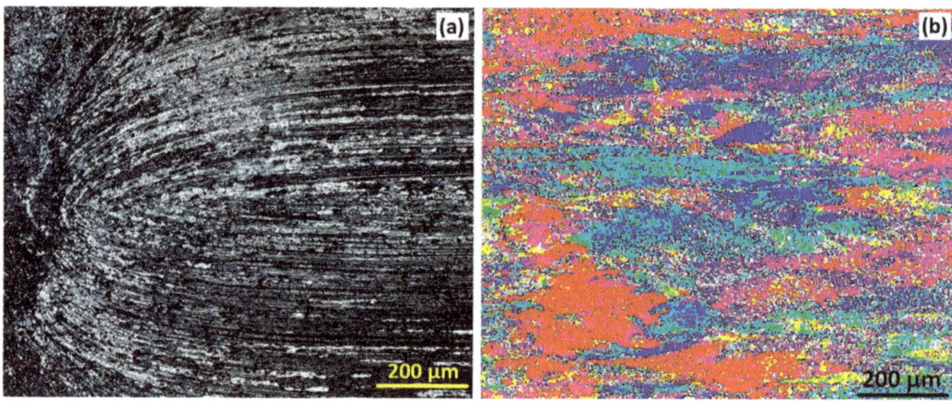

Figure 9. The microstructure of the RS Hourglass border, (**a**) Optical microscopy and (**b**) EBSD colour map.

3.9. LAGBs and HAGBs (in the Weld Region)

The grain boundary mapping for different regions of the weld are shown in Figure 10. The microscopic observations in Figure 10 were processed based on the colour identification for the LAGBs and HAGBs, identified as blue colours (Figure 10) and red colour (Figure 11), respectively.

In the stir zone of Figure 10e the density of the blue colour maps is high, indicating the dominance of LAGBs. Our interpretation is that the LAGBs are representative of the position of the sub-grain boundaries or precipitates, which are a direct result of the DRX.

The comparison between the grain orientation maps and grain boundary maps indicates that the position of the LAGBs is aligned with the position of the crystallographic misorientation inside the grains. This is more evident in the AS of the weld or the mid-SZ region where the texture experiences severe plastic deformation compared to the rest of the weld region.

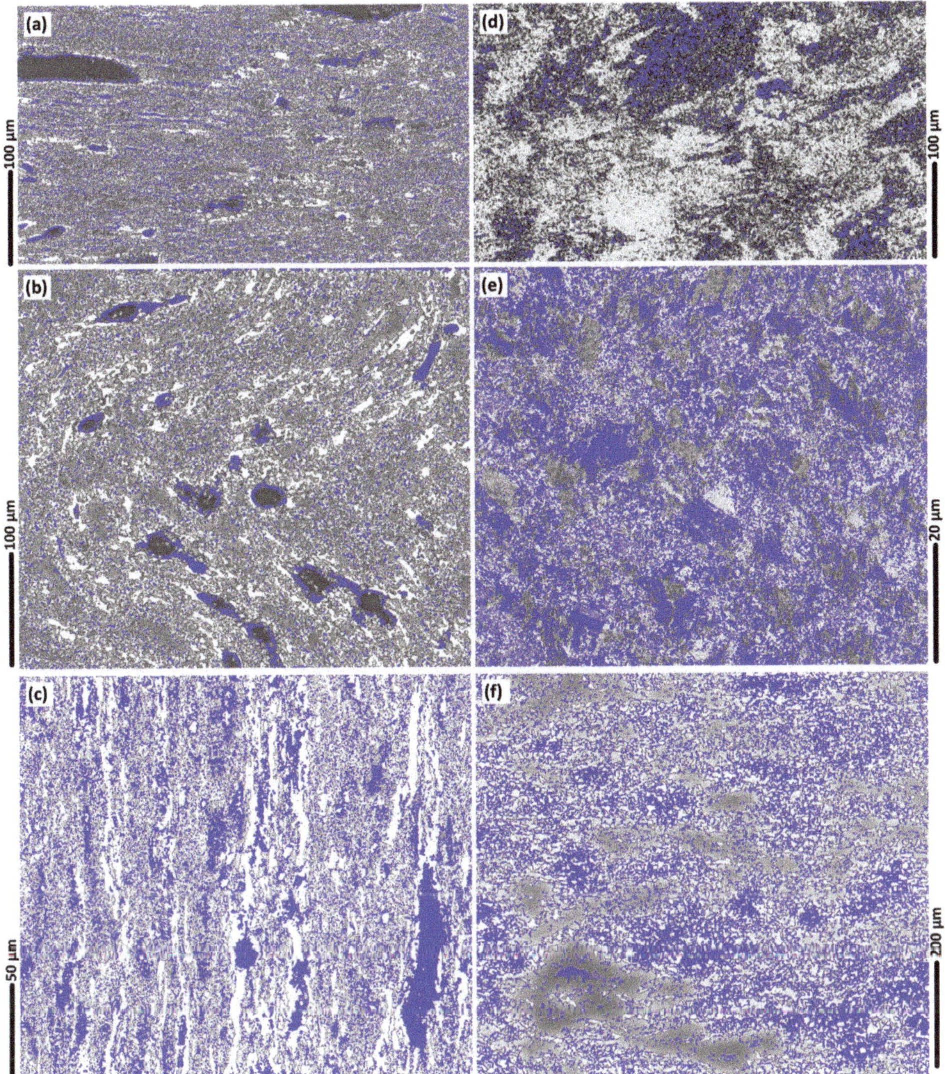

Figure 10. The micropatterns of the presence of low-angle grain boundaries (LAGBs, for 2°–10°) (**a–f**) in blue colour, for different regions of the weld; Base Metal (BM) (**a**); AS Hourglass border (**b**); Flow-arms (**c**); Sub-shoulder region (**d**); Mid SZ (**e**); RS Hourglass border (**f**).

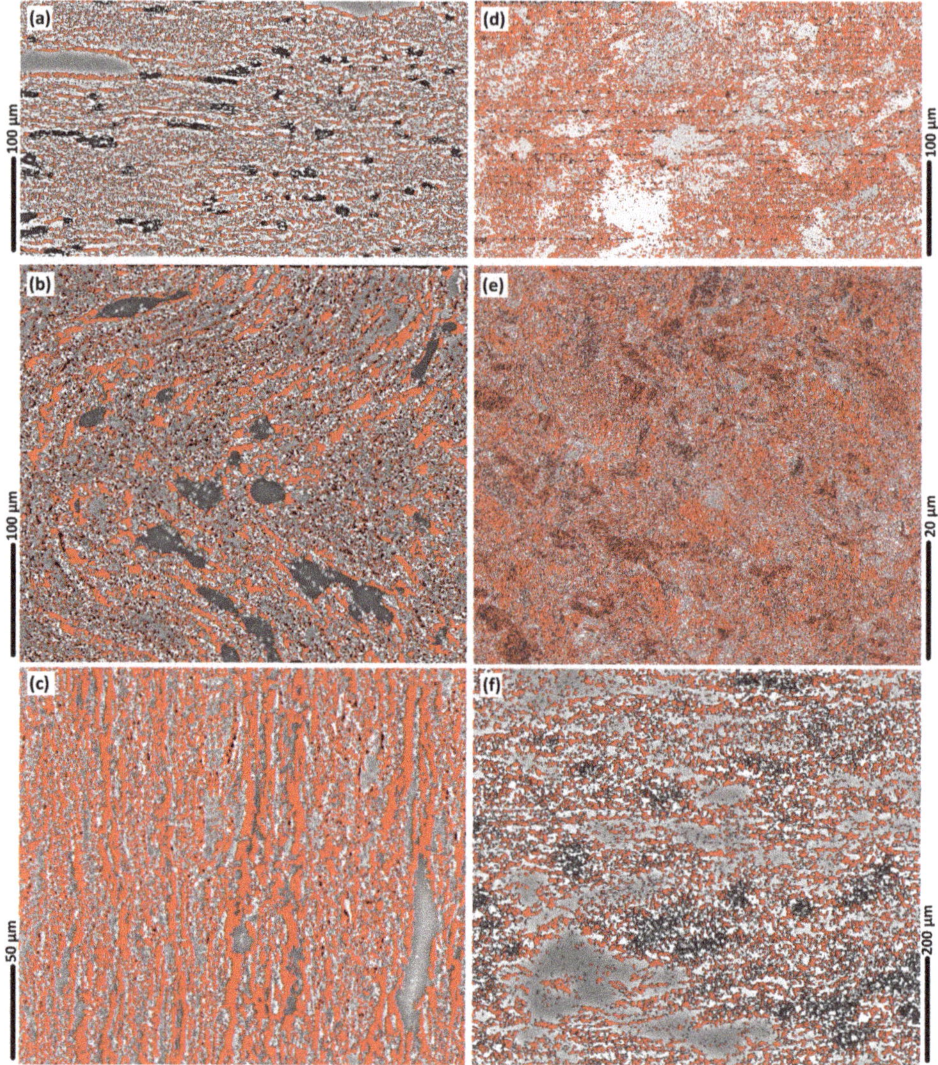

Figure 11. The micropatterns of the presence of high-angle grain boundaries (HAGBs, more than 10°) (**a**–**f**) in red colour, for different regions of the weld; BM (**a**); AS Hourglass border (**b**); Flow-arms (**c**); Sub-shoulder region (**d**); Mid-SZ (**e**); RS Hourglass border (**f**).

For a better understanding of the BFSW weld texture, the EBSD Euler contrast maps were constructed by post-processing of the EBSD data to denote the overlaid grain boundaries with a transition misorientation angle of 15°.

Assuming a 15° threshold for grain boundary elucidation, gives the results shown in Figure 12. Results emphasise the display of the HAGBs in the Euler orientation map. Precipitates are more prominent in these views compared to the previous IPF maps. Otherwise the morphology and shear bands show the same outcomes as before, indicating that the results are insensitive to choice of threshold angle. Moreover, compared to IPF maps, the Euler maps are more limited in ability to detect small orientation changes of the LAGBs, hence these are not distinctly observable.

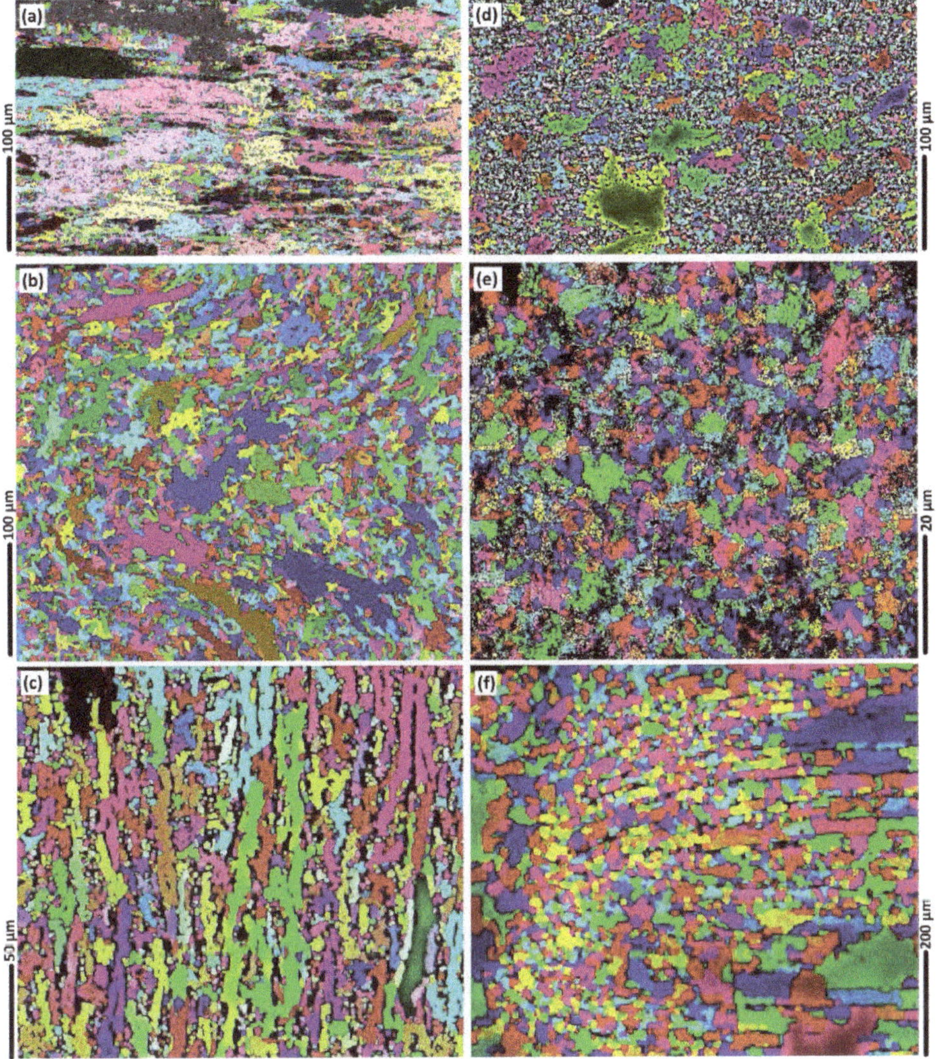

Figure 12. The Euler map micropatterns of the presence of HAGBs with post processing using a 15° angle, for different regions of the weld; BM (**a**); AS Hourglass border (**b**); Flow-arms (**c**); Sub-shoulder region (**d**); Mid-SZ (**e**); RS Hourglass border (**f**).

4. Discussion

4.1. Comparison between Methods

When applied to the BFSW welding case, the main benefit of optical microscopy (with etchant) is that it shows the size of the grains, i.e., the morphology. It does so at relatively low cost. It should be noted that the optical microscopy micrographs delineate the grains by the same colour without any information regarding the distribution of the crystallographic orientation or lattice structure. Moreover, the AA6082-T6 is difficult to obtain sufficient contrast in OM, and depends on suitable reagents and etching procedures.

A better understanding of weld texture arises from application of EBSD, which shows the grain orientation. This is useful because it shows that adjacent grains often have very different orientations. This is attributed to the stirring action. The EBSD map can show the grain orientation distribution of the sample, illustrating the morphology and size of grains with a precise accuracy constructed with the spatial digital pixel derivation. Based on the EBSD mapping analysis, the neighbouring pixels with similar crystallographic orientations are representative of a grain region, identifying by the same colour.

The study verifies that the method of optical microscopy with etchant gives results that are consistent with EBSD, though naturally the type of information available is different. There are advantages and disadvantages to the two methods (see Table 3).

Table 3. Comparative details of the metallographic measurements performed for the bobbin friction stir welding (BFSW) A6082-T6 weld structure by optical microscopy (OM) and EBSD.

Metallographic Measurement	Pros	Cons
Optical microscopy (with etchant)	Grain boundaries visible (but orientation not)	Precipitation not evident
EBSD	Crystal orientation visible. Misorientation between grains is evident	Precipitation not evident at this level of magnification
Combination of both methods	Characterised microscopic features of the BFSW weld by OM, was validated by EBSD. Further details of the shear texture in different regions of the weld were evaluated by EBSD	Due to repolishing, the measurements are time-consuming and it is not possible to repeat the exact position of the microscopic features

Another powerful instrument for characterisation of the DRX detail and grain growth structures within the weld region is the transmission electron microscope (TEM) [32]. Specifically, TEM also enables detection of precipitation phenomena, often intermixed with the dislocation network and the grain boundaries structure [33,34]. This might elucidate the solid-state plastic flow mechanism in the BFSW processed structure of the AA6082-T6 in correlation with the microstructural effect of the T6 temper. This is left for future research, as a TEM study is beyond the scope of this paper.

4.2. *Microstructure of Welded AA6086-T6*

In this work, we present a metallographic measurement (by OM and EBSD methods) to elucidate some microstructural performance of the AA6086-T6 weld texture, dominant by the thermomechanical nature of the BFSW process.

4.2.1. Shear Bands

From the EBSD colour orientation mapping, it can be concluded that the (100) red and {111} <101> turquoise are the dominant orientations at the hourglass borders, indicated by a layered/fibrous texture.

After the severe shearing, the polycrystalline lattice transforms to the zero-distortion planes by the displacement. The most plausible shear plane with the minimum stacking fault energy in FCC lattice is (100)//WD. Moreover, during the DRX, the slip system of {110} <111>//TD activates the rearrangement of orientation which is revealed as turquoise colour. This is a diffusive transformation to stabilise the interface energy between the shear texture of (100) plane with the shear strained region at its immediate proximity to obtain a coherent interface.

The grain orientation maps of the BFSW sample confirmed that the shear texture at the AS interface of the weld region indicates a more abrupt transition compared to the RS. The occurrence of the discontinuity defects (e.g., tunnel void) occurs at the AS, rather than the RS. Based on this observation, our interpretation is that the texture characteristics in BSW weld correlate with a location-dependency microscopic evolution, where the shear-induced locations (due to the rotation and advancement of the tool) can increase the potential of failure within the weld structure.

The relative motion between the directions of the rotation and advancement of the tool are the same at the AS, which results in an elevated rate of flow (directed forward) and hence plastic deformation. On the other hand, because of the opposite directions of the rotation and advancement of the tool at the RS, the stirred mass is compacted at the trailing edge of the tool, leading to the mixing of the layers (leading to lower level of distinction between the layers in metallographic observations), and refilling of the possible discontinuities (no void formation observed in the RS).

The elongated-shape texture at the hourglass border was studied in TD-ND plane of the EBSD plots. However, because of the dependency of this texture to the (100) shear plane, it needs to be further studied in ND-WD plane through the crystal direction aligned to both ND and RD. These directions may better identify the shear at the circumferential of the tool for the deposited layers along the AS.

The crystallographic nature of these shear band textures may be worth further study. The thermomechanical behaviour of the BFSW and semi-solid plastic deformation are possible explanations for the orientation evolution of the shear texture at the hourglass border interface.

The (100) shear direction in the FCC lattice can be activated when the uniaxial hot deformation compression happens to the texture. The microscopic observations suggest that the high mobility of the grain-boundaries during the deformation results in formation of elongated grains within the (100) shear texture.

In the growth competition between different crystallographic directions, the (100) planes provide a preferential axis for the faster grain-boundary migration. Hence the red (001) material is frequently surrounded by turquoise material {111} <110>. Turquoise is also sheared material, but in the orthogonal orientation. This indicates that there is an element of secondary alignment of grains alongside the main (red) shear bands. Some possible interpretations are that the turquoise material represents material that is compressed between the shear bands. There are also pink and yellow regions, but blue and green are sparse.

During the recovery and DRX after the process, it is assumed that the stored strain and the temperature can provide the driving force to activate other crystallographic systems to change the texture. However, because the elongated grains from the (100) system were formed already during the mechanical stirring process, the similar structures in their proximity has more preference. In the case of FCC metals, the crystallographic system of {111} <110> is a stable orientation for the recovery of the plane-strain compression deformed texture during the recrystallisation. This is the turquoise colour regions in EBSD plots, which is situated between the (111) and (101). In OM these shear bands are evident as dark lines. These lines comprise layers that are sheared and stacked in directions aligned with the shear flow changes.

Shear causes the grains to be compacted (as evidenced in the red (001) orientation). By definition this means fewer stacking faults, and fewer dislocations. The EBSD shows that the shear effect is layered—there is no uniform shear across the section, though there is similarity of orientation between adjacent sheared areas. This has potentially significant implications for the modelling of internal flows via computational fluid dynamics (CFD) modelling, where it is assumed that the flow is locally consistent. In the BFSW samples under examination, the shear occurs in discrete layers. This may have something to do with the transport of material around the pin, in a type of packet flow.

4.2.2. Internal Flow

The homogeneity of the grain structure and the random distributed grain refinement took placed in the SZ (as the middle of the weld region) represent a simultaneous grain fragmentation and severe plastic deformation during the FSW process and the subsequent DRX taking place by the re-cooling, respectively.

The microscopic comparison between the sharp tapered-shape boundary at the AS and the ellipse-shape boundary at the RS may represent a difference in rate of the torsion for the two zones.

This can be also related to the different directions of the tool motion and the mass transportation, by which causing different strain rates and the subsequent shearing patterns. Furthermore, different temperature gradient in the AS and RS can directly influence the DRX mechanism subjected to the frictional heating.

The elongated grains as the flow-arm region was characterised as a periodic deformed structure caused by both deformation and thermal load induced by the pin-driven stirring action. Additionally, underneath the shoulder, a continuous plastic flow appears which forms the sub-shoulder region. The microstructure in the sub-shoulder region is slightly coarsened compared to the internal parts of the SZ. This may be related to the cooling conditions experienced by the surface of the workpiece in free contact with the air as the coolant.

The triple flow junction at the middle of the hourglass-border is attributed to the two different characteristics of the flow there. The middle part of the weld region experiences severe thermo-plastic flow regime driven by the stirring action of the pin, which appears to provide two flow path ways towards the upper and lower parts of the cross-section.

4.3. Implications: Towards an Interpretation of the Interaction between Physical Metallurgy and Flow

Based on the theory of FSW, the alteration of the texture in different region of the weld (SZ, HAZ, TMAZ) was expected. However, the present work elaborates on the structures and DRX mechanisms for the BFSW case. Shear bands features were identified in different microscopic details through the hourglass-borders and the sub-shoulder region. These can be explained by the induced strain and thermal history.

In general, the microscopic observations confirmed that the deformation induced by the rotating BFSW tool is typically in the form of shear. However, due to the physical rotation of the tool, the shear texture shows different local orientation throughout the weld texture. Moreover, it should be noted that the shear flow layers within the weld are affected by the depth of the material as the position of the pin and shoulders can induce different local shearing modes. The shear layers are not necessarily aligned with the weld direction. This can be due to the rotating nature of the tool. For example, even in the middle of the SZ, the equiaxed grain texture shows the minimum of anisotropy within the sample, representative of the uniform distribution of the shear. For the processing of the aluminium alloys, the EBSD texture observations should be compared to the possible shear planes reported in the literature for the face-cantered cubic (FCC) crystallographic unit: red for (001).

The absence of the precipitate phase in the microstructure observations for both of OM and EBSD results implies that the BFSW welding happened without precipitation. However, this is not usual for a dynamic recrystallised texture that experienced a severe plastic deformation accompanied with the frictional heat generation. Possibly this may be related to the scanning mode of the microscope, and hence this may need to be checked in future studies.

The formation of the shear bands is related to the details of the mechanical stirring action. Our explanation is as follows. By the contact between the rotating tool and the workpiece, the primary grains located at the advancing edge of the tool undergo fragmentation and subsequent plasticising by the frictional heat and deformation. The simultaneous rotation and advancement of the tool squeeze the softened grains to the trailing edge of the tool. This causes a bending and elongation yielding through the morphology of the grains which are deposited at the AS and RS sides of the tool, where the shear is maximum. However, at the RS because of the opposite directions of the rotation and translation, the flow layers are mixing together, and the elongated grain structure is less clear. Therefore, the hourglass interface between the TMAZ and the SZ on the AS is more observable rather than the RS. In other words, the plastic flow on the AS swept in a smooth narrow shear band at the proximity of the pin, while the irregular transformation of the flow at the hourglass-interface on the RS cause a kind of mixing of the shear-bands. At the AS-RS region behind the tool, the stirring action breaks up the transformed grains into the ultrafine equiaxed grains.

The coarsened and elongated grain textures at the AS and RS compared to the mid-SZ equiaxed grains can be attributed to the multi-directional strain-induced flow fields, caused by the top and bottom shoulders. While in mid-SZ region, pin-induced shear appears to dominate the plastic flow, and the subsequent stored-strain becomes the driving force of the DRX. Therefore, the fully DRX in mid-SZ produce an ultrafine grain structure compared to the AS and RS interface regions, because of a higher rate of stored-strain driven by the direct action of the stirring pin. Moreover, at the sub-shoulder region, the ultrafine DRX grains are intensified by the severe frictional effect of the shoulder and the subsequent thermal dissipation due to the cooling by the free air surrounding the weld surface. Therefore, the sub-shoulder region shows a different flow pattern compared to the mid-SZ region and the hourglass-borders.

Based on the ultrafine equiaxed grain structure, it is proposed that the mid-SZ grains have undergone a severe DRX. In contrast the elongated grains in the transition region at the hourglass interface of the weld region indicate a different morphology implying a local deformation and recovery with partial DRX.

The literature explains that the stirring action is directly responsible of the formation of the shear texture. In the case of BFSW the shear deformation around the pin is non-uniform with a thermal and strain gradient across the weld. Hence in BFSW welds there is an inherent flow mechanism (induced from the pin and shoulder features), which transports the stirred mass around the weld in discrete packets. This induced shear, which was followed by a DRX process (partial or full depending on the shear), which led to the observed polycrystalline structures (and grain boundaries).

Regarding the texture evolution it should be noted that beside the shearing deformation, the frictional heat input (generated by the interaction between the tool-workpiece) has a key role in microstructural alteration during DRX. Similar to the induced strain, the variation in thermal flux from the position of the pin (mid-SZ) to the top and bottom shoulders, causes local heating on the deformed material. Beside the stored strain, this heat input contributes to accomplish a fully-DRX through the mid-SZ. Other regions of the FSW weld, such as hourglass-borders, TMAZ and HAZ, also are affected by heat, but to a lesser extent. Hence the thermal contribution to the DRX varies according to the location in the weld. Temperature also has a direct effect in formation of LAGBs/HAGBs and precipitation during DRX in different regions of the weld.

Welding process settings (workpiece thickness, tool geometry, welding speeds) are expected to affect the microscopic details of the weld texture. Increase in the thickness of the workpiece may reduce the generated heat and shrink the thermal flux, hence decreasing the DRX process. However, this might possibly be compensated by more complexity in tool geometry to induce more flow through the stirring action. Similarly, increase in welding speeds (ω, V) is expected to increase the frictional heat at the position of the tool-material, resulting in a more severe plastic deformation. Therefore, the DRX mechanism can potentially be intensified by suitable optimisation of the welding parameters. This might be worth investigating via microscopic observation to identify the optimum settings.

4.4. Future Work

The 6-series aluminium alloys (Al-Mg-Si) such as AA6082-T6 as here, are generally problematic in FSW processing. It is already known that precipitation occurs during the process [7]. The present study identified that the texture includes shear bands with fibrous morphology in the grain orientation. One of the possible reasons can be work hardening. This is a plausible mechanism based on our observation of the precipitation in Euler maps. However, the characterisation ideally needs to be developed by TEM and in higher magnification to identify the origins of the phenomenon and elucidate the origins of the precipitation mechanism. Hence, we suggest that future work could consider investigating work hardening and anisotropic behaviour of the texture during stirring and DRX.

5. Conclusions

The study verifies that the method of optical microscopy of AA6082-T6 BFSW weld with etchant per [5] gives results that are consistent with EBSD, though naturally the type of information available is different. The EBSD results, including grain orientation mapping and grain boundary mapping, reveal more details of the weld texture. The EBSD colour mapping demonstrated the grain misorientation in different regions of the weld, caused by the shear during stirring action and the subsequent dynamic recrystallisation (DRX). The grain boundary mapping, based on the Taylor factor, distinguished low-angle grain boundaries (LAGBs) and high-angle grain boundaries (HAGBs) through the weld texture.

It was evident that different grain orientations were formed by the shear strain induced by the tool performance in AS and RS of the weld. The hourglass-border as a direct outcome of the shearing was formed as the aligned bands at the AS, while the complex action of the tool caused the mixing of shear bands at the RS.

The grain structures that occur in AA6082-T6 BFSW welds are due to a complex set of factors. The motion of the pin and shoulder features transports material around the weld, which induces shear. The shear deformation around the pin is non-uniform with a thermal and strain gradient across the weld, and hence the DRX processes are also variable, giving a range of polycrystalline and grain boundary structures.

This micro-pattern is interpreted as DRX features of the FSW process whereby the stored strain induced by plastic deformation affects the grain structure of the TMAZ by grain misorientation. This can be detected as the microscopic change in etching response in the optical microscopy (Figure 8a), and the grain misorientation alteration in the EBSD colour mapping.

This paper makes the original contribution of identifying the content of the flow lines that are evident in the macro-structure. These lines are evident in optical microscopy, but the images in the literature are typically diffuse due to the difficult of etching this material. As shown here, with suitable etchants OM shows the grain boundary features more clearly. More accuracy is evident with EBSD, which also avoids risks of introducing artefacts due to over-etching. The results show the OM and EBSD results are consistent. The overall interpretation is that the welded region contains fine discrete layers of material, of approximately 10 μm thickness. In these the shear has caused dynamic recrystallisation, which has resulted in a string of grains with similar orientation. When etched and observed with OM, these appear as dark and light lines.

Furthermore, these shear bands are found to have different characteristics across the weld section. In turn this is attributed to the different flow and thermal regimes in these locations. A homogeneous continuous fluid would be expected to have a smooth shear field. The actual observer shear occurs in discrete bands. This, plus the abrupt changes in the lines, suggests that the internal flow regime is more complex than might be expected.

Author Contributions: Methodology and Formal analysis, A.T., D.J.P.; Supervision, D.J.P., D.C., K.S.; Validation, A.T., D.J.P.; Writing—original draft, A.T.; Writing—review and editing, A.T., D.J.P.

Funding: This research received no external funding.

Acknowledgments: Thanks are extended to Mike Flaws and Kevin Stobbs for assistance with electron and optical microscopy, respectively.

Conflicts of Interest: The authors declare no conflict of interest.

References

1. Thomas, W.; Nicholas, E.; Needham, J.; Murch, M.; Temple-Smith, P.; Dawes, C. Friction Stir Butt Welding. GB Patent Application No 9125978.8, 6 December 1991.
2. Thomas, W.; Nicholas, E. Friction stir welding for the transportation industries. *Mater. Des.* **1997**, *18*, 269–273. [CrossRef]

3. Threadgill, P.; Leonard, A.; Shercliff, H.; Withers, P. Friction stir welding of aluminium alloys. *Int. Mater. Rev.* **2009**, *54*, 49–93. [CrossRef]
4. Sued, M.; Tamadon, A.; Pons, D. Material flow visualization in bobbin friction stir welding by analogue model. In Proceedings of the Mechanical Engineering Research Day 2017, Melaka, Malaysia, 30 March 2017; Volume 2017, pp. 1–2.
5. Tamadon, A.; Pons, D.; Sued, K.; Clucas, D. Development of metallographic etchants for the microstructure evolution of a6082-t6 bfsw welds. *Metals* **2017**, *7*, 423. [CrossRef]
6. Tamadon, A.; Pons, D.; Sued, K.; Clucas, D. Formation mechanisms for entry and exit defects in bobbin friction stir welding. *Metals* **2018**, *8*, 33. [CrossRef]
7. Tamadon, A.; Pons, D.; Sued, K.; Clucas, D. Thermomechanical grain refinement in aa6082-t6 thin plates under bobbin friction stir welding. *Metals* **2018**, *8*, 375. [CrossRef]
8. Sued, M.K. Fixed Bobbin Friction Stir Welding of Marine Grade Aluminium. Ph.D. Thesis, University of Canterbury, Christchurch, New Zealand, 2015.
9. Avallone, E.A.; Baumeister III, T. *Marks' Standard Handbook for Mechanical Engineers*; McGraw-Hill Education: New York, NY, USA, 1986.
10. Aginagalde, A.; Gomez, X.; Galdos, L.; García, C. Heat treatment selection and forming strategies for 6082 aluminum alloy. *J. Eng. Mater. Technol.* **2009**, *131*, 044501. [CrossRef]
11. Mohamed, A.; Samuel, F. A review on the heat treatment of al-si-cu/mg casting alloys. In *Heat Treatment–Conventional and Novel Applications*; IntechOpen: London, UK, 2012; pp. 55–72.
12. Sued, M.; Pons, D.; Lavroff, J.; Wong, E.-H. Design features for bobbin friction stir welding tools: Development of a conceptual model linking the underlying physics to the production process. *Mater. Des.* **2014**, *54*, 632–643. [CrossRef]
13. Davies, P.; Wynne, B.; Rainforth, W.; Thomas, M.; Threadgill, P. Development of microstructure and crystallographic texture during stationary shoulder friction stir welding of ti-6al-4v. *Metall. Mater. Trans. A* **2011**, *42*, 2278–2289. [CrossRef]
14. Tayon, W.A.; Domack, M.S.; Hoffman, E.K.; Hales, S.J. Texture evolution within the thermomechanically affected zone of an al-li alloy 2195 friction stir weld. *Metall. Mater. Trans. A* **2013**, *44*, 4906–4913. [CrossRef]
15. Davis, J.R. *Aluminum and Aluminum Alloys*; ASM International: West Conshohocken, PA, USA, 1993.
16. Fonda, R.; Bingert, J.; Colligan, K. Development of grain structure during friction stir welding. *Scr. Mater.* **2004**, *51*, 243–248. [CrossRef]
17. Fonda, R.; Bingert, J. Texture variations in an aluminum friction stir weld. *Scr. Mater.* **2007**, *57*, 1052–1055. [CrossRef]
18. Fonda, R.; Knipling, K. Texture development in friction stir welds. *Sci. Technol. Weld. Join.* **2011**, *16*, 288–294. [CrossRef]
19. Fonda, R.; Knipling, K.; Bingert, J. Microstructural evolution ahead of the tool in aluminum friction stir welds. *Scr. Mater.* **2008**, *58*, 343–348. [CrossRef]
20. Fonda, R.; Reynolds, A.; Feng, C.; Knipling, K.; Rowenhorst, D. Material flow in friction stir welds. *Metall. Mater. Trans. A* **2013**, *44*, 337–344. [CrossRef]
21. Prangnell, P.; Heason, C. Grain structure formation during friction stir welding observed by the 'stop action technique'. *Acta Mater.* **2005**, *53*, 3179–3192. [CrossRef]
22. Coelho, R.S.; Kostka, A.; Dos Santos, J.; Pyzalla, A.R. Ebsd technique visualization of material flow in aluminum to steel friction-stir dissimilar welding. *Adv. Eng. Mater.* **2008**, *10*, 1127–1133. [CrossRef]
23. Hilgert, J.; Schmidt, H.; Dos Santos, J.; Huber, N. Thermal models for bobbin tool friction stir welding. *J. Mater. Process. Technol.* **2011**, *211*, 197–204. [CrossRef]
24. Hilgert, J.; Hütsch, L.L.; dos Santos, J.; Huber, N. Material Flow Around a Bobbin Tool for Friction Stir Welding. In Proceedings of the COMSOL Conference, Paris, France, 17–19 November 2010.
25. Hilgert, J.; Dos Santos, J.; Huber, N. Shear layer modelling for bobbin tool friction stir welding. *Sci. Technol. Weld. Join.* **2012**, *17*, 454–459. [CrossRef]
26. Tamadon, A.; Pons, D.; Sued, M.; Clucas, D.; Wong, E. Preparation of plasticine material for analogue modelling. In Proceedings of the International Conference on Innovative Design and Manufacturing (ICIDM2016), Auckland, New Zealand, 24–26 January 2016.
27. Beardsley, A.; Bishop, C.; Kral, M. Ebsd characterization of pilgered alloy 800 h after heat treatment. *Mater. Perform. Charact.* **2016**, *5*, 717–739.

28. Jackson, M.A.; Groeber, M.A.; Uchic, M.D.; Rowenhorst, D.J.; De Graef, M. H5ebsd: An archival data format for electron back-scatter diffraction data sets. *Integr. Mater. Manuf. Innov.* **2014**, *3*, 44–55. [CrossRef]
29. Paul, H.; Driver, J.; Tarasek, A.; Wajda, W.; Miszczyk, M. Mechanism of macroscopic shear band formation in plane strain compressed fine-grained aluminium. *Mater. Sci. Eng. A* **2015**, *642*, 167–180. [CrossRef]
30. Li, R.; Xie, Q.; Wang, Y.-D.; Liu, W.; Wang, M.; Wu, G.; Li, X.; Zhang, M.; Lu, Z.; Geng, C. Unraveling submicron-scale mechanical heterogeneity by three-dimensional x-ray microdiffraction. *Proc. Natl. Acad. Sci. USA* **2018**, *115*, 483–488. [CrossRef] [PubMed]
31. Jeong, H.T.; Park, S.D.; Ha, T.K. Evolution of shear texture according to shear strain ratio in rolled fcc metal sheets. *Met. Mater. Int.* **2006**, *12*, 21–26. [CrossRef]
32. Cabibbo, M.; Meccia, E.; Evangelista, E. Tem analysis of a friction stir-welded butt joint of al–si–mg alloys. *Mater. Chem. Phys.* **2003**, *81*, 289–292. [CrossRef]
33. Murr, L.; Liu, G.; McClure, J. A tem study of precipitation and related microstructures in friction-stir-welded 6061 aluminium. *J. Mater. Sci.* **1998**, *33*, 1243–1251. [CrossRef]
34. Lityńska, L.; Braun, R.; Staniek, G.; Dalle Donne, C.; Dutkiewicz, J. Tem study of the microstructure evolution in a friction stir-welded alcumgag alloy. *Mater. Chem. Phys.* **2003**, *81*, 293–295. [CrossRef]

Article

Nugget Formation and Mechanical Behaviour of Friction Stir Welds of Three Dissimilar Aluminum Alloys

Neves Manuel [1,2,*], Ivan Galvão [1,3], Rui M. Leal [1,4], José D. Costa [1] and Altino Loureiro [1]

1 CEMMPRE, Departamento de Engenharia Mecânica, Universidade de Coimbra, Rua Luís Reis Santos, 3030-788 Coimbra, Portugal; ivan.galvao@dem.uc.pt (I.G.); rui.leal@dem.uc.pt (R.M.L.); jose.domingos@dem.uc.pt (J.D.C.); altino.loureiro@dem.uc.pt (A.L.)
2 Escola Superior Politécnica do Namibe, Rua Amílcar Cabral, Moçâmedes 201, Angola
3 ISEL, Departamento de Engenharia Mecânica, Instituto Politécnico de Lisboa, Rua Conselheiro Emídio Navarro 1, 1959-007 Lisboa, Portugal
4 LIDA-ESAD.CR, Instituto Politécnico de Leiria, Rua Isidoro Inácio Alves de Carvalho, 2500-321 Caldas da Rainha, Portugal
* Correspondence: uc2013112368@student.uc.pt; Tel.: +351-239-790-700

Received: 14 May 2020; Accepted: 9 June 2020; Published: 11 June 2020

Abstract: The aim of this research was to investigate the influence of the properties of the base materials and welding speed on the morphology and mechanical behavior of the friction stir welds of three dissimilar aluminum alloys in a T-joint configuration. The base materials were the AA2017-T4, AA5083-H111, and AA6082-T6 alloys in 3 mm-thick sheets. The AA6082-T6 alloy was the stringer, and the other alloys were located either on the advancing or retreating sides of the skin. All the T-joint welds were produced with a constant tool rotation speed but with different welding speeds. The microstructures of the welds were analyzed using optical microscopy, scanning electron microscopy with energy dispersive spectroscopy, and the electron backscatter diffraction technique. The mechanical properties were assessed according to micro-hardness, tensile, and fatigue testing. Good quality welds of the three dissimilar aluminum alloys could be achieved with friction stir welding, but a high ratio between the tool's rotational and traverse speeds was required. The welding speed influenced the weld morphology and fatigue strength. The positioning of the skin materials influenced the nugget morphology and the mechanical behavior of the joints. The joints in which the AA2017 alloy was positioned on the advancing side presented the best tensile properties and fatigue strength.

Keywords: friction stir welding; three dissimilar aluminum alloys; welding speed; T-joints; microstructure; mechanical properties

---

1. Introduction

Aluminum alloys of the 5xxx, 6xxx, and 2xxx series are widely used in various industrial sectors, such as shipbuilding, aerospace, building structures, bridge structures, and even military vehicles, due to their lightness, mechanical strength, and resistance to corrosion [1,2]. T-joints are currently used in these industries to increase the stiffness of thin plates.

Friction stir welding (FSW) is a solid-state welding process that prevents porosity and cracking, but the formation of defects is influenced by the flow of materials around a tool. The flow of material in this zone depends on the tool's geometry, the tool's rotation speed and welding speed, the tool's axial force or displacement, and even the tool's tilt angle, whether in similar or dissimilar material welds [3,4]. Fratini et al. [5], when comparing similar welds for the AA6082-T6 and AA2024-T4 alloys in a T-joint configuration, concluded that the temperature fields and strain rate influence the flow of a material in the stir zone and, hence, the integrity of welds.

In welds between dissimilar materials of different families, the properties of the base materials have to be considered because of the formation of intermetallic compounds that have very different physical and mechanical properties from the base materials and influence the flow of materials in the stir zone [6,7]. However, even for dissimilar welds of materials within the same family, the influence of the properties of the materials on the formation of the weld can be significant. Silva et al. [8] stated that in the dissimilar butt welds of AA7075-T6 to AA2024-T3, the mixing of the two materials in the stir zone was greatly influenced by the geometry and rotational speed of the tool, only obtaining satisfactory mixing at high rotational speeds.

Dinaharan et al. [9] stated that the material that is on the advancing side occupies most of the stir zone and influences the strength of butt weld in aluminum 6061 in rolled and cast plates. Barbini et al. [10] showed that, for butt welds between AA2024-T3 and AA7050-T7651, a better material flow in the stir zone is obtained when the AA2024-T3 is located on the advancing side. Better material flow in the stir zone was also observed for butt welds between AA2024-T6 and AA6061-T6, when the latter material was placed on the advancing side [11]. The results mentioned above suggest that the flow of materials in the stir zone depends on the mechanical properties of the materials on the advancing and retreating sides. Manuel et al. [12] also stated that for T-joints between AA6082 and AA5083, material flow and defect formation are greatly influenced by the joint type, tool geometry, and process parameters, as well as the materials' properties. The investigations found in the literature about three dissimilar aluminum alloys only concern lap joints [13,14].

The FSW of three dissimilar aluminum alloys has scientific and industrial interest, since there is a need for new combinations of materials and there is no clear understanding of how material properties influence weld formation. The aim of this study was to analyze the influence of the properties of the base materials and the variation of the welding speed on the morphology and mechanical properties of welds of three dissimilar aluminum alloys arranged in a T-joint configuration.

2. Materials and Methods

The experimental tests were performed on 3 mm-thick sheets of AA2017-T4, AA5083-H111, and AA6082-T6 aluminum alloys. Alloys that require plastic deformation during construction are used without much hardening, as in the case the alloys 5083-H111 and 2017-T4, while reinforcements are made of hardened alloys, as is the case for 6082-T6. The chemical composition and mechanical properties of these alloys are listed in Tables 1 and 2, respectively.

Table 1. Chemical composition of the base materials (wt.%).

Alloy	Cu	Mg	Mn	Si	Cr	Others
AA2017-T4	4.5	0.8	0.7	0.8	≤0.1	Bal.
AA5083-H111	0.025	4.5	0.57	0.09	0.25	Bal.
AA6082-T6	0.09	0.7	1.0	0.53	<0.25	Bal.

Table 2. Mechanical properties of the tested aluminum alloys.

Properties	AA2017-T4	AA5083-H111	AA6082-T6
Ultimate Tensile Strength (MPa)	416.8 ± 3	317.5 ± 5.8	344.6 ± 5.7
Tensile Yield Strength (MPa)	293 ± 9.2	145 ± 4.1	286 ± 13.4
Elongation at Break (%)	18 ± 3.9	22.7 ± 1.4	18.4 ± 1.6
Vickers Hardness ($HV_{0.2}$)	116.7 ± 4.2	82.3 ± 1.4	115 ± 0.8

A tool of H13 quenched and tempered steel, composed of a progressive (cylindrical and conical) threaded pin of 5.2 mm in length and a shoulder of 18 mm in diameter with a concavity of 5°

(see Figure 1a) was used. Previous tests have shown that this tool geometry has obtained defect-free welds in dissimilar aluminum alloys [12].

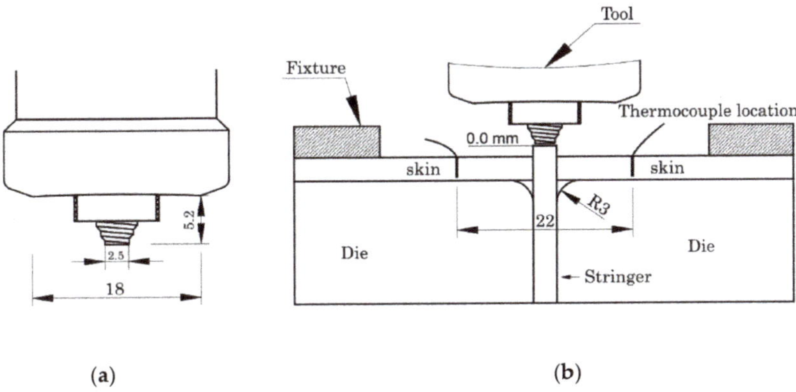

(a) (b)

Figure 1. (a) Geometry of the tool; (b) welding setup and starting point to measure tool penetration.

The setup used to perform the welds is shown in Figure 1b. The stringer protruded 1.4 mm in order to provide enough material to fill the empty volumes between plates and dies in the fillets. The AA6082 alloy was the stringer for all the weld series. The AA5083 and AA2017 alloys were positioned on the advancing and retreating sides of the skin, respectively, in some series—they were designated as 562 and the opposite in another series, designated as 265. The dimensions of the stringer and skin plates were 330 × 37.4 × 3 and 330 × 80 × 3 mm, respectively. The oxides were removed from the interfaces by sanding, and the plates were cleaned with alcohol just before the welding.

The welds were performed in position control using a Cincinnati Milacron 207 Mk milling machine, and the tool's rotational speed (w-500 rpm), plunge depth (7.1 mm), and tilt angle (3°) were maintained constant for all series. The tool's welding speed (v) was changed according to Table 3. The weld series designation consisted of the designation of the sequence of the materials, followed by the welding speed. The parameters were chosen based on previous experience. The (w/v) ratio is also indicated in Table 3 due to its relationship with the heat input, which increases with the ratio [15].

Table 3. Welding parameters used to make the three dissimilar T-joints.

Material Position	Series	V (mm/min)	w/v (r/mm)
562	562-30	30	16.7
	562-120	120	4.2
	562-280	280	1.8
265	265-30	30	16.7
	265-120	120	4.2
	265-230	230	2.2

During welding, the thermal cycles were measured by k-type thermocouples embedded in small holes close to the shoulder's trajectory, on the advancing and retreating sides; see Figure 1b. A data translation device with an acquisition rate of 75 Hz and cold junction compensation was used to record the thermal cycles.

After welding, 63 × 25 × 25 mm specimens were transversely removed to the welding direction of all the series, ground down with sandpaper P2500, and then polished using 3 and 1 μm diamond suspensions. Modified Keller's reagent was effective in revealing the grain boundary of AA2017 and Weck's reagent was effective for such for AA6082, but no effective reagent was found for AA5083. The grain size was determined by the Heyn intercept method. A Leica DM4000M LED optical microscope was used to analyze the samples. In order to better identify which of the alloys was present in each zone of the nugget, semi-quantitative chemical analyses were performed by scanning electron microscopy/energy dispersive spectroscopy (Zeiss, MERLIN, Field Emission Scanning Electron Microscope-Gemini II/Oxford Instruments, X-MAXN) at 10 kV, bearing in mind that the AA2017 alloy was Cu-rich, the AA5083 alloy was Mg-rich and the AA6082 alloy had intermediate contents of Mg and Si. The samples for electron backscatter diffraction (EBSD) analysis were grinded and polished with 3 and 1 μm diamond suspensions and finished with colloidal silica. Finally, an electrolytic polishing was performed by using a solution of nitric acid and methanol at −10 °C and 15–20 V for 30 s. EBSD was performed with an FEI QUANTA 400F scanning electron microscope provided with an EBSD detector system using a scan step size of 0.65 μm.

The Vickers microhardness profiles of the welds were determined using an HMV-G SHIMADZU tester on the weld's cross-section, using 200 g for 15 s. The distance between the test points was 0.5 mm in the nugget zone and 1 mm in the heat-affected zone (HAZ) and base materials.

The base materials were tensile tested at room temperature and at high temperatures (320 and 450 °C) using loading speeds of 2 and 72 mm/min, respectively on an Instron 4206 machine provided with a three-stage oven. Three specimens were tested for each base material.

The welded specimens for the tensile and fatigue tests were cut transversely to the direction of the weld with the dimensions of 180 × 20 mm (length × width). The shape and size of the fatigue specimens are shown in Figure 2. The edges of the specimens were rounded and polished to avoid a concentration of surface stresses and the initiation of cracks. The tensile tests were performed in accordance with the ASTM E8 standard for testing metallic materials [16], and the loads were applied in the direction of the skin. Three tensile specimens were tested for each weld series. The local strain fields were recorded with an ARAMIS 3D 5 M optical extensometer from GOM GmbH with digital image correlation (DIC). This is a real time system of measurement of 3D surface strains, based on triangulation, that employs image registration and the tracking of changes in images over time. The specimens were prepared by applying a random black speckle pattern over the previously mat white-painted side surface.

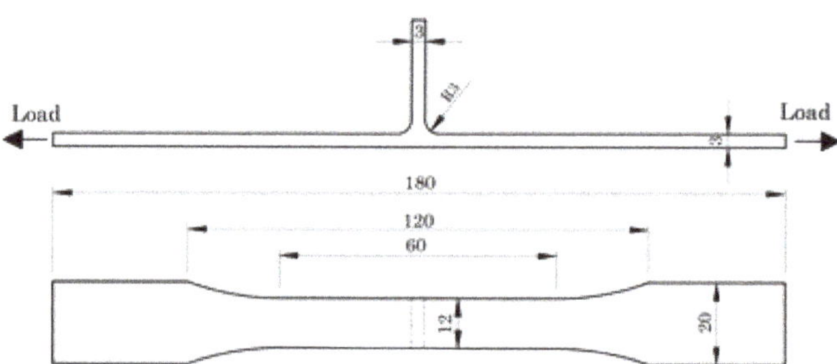

Figure 2. Fatigue specimen—shape and size.

The fatigue tests were carried out using an Instron servo-hydraulic machine coupled to an Instron Fast Track 8800 acquisition and control system. The range of stresses varied between 150 and 200 MPa with a frequency of 15–25 Hz; the frequency decreased as the maximum load applied increased, and the stress ratio was set to 0.02. Two test pieces were used for each test condition. In cases where there was greater dispersion, a third trial was carried out. The fracture surface of the fatigue specimens was studied using a Zeiss, MERLIN, field emission scanning electron microscope.

3. Results and Discussion

3.1. Thermal Cycles in the Welds

The formation of defects is currently attributed to insufficient heat generation in the weld, inadequate material flow around the pin, and the insufficient consolidation of the deformed material at the back of the pin, all of which are factors controlled by welding parameters [17]. For this reason, the influence of some process parameters on the thermal cycles induced in the welds was analyzed.

Figure 3 shows the thermal cycles measured on the advancing side for the 562 series. As the welding speed of the series increased, the peak temperature and the cooling cycle time decreased due to the lower ratios (*w/v*); see Table 3.

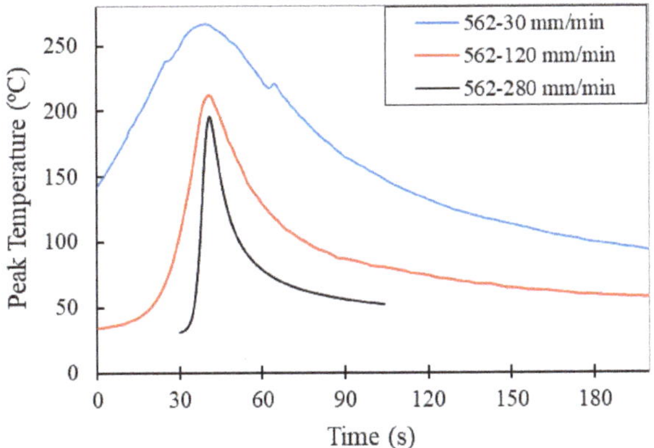

Figure 3. The thermal cycles measured in the 562 series, as performed using different welding speeds.

Though these thermal cycles were recorded far from the center of the nugget (11 mm), the curves illustrate the influence of the welding speed on the heat input during the process. On the other hand, the maximum temperature reached in the welds was higher on the advancing side than on the retreating side by about 27 °C, as illustrated in Figure 4a. This was due to the asymmetry of heat generation around the tool, as has been shown by other researchers [18,19]. The induced thermal cycle was also greatly influenced by the materials that were located on the advancing and retreating sides. Figure 4b shows that the measured peak temperature was higher when AA2017 was on the advancing side than when it was on the retreating side. This means that there was a larger energy consumption to plastically deform this material than AA5083, so the material position had a more significant effect on the heat generated than the asymmetry of the process.

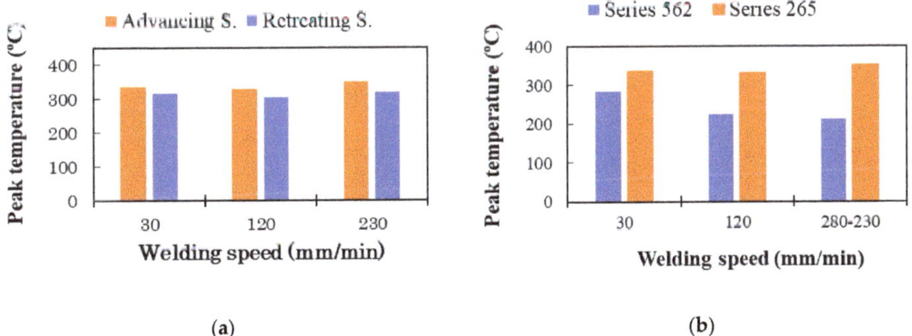

Figure 4. Peak temperature measured (**a**) on the advancing and retreating sides of the weld series 265 and (**b**) on the advancing side of the 562 and 265 weld series.

Figure 4b also shows that the peak temperature observed for the 265-230 series was higher than that obtained in the 265-120 series; this was due to a slight offset of the tool closer to the thermocouple on the advancing side in the 265-230 series.

3.2. Morphology of the Welds

All welds showed good surface appearance; however, the differences mentioned in the thermal cycles should have influenced the formation of the weld nugget. Figure 5 illustrates the cross-sectional macrographs of the 562 and 265 weld series. This image shows that all the welds had well-defined radii of the fillets, which indicates the effectiveness of the welding parameters and the adopted T-joint configuration. The points where chemical analysis (EDS) was performed are marked with numbers in these macrographs, e.g., Z1, Z2, and Z3, to better aid the interpretation of the flow of the three materials in the nugget. The chemical composition of the different zones is indicated in Table 4.

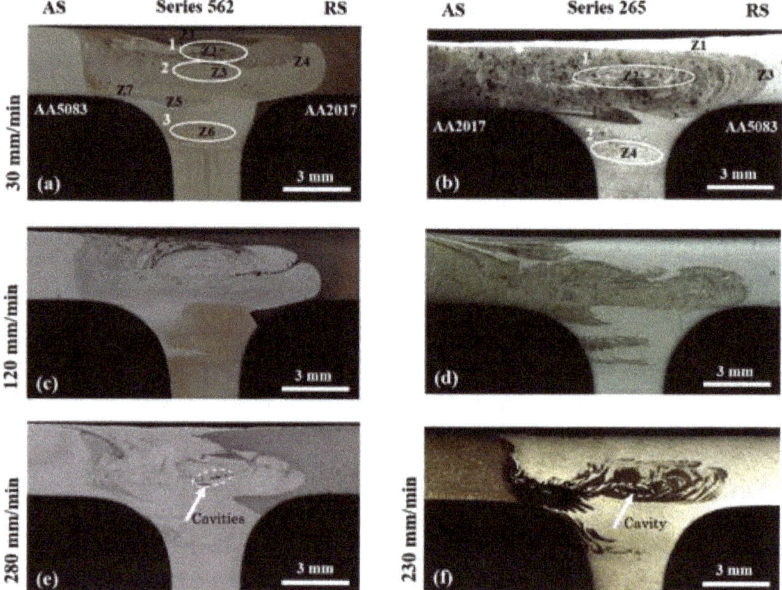

Figure 5. Cross-section macrographs of the three dissimilar materials weld series: (**a**) 562-30, (**b**) 265-30, (**c**) 562-120, (**d**) 265-120, (**e**) 562-280, and (**f**) 265-230.

Table 4. Chemical composition in the various zones of both the 562-30 and 265-30 series (wt.%).

Welds Series	Zone	Mg	Si	Cu	Material Composition
562-30	Z1	0.66	0.45	1.75	2017
	Z2	0.81	0.79	0.14	2017/5083/6082
	Z3	2.3	0.57	0.4	2017/5083/6082
	Z4	3.32	0.32	0.37	5083/2017
	Z5	4.85	...	...	5083
	Z6	0.72	0.64	0.4	2017/6082
	Z7	1.29	0.51	0.6	2017/5083/6082
265-30	Z1	0.6	0.7	0	6082
	Z2	4.3	0.6	4.7	2017/5083/6082
	Z3	1.2	...	3	2017/5083
	Z4	0.6	0.8	1.6	2017/6082

An analysis of the macrographs revealed differences in nugget morphologies that varied with increasing welding speeds, as well as with the position of the base materials on the advancing or retreating side. It was possible to observe a great asymmetry in the flow of materials in all the macrographs in relation to the welding center line in both the 562 and 265 series.

The 562-30 series featured three onion ring structures—two located in the skin and one in the stringer fillet zone (marked with ellipses 1, 2, and 3, respectively)—and other zones with different colors; see Figure 5a. Zone 1 essentially consisted of AA2017 positioned on the rear side, which was dragged by the shoulder and remained at the top of the nugget, as seen in Table 4. This table shows only some of the measurements in the 562-30 and 265-30 series in order to illustrate the main differences and reduce the size of the article. The onion ring structure just below ellipse 1—zone 2—was essentially composed of interspersed layers of the three alloys; the dark layers consisted of AA2017. The second onion ring structure indicated by ellipse 2—zone 3—had interleaved layers composed of the three alloys, varying in composition with the location, but with less contribution from AA2017; see Table 4. In zone 4, there was a really good contribution from AA5083. Zone 5 consisted of AA5083 coming from the advancing side. While the material flow on top was a shoulder-driven flow, the last flows were, in fact, pin-driven flows. The onion ring structure was composed of the three materials on the advancing side, as illustrated by zone 7. This showed a good mix of the three materials in the stir zone, contrary to what has been suggested for dissimilar welds [20].

Zone 6 corresponded to the stringer region next to the advancing side fillet, influenced by the action of the pin tip. There was the formation of an onion ring structure, resulting from the contribution of only AA6082 and AA2017. This showed that only one material of the skin, the AA2017, flowed downward into the fillet zone between the tool and dies, which is contrary to what was suggested by Manuel et al. [12] for dissimilar welds.

For the 265-30 series, only one onion ring was formed on the skin and another was formed on the stringer fillet zone, as indicated by ellipses 1 and 2 in Figure 5b. The top layer in this weld consisted of AA6082 (see zone 1), as opposed to the 562-30 series, which had AA2017 on the top; see Table 4. Zone 2 consisted of interspersed layers of the skin materials AA2017 and AA5083. The AA6082 alloy also probably coexisted in this zone, but it was difficult to distinguish via the chemical composition analysis. The peripheral zone of the nugget on the retreating side consisted of layers of only AA5083 and AA2017, according to zone 3. The onion ring structure in the fillet (zone 4) consisted of interleaved layers composed of AA6082 and AA2017, with no AA5083.

As the welding speed increased, in the 562-120 series, the onion ring structure on the fillet tended to disappear, although the material flow remained in this zone; see Figure 5c. In addition, the two onion ring structures in the skin tended to separate, with material from the retreating side entering between the onion rings. This is more visible in Figure 6, where the morphology of the material flow from the 562-30 and 562-120 weld series is compared at a higher magnification. Figure 6a,c compares the advancing sides and the retreating sides (Figure 6b,d).

For the 265-120 series (Figure 5d), the onion ring structure tended to fade, revealing a lack of time or an inability for an orderly flow of layered materials but with increased participation of the AA2017 alloy. This suggested that this alloy had a greater capacity for hot plastic flow and a greater need for time and/or temperature for the AA5083 alloy to flow in layers when it was located on the retreating side.

The onion ring structures disappeared in the 562-280 and 265-230 series, and the stir zone presented a more chaotic appearance with the formation of small internal cavities, either in the center or in the advancing side, as marked with arrows in Figure 5e,f. This was because the heat input decreased as the w/v ratio decreased, so less material was dragged by the tool and there was less time and temperature for the stable and periodic flow of materials (as suggested by Yoon [21]), thus causing the formation of defects.

Furthermore, it was found that the AA5083 alloy never went downwards to the stringer, regardless of its advancing or retreating side position. This showed that the formation of the nugget depended not only on the position of the materials and process parameters but also on their intrinsic ability to deform at a high temperature.

Figure 7a shows the engineering tensile stress–strain curves of the three base materials obtained at temperatures of 320 and 450 °C, as well as the variation of the yield stress with temperature;

see Figure 7b. The alloys still had very different yield stresses and tensile strengths at 320 °C, and AA2017 had the highest values. The AA2017 and AA6082 alloys showed significant work softening soon after reaching yield stress, unlike the AA5083 alloy that hardened until a strain of 5% before starting to soften. In turn, the AA5083 alloy exhibited much greater plastic strain at fracture than the AA2017 and AA6082 alloys, probably due to dynamic recovery, as suggested by Shi et al. [22]. At 450 °C, a noticeable loss of mechanical strength was observed for all the alloys, as was an obvious increase in strain at fracture. Some dissolution and coarsening of strengthening precipitates and partial recrystallization may explain the loss of strength and the increased ductility of the heat-treatable alloys (AA6082 and AA2017) [22,23]. In addition to the loss of strength, the AA5083 alloy had a large steady state flow due to dynamic recrystallization [24]. For this temperature, the yield stress of this alloy was higher than that of AA6082; see Figure 7b.

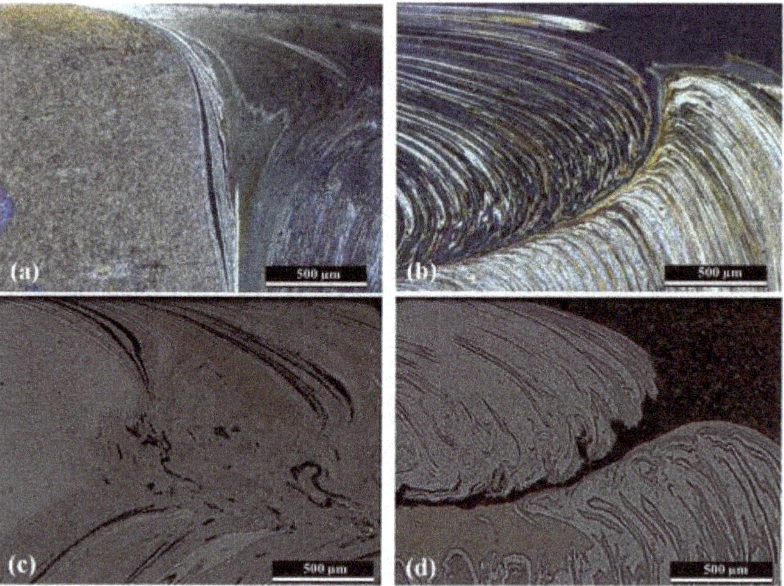

Figure 6. Comparison of material flow between the advancing and retreating sides of 562 welds series: (**a**) Advancing Side-562-30, (**b**) Retreating Side-562-30, (**c**) Advancing Side-562-120, and (**d**) Retreating Side-562-120.

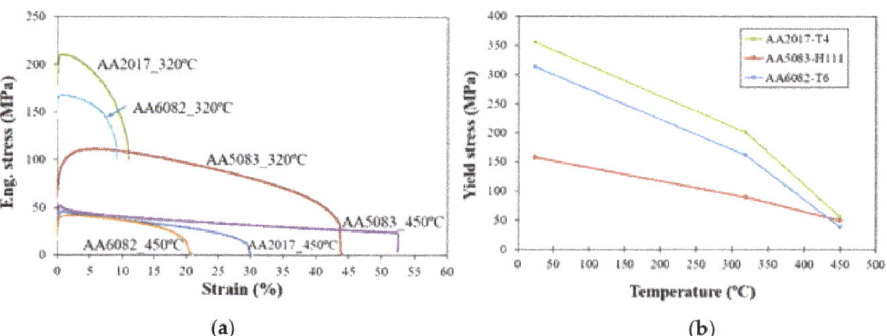

Figure 7. Variation of the mechanical properties of base materials with temperature: (**a**) Engineering tensile stress–strain curves and (**b**) yield stress curves.

The heated material was mainly exposed to compression and shear stresses in the stir zone due to its complex interaction with the tool and the colder surrounding material [25]. Though the tensile behavior of the materials at high temperatures did not match what happens in welding, it helped to understand the material flow and was easier to obtain. In welding, temperature decreases rapidly as the distance to the tool increases, so, for the AA5083 alloy, it should be difficult for the tool to drag the material in this zone because some work hardening occurs at a low temperature. Therefore, a little volume of material is dragged around the pin, which results in the poor weldability of the alloy. This is compatible with the difficulty for the AA5083 alloy to flow downwards from the skin to the stringer, which was observed in the welds presented above. The AA6082 and AA2017 alloys experienced significant temperature softening, which allowed the AA2017 to flow downwards from the skin to the stringer, as illustrated above. Similar behavior was observed by Leitão et al. [26] for the AA6082 and AA5083 alloys used in butt welds. The results presented above suggest that, for this welding process, the AA2017 and AA6082 alloys show better weldability than the AA5083.

3.3. Microstructure

The microstructures in the nugget of the welds between the three dissimilar Al alloys were very complex and difficult to uncover, especially when taking the definition of the grain boundary into consideration. The etchants for each material were different and did not always work when the materials were together. Figure 8a shows a micrograph of the AA6082 base material etched with Weck's reagent. The material was composed of grains elongated in a rolling direction with an average grain size of 59 × 26.3 µm. The distribution of each material in the nugget was very complex, as is illustrated in Figure 8b, for a 562-120 weld. This image shows, at a smaller magnification, a refined grain structure in the areas where AA6082 was present. In the surrounding areas, which were made up of the other materials, the grain boundary was not revealed with this etchant. Figure 8c,d illustrates the microstructure of the same alloy in the nuggets of the 562-30 and 265-30 weld series, respectively. When comparing these last two figures with Figure 8a, a marked refinement of the grain in the nugget is visible. The difference in the average grain size in the nugget between the 562-30 and 265-30 series was marginal, as the grain sizes were about 5.1 ± 1.5 and 5.2 ± 1.6 µm, respectively. A large standard deviation was observed, because the grain size varied with the location. According to the thermal cycles illustrated in Section 3.1, a larger difference in grain size from both weld series would be expected. A possible cause of this similarity was that most of the grain sizes were measured in the central nugget or even on the retreating side, where the grains were most visible.

Increasing the welding speed—from 30 to 280 mm/min, for instance—had a very small effect on the nugget grain size, with it remaining within the range of 4–5 µm. For AA2017, although its grain size (20.2 × 9.5 µm) was smaller than that of AA6082, the nugget grain size was also in the 4–5 µm range for all the welding series. The significant plastic deformation suffered by the materials in the stir zone and the recrystallization occurring in small and very confined areas, as seen in Figure 8b, could penalize grain growth with the ephemeral increase in heat input. The authors believe that the coexistence of the three alloys in the nugget cancelled the effect that the increase in welding speed had on the decrease in grain size, contrary to that suggested by Kalemba-Rec et al. [27]. A study by Ahmed et al. [28] found that grain size decreases with the increase in welding speed for similar welds, but the same effect does not occur for dissimilar welds.

The nuggets of both weld series were analyzed by EBSD to study the weld microstructure further, as shown in Figures 9 and 10 for the 562 and 265 welds, respectively. The regions under analysis are indicated by dots in Figures 9a and 10a to make them more discernible in the weld macrographs, although they are small vertical or horizontal scanning lines. It can be observed from the figures that the onion-ring regions of both welds tended to present a refined microstructure, which agreed well with the metallographic study, but the grain size was not uniform in the nugget. A gradient in grain structure is visible in these maps, with the grains composing the onion-ring regions presenting a smaller grain size and being more equiaxed than the grains in neighboring areas; see Figure 9b,c and Figure 10b–d. Plastically deformed grains are visible both in the onion-ring regions and the neighboring areas (see Figures 9b and 10c), which agrees well with the flow features characterized above. A further refined microstructure (2 µm) is also observed in Figure 9d, and this corresponds to a region in the stringer that was under the action of the pin tip where little heat was generated. The great microstructural heterogeneity of the weld nugget was due not only to the differences in local heat generated but also to the very complex flow features of the different alloys that made it up.

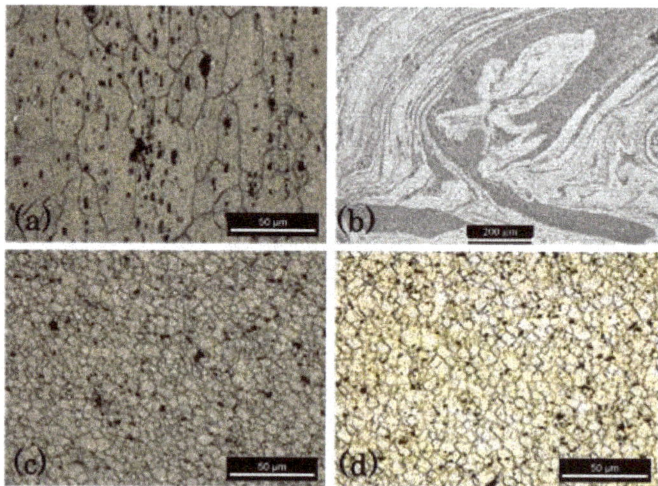

Figure 8. Microstructure of the: (**a**) base material AA 6082, (**b**) nugget of the 562-120 series, (**c**) nugget of the 562-30 series, and (**d**) nugget of the 265-30 series.

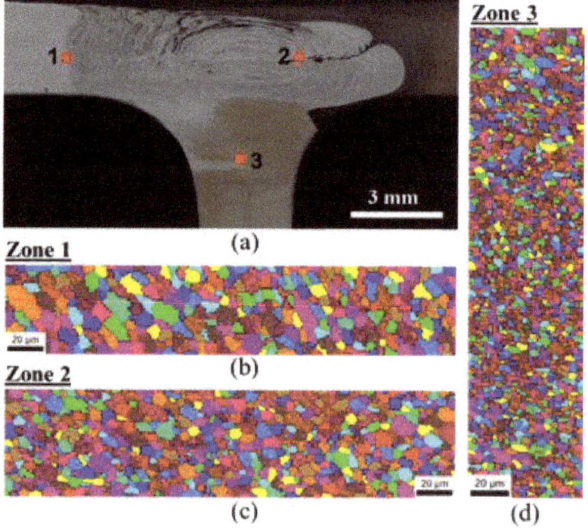

Figure 9. Electron backscatter electron backscatter diffraction (EBSD) analysis conducted for the 562-120 weld: (**a**) zones analyzed and EBSD maps registered in zones 1 (**b**), 2 (**c**), and 3 (**d**).

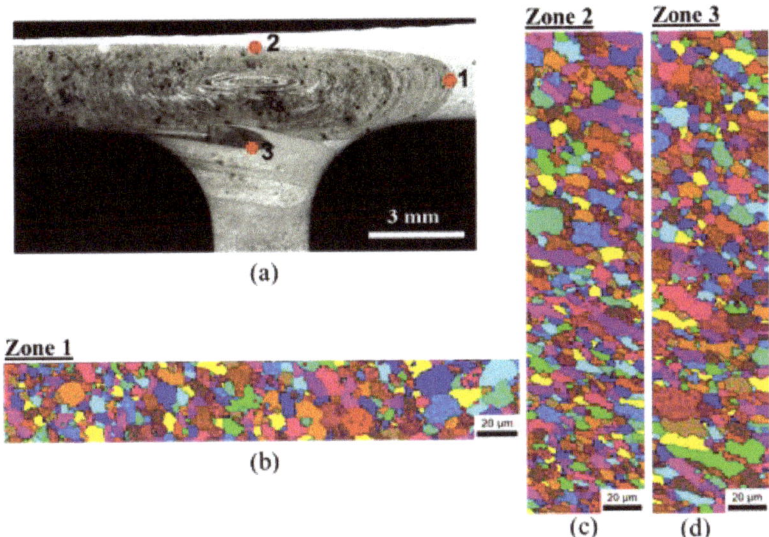

Figure 10. EBSD analysis conducted for the 265-30 weld: (**a**) zones analyzed and EBSD maps registered in zones 1 (**b**), 2 (**c**), and 3 (**d**).

3.4. Hardness and Tensile Behavior

The skin hardness profiles of the 562 and 265 weld series are shown in Figure 11a for defect-free welds. The hardness of the base materials is also shown in the same figure using dashed (AA2017) and dotted (AA5083) lines. It appears that the increase in welding speed did not significantly change the hardness in the HAZ, either on the AA5083 or AA2017 sides. However, there was a slight loss in hardness in the HAZ close to the tool path, mainly when AA2017 was located on the advancing side for the lowest welding speed. This could be attributed to the dissolution of hardening precipitates, as stated by Dong et al. [29]. The stir zone had an irregular hardness pattern with some peaks in the current welds as a result of the non-homogeneous mixing of the three different base materials during the process, as illustrated in Section 3.2.

The stringer hardness profiles for the 562 and 265 series are shown in Figure 11b. In all the series, there was a significant reduction in hardness in the thermomechanically-affected zone (TMAZ) and HAZ, which was generally attributed to the dissolution and coarsening, respectively, of the hardening precipitates [29]. This reduction in hardness was higher for the lowest welding speed series due to the higher and longer thermal cycles, as per Section 3.1.

Figure 12 shows the tensile curves of the skin of defect-free specimens of the 562 and 265 series, as well as of the three base materials for comparison. Table 5 summarizes the mean values of tensile strength and strain at failure of each series, as well as the zones where the failure occurred. The effect of welding speed on weld strength is well-illustrated in Figure 12, where the welds performed at the highest speed were slightly above the others but just below the AA5083 alloy.

Table 5 shows that specimens from both series (562 and 265) made at the lowest welding speed (30 mm/min) had the lowest efficiency values (losses of about 20%) and broke in the HAZ, close to the stir zone, on the AA5083 side. The efficiency is defined as the ratio between the maximum tensile strength of each weld series and the tensile strength of the least resistant base material, in this case being AA5083. Specimens from both series performed at 120 mm/min had the highest efficiency values, as suggested by the hardness results, although the difference was small; see Figure 11a. In addition, these test specimens broke in the HAZ on the AA5083 side but far from the stir zone. This is illustrated in Figure 13, which represents the distribution of the strain fields, obtained by an optical extensometer in two specimens produced with different welding speeds, and at an instant close to failure. Figure 13a illustrates a specimen from the 562-30 series whose highest plastic deformation, the red zone, and failure occurred in the HAZ next to the stir zone, while Figure 13b illustrates a specimen form the 562-120 series that also broke in the HAZ but far from the stir zone.

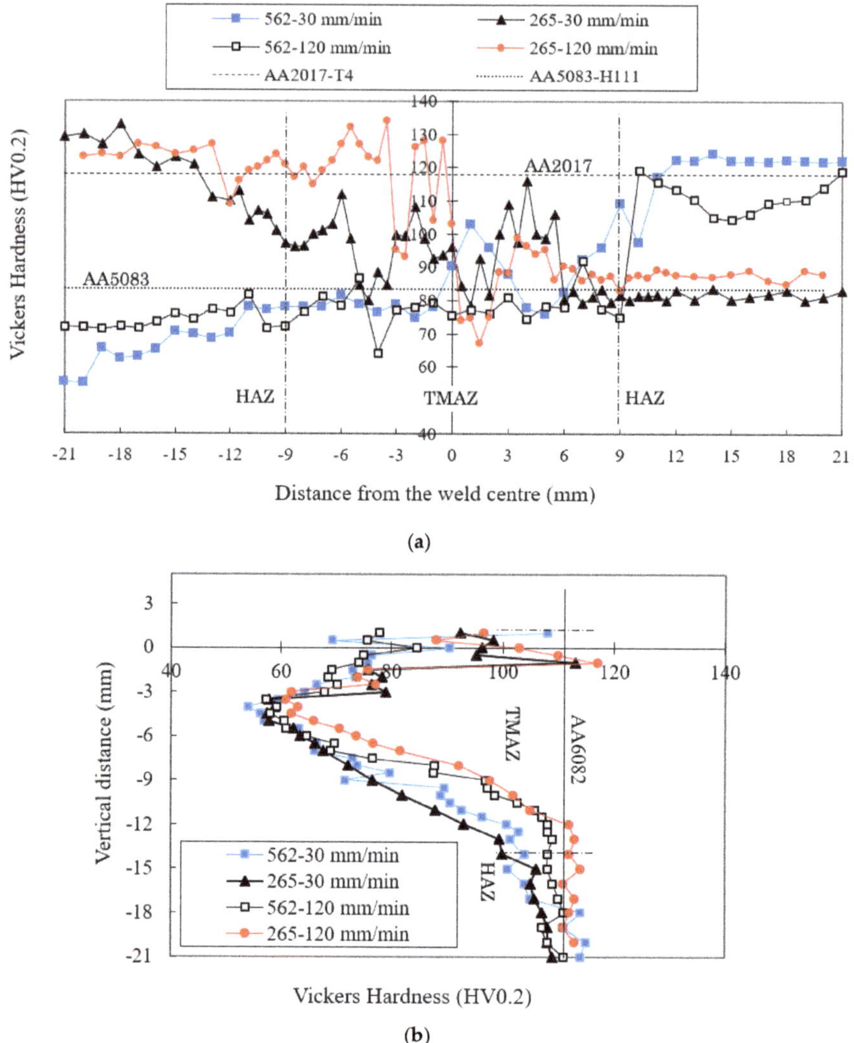

Figure 11. Effect of variation of the welding speed on the hardness profile in the (**a**) skin and (**b**) stringer.

Table 5. Average tensile results of defect-free welded specimens.

Weld Series		Ultimate Tensile Strength (MPa)	Efficiency (%)	Strain (%)	Fracture Zone
562	562-30	273.3 ± 0.5	86.0	10.6 ± 1.8	HAZ
	562-120	277.4 ± 0.4	87.4	8.3 ± 2.6	HAZ
265	265-30	271.7 ± 0.4	85.6	8.9 ± 0.4	HAZ
	265-120	296.2 ± 4.2	93.3	8.2 ± 2.5	HAZ

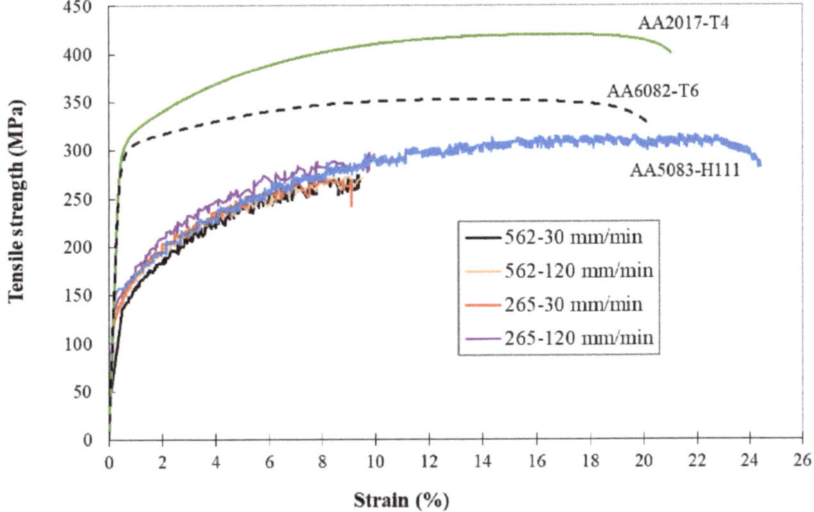

Figure 12. Tensile stress–strain curves of the 562 and 265 weld series and the base materials.

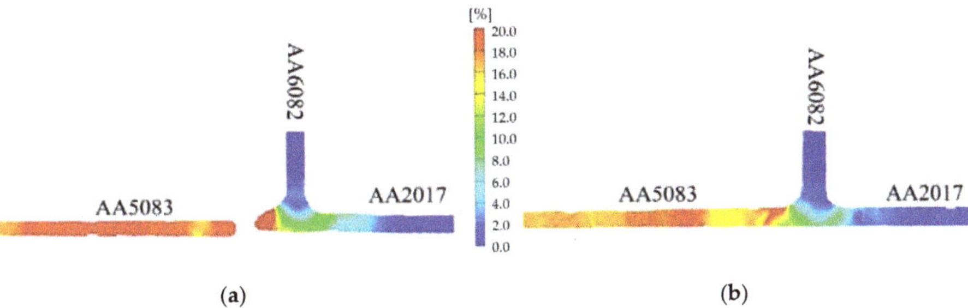

Figure 13. Strain distribution in tensile test specimens close to failure from the (**a**) 562-30 and (**b**) 562-120 series.

Figure 14 illustrates the influence of the welding speed on the fracture surface morphology of test pieces from series 562-30 and 562-280 performed at speeds of 30 and 280 mm/min, respectively. Figure 14a shows a ductile fracture surface of a 562-30 specimen with thin dimples and some larger dimples, as shown in the enlargement (5000x) of a zone in the lower right corner of the image. The fractures of the remaining series had similar morphologies because they occurred in the HAZ of the same alloy. Figure 14b shows a ductile fracture surface of a 562-280 series specimen, but the fracture was caused by a defect here, already mentioned in Section 3.2 and indicated by an arrow in the image.

3.5. Fatigue Strength

Figure 15 shows the S–N (stress range/number of cycles to failure) curves for the 562 and 265 series made at 30 and 120 mm/min, as well as the curve for the AA5083 alloy (the least resistant of the base materials). The fatigue test specimens that did not break for more than one and a half million cycles are shown with a horizontal arrow.

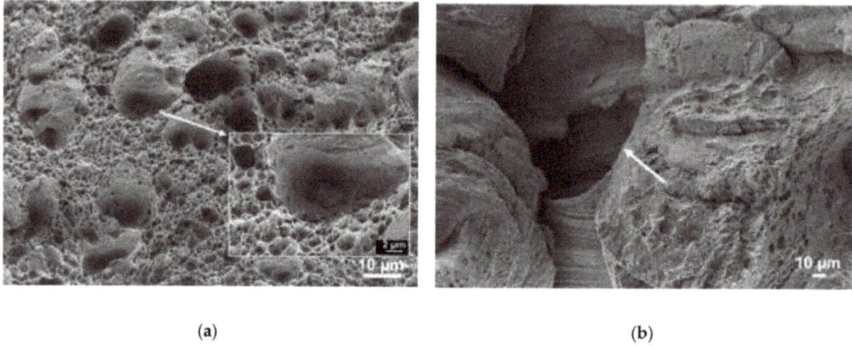

(**a**) (**b**)

Figure 14. Fracture surface morphology of a specimen in the (**a**) 562-30 and (**b**) 562-280 series.

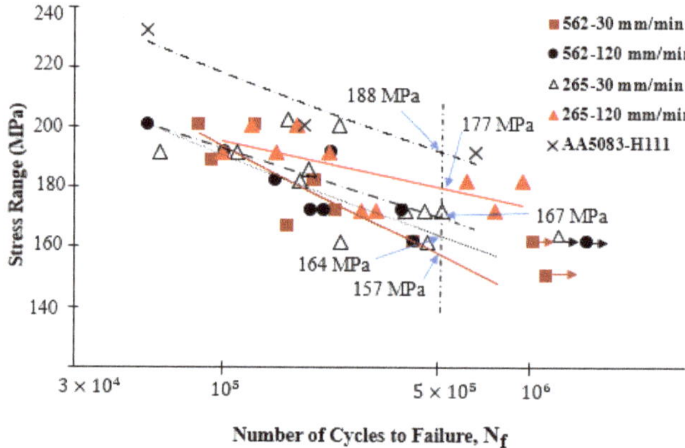

Figure 15. Influence of welding speed on the fatigue strength of the 562 and 265 series.

This figure shows that all the weld series presented a lower fatigue strength than the base material, which indicates that the welding process reduced fatigue strength regardless of the welding parameters used. However, the 562-120 series, produced at the advancing speed of 120 mm/min, presented better resistance to fatigue compared to the 562-30 series, performed at a lower speed of 30 mm/min but with an increase in the fatigue strength of only 4.5% at 5×10^5 cycles.

A similar increase in the fatigue strength with the increase in the welding speed could be observed in the 265 series for identical welding speeds of about 5.6%. Ericsson and Sandstrom [30] observed that for low welding speeds, their welds using the AA6082 alloy showed a slight increase in mechanical strength and fatigue compared to the higher welding speed due to the increased amount of heat generated in welding per unit length. In the current study, the opposite behavior was observed, which can be explained by the increase in mechanical strength with the increase in the welding speed, as shown in Table 5.

Figure 15 further shows that the 265 series had a slightly higher fatigue strength than the 562 series, while the 265-120 series had the highest strength. This may also have been related to the tensile properties of the welds, as specified in Table 5.

One factor that influenced the fatigue strength had to do with the surface finish of the samples tested, as the presence of small surface defects made crack initiation and propagation more probable, thus leading to premature failures [31]. The specimens were polished before testing in the current work; however, the welding crown was more difficult to polish without significantly reducing the thickness of the skin, which introduced some variability in the results. The specimens broke mostly at the TMAZ or close to the HAZ in the AA5083 alloy, either in the higher or lower stress ranges.

The fracture surface of the fatigue specimens had identical morphology. Figure 16 illustrates the fracture surface of the specimen from the 562-30 series subjected to a stress range of 180 MPa that broke after 193,700 cycles. Figure 16a shows a general view of the fracture surface of the specimen, where the fracture zones analyzed in more detail are indicated with rectangles. The crack started on the welding surface marked with an arrow and was propagated by fatigue through the thickness of the skin. The orientation of the characteristic fatigue striations was perpendicular to the growth direction of the crack, as shown in Figure 16b,c in more detail. Figure 16c shows the detail marked with a rectangle in Figure 16b in greater magnification. In the final part of the crack propagation phase, a ductile fracture occurred, as evidenced by the presence of small dimples and some larger dimples in the transition zone, as seen in Figure 16d. The fracture of this test specimen occurred in the TMAZ on the advancing side of the AA5083 alloy.

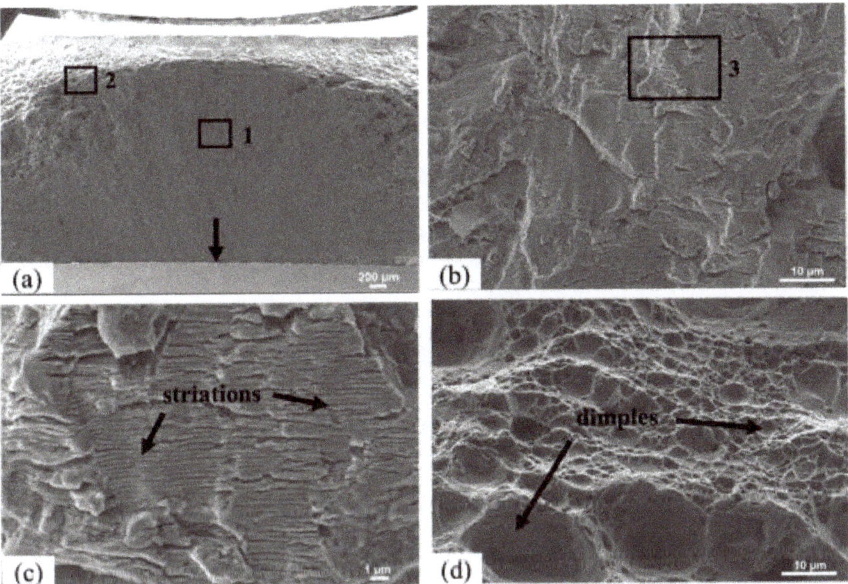

Figure 16. Fracture surfaces of a sample from the 562-30 series: (**a**) general fracture surface, (**b**) detail from location 1, (**c**) detail from location 3, and (**d**) detail from location 2.

4. Conclusions

The carried out research led to the following conclusions:

- It is feasible to achieve good quality FSWs with fillets between three dissimilar aluminum alloys.
- The nugget formation of FS welds between three dissimilar aluminum alloys is greatly influenced by the welding speed, mechanical properties, and location of the alloys, either on the advancing or retreating sides.
- A very low tool rotational to welding speed ratio (*w/v*) leads to the formation of welding defects.
- The weld nugget has a large dispersion of grain sizes, but the welding speed does not affect the grain size in the nugget.
- Increasing the welding speed increases the static and fatigue resistance of the welded joints.

- Placing the more resistant alloy (AA2017) on the advancing side rather than on the retreating side generates higher local weld temperature and provides stronger joints and with better fatigue behavior.

Author Contributions: The conceptualization and revision of the article, A.L. and I.G.; methodology, data treatment, and writing—review and editing of the manuscript, N.M. and R.M.L.; supervision and formal analysis of the results of the fatigue tests, J.D.C. All authors have read and agreed to the published version of the manuscript.

Funding: This research was funded by FEDER and FCT, as indicated in the acknowledgement.

Acknowledgments: This research is sponsored by FEDER funds through the program COMPETE–Programa Operacional Factores de Competitividade–and by national funds through FCT–Fundação para a Ciência e a Tecnologia –, under the project UIDB/00285/2020. The author N. Manuel is supported by Ministério das Pescas from Angola through a fellowship. All support is gratefully acknowledged.

Conflicts of Interest: The authors declare no conflict of interest.

References

1. Burcu, E.; Kumruoglu, L.C. 5083 type Al-Mg and 6082 type Al-Mg-Si alloys for ship building. *Am. J. Eng. Res.* **2015**, *4*, 146–150.
2. Doherty, K.; Squillacioti, R.; Cheeseman, B.; Placzankis, B.; Gallardy, D. Expanding the Availability of Lightweight Aluminum Alloy Armor Plate Procured from Detailed Military Specifications. In Proceedings of the 13th International Conference on Aluminum Alloys, Pittsburgh, Russia, 3–7 June 2012; pp. 541–546.
3. Kim, Y.; Fujii, H.; Tsumura, T.; Komazaki, T.; Nakata, K. Three defect types in friction stir welding of aluminum die casting alloy. *Mater. Sci. Eng. A* **2006**, *415*, 250–254. [CrossRef]
4. Hasan, M.M.; Ishak, M.; Rejab, M.R.M. Effect of pin tool flute radius on the material flow and tensile properties of dissimilar friction stir welded aluminum alloys. *Int. J. Adv. Manuf. Technol.* **2018**, *98*, 2747–2758. [CrossRef]
5. Fratini, L.; Buffa, G.; Shivpuri, R. Influence of material characteristics on plastomechanics of the FSW process for T-joints. *Mater. Des.* **2009**, *30*, 2435–2445. [CrossRef]
6. Bhattacharya, T.; Das, H.; Pal, T. Influence of welding parameters on material flow, mechanical property and intermetallic characterization of friction stir welded AA6063 to HCP copper dissimilar butt joint without offset. *Trans. Nonferrous Met. Soc. China* **2015**, *25*, 2833–2846. [CrossRef]
7. Choi, J.-W.; Liu, H.; Fujii, H. Dissimilar friction stir welding of pure Ti and pure Al. *Mater. Sci. Eng. A* **2018**, *730*, 168–176. [CrossRef]
8. Da Silva, A.; Arruti, E.; Janeiro, G.; Aldanondo, E.; Alvarez, P.; Echeverria, A. Material flow and mechanical behaviour of dissimilar AA2024-T3 and AA7075-T6 aluminium alloys friction stir welds. *Mater. Des.* **2011**, *32*, 2021–2027. [CrossRef]
9. Dinaharan, I.; Kalaiselvan, K.; Vijay, S.J.; Raja, P. Effect of material location and tool rotational speed on microstructure and tensile strength of dissimilar friction stir welded aluminum alloys. *Arch. Civ. Mech. Eng.* **2012**, *12*, 446–454. [CrossRef]
10. Barbini, A.; Carstensen, J.; Dos Santos, J.F. Influence of Alloys Position, Rolling and Welding Directions on Properties of AA2024/AA7050 Dissimilar Butt Weld Obtained by Friction Stir Welding. *Metals* **2018**, *8*, 202. [CrossRef]
11. Pabandi, H.K.; Jashnani, H.R.; Paidar, M. Effect of precipitation hardening heat treatment on mechanical and microstructure features of dissimilar friction stir welded AA2024-T6 and AA6061-T6 alloys. *J. Manuf. Process.* **2018**, *31*, 214–220. [CrossRef]
12. Manuel, N.; Silva, C.; Da Costa, J.M.D.; Loureiro, A. Friction stir welding of T-joints in dissimilar materials: Influence of tool geometry and materials properties. *Mater. Res. Express* **2019**, *6*, 106528. [CrossRef]
13. Boșneag, A.; Constantin, M.A.; Nitu, E.; Iordache, M. Friction Stir Welding of three dissimilar aluminium alloy: AA2024, AA6061 and AA7075. *IOP Conf. Ser. Mater. Sci. Eng.* **2018**, *400*, 022013. [CrossRef]
14. Boşneag, A.; Constantin, M.A.; Nitu, E.L.; Iordache, M. Friction Stir Welding of three dissimilar aluminium alloy used in aeronautics industry. *IOP Conf. Ser. Mater. Sci. Eng.* **2017**, *252*, 12041. [CrossRef]

15. Abnar, B.; Kazeminezhad, M.; Kokabi, A. Effects of heat input in friction stir welding on microstructure and mechanical properties of AA3003-H18 plates. *Trans. Nonferrous Met. Soc. China* **2015**, *25*, 2147–2155. [CrossRef]
16. ASTM. *E8/E8M—13a—Standard Test Methods for Tension Testing of Metallic Materials*; ASTM: West Conshohocken, PA, USA, 2013.
17. Khan, N.Z.; Khan, Z.A.; Siddiquee, A.N.; Al-Ahmari, A.; Abidi, M.H. Analysis of defects in clean fabrication process of friction stir welding. *Trans. Nonferrous Met. Soc. China* **2017**, *27*, 1507–1516. [CrossRef]
18. Aval, H.J.; Serajzadeh, S.; Kokabi, A.H.; Loureiro, A. Effect of tool geometry on mechanical and microstructural behaviours in dissimilar friction stir welding of AA 5086–AA 6061. *Sci. Technol. Weld. Join.* **2011**, *16*, 597–604. [CrossRef]
19. Nandan, R.; Roy, G.G.; Debroy, T. Numerical simulation of three-dimensional heat transfer and plastic flow during friction stir welding. *Met. Mater. Trans. A* **2006**, *37*, 1247–1259. [CrossRef]
20. Alvarez, P.; Janeiro, G.; Da Silva, A.A.M.; Aldanondo, E.; Echeverria, A. Material flow and mixing patterns during dissimilar FSW. *Sci. Technol. Weld. Join.* **2010**, *15*, 648–653. [CrossRef]
21. Yoon, T.-J.; Yun, J.-G.; Kang, C.-Y. Formation mechanism of typical onion ring structures and void defects in friction stir lap welded dissimilar aluminum alloys. *Mater. Des.* **2016**, *90*, 568–578. [CrossRef]
22. Shi, C.; Lai, J.; Chen, X.-G. Microstructural Evolution and Dynamic Softening Mechanisms of Al-Zn-Mg-Cu Alloy during Hot Compressive Deformation. *Materials* **2014**, *7*, 244–264. [CrossRef]
23. Ye, T.; Wu, Y.; Liu, W.; Deng, B.; Liu, A.; Li, L. Dynamic Mechanical Behavior and Microstructure Evolution of an Extruded 6013-T4 Alloy at Elevated Temperatures. *Metals* **2019**, *9*, 629. [CrossRef]
24. Agarwal, S.; Krajewski, P.; Briant, C.L. Dynamic Recrystallization of AA5083 at 450 °C: The Effects of Strain Rate and Particle Size. *Met. Mater. Trans. A* **2008**, *39*, 1277–1289. [CrossRef]
25. Imam, M.; Sun, Y.; Fujii, H.; Ma, N.; Tsutsumi, S.; Ahmed, S.; Chintapenta, V.; Murakawa, H. Deformation characteristics and microstructural evolution in friction stir welding of thick 5083 aluminum alloy. *Int. J. Adv. Manuf. Technol.* **2018**, *99*, 663–681. [CrossRef]
26. Leitão, C.; Louro, R.; Rodrigues, D. Analysis of high temperature plastic behaviour and its relation with weldability in friction stir welding for aluminium alloys AA5083-H111 and AA6082-T6. *Mater. Des.* **2012**, *37*, 402–409. [CrossRef]
27. Kalemba-Rec, I.; Kopyściański, M.; Miara, D.; Krasnowski, K. Effect of process parameters on mechanical properties of friction stir welded dissimilar 7075-T651 and 5083-H111 aluminum alloys. *Int. J. Adv. Manuf. Technol.* **2018**, *97*, 2767–2779. [CrossRef]
28. Ahmed, M.; Ataia, S.; Seleman, M.E.-S.; Ammar, H.; Ahmed, E. Friction stir welding of similar and dissimilar AA7075 and AA5083. *J. Mater. Process. Technol.* **2017**, *242*, 77–91. [CrossRef]
29. Dong, P.; Li, H.; Sun, D.; Gong, W.; Liu, J. Effects of welding speed on the microstructure and hardness in friction stir welding joints of 6005A-T6 aluminum alloy. *Mater. Des.* **2013**, *45*, 524–531. [CrossRef]
30. Ericsson, M. Influence of welding speed on the fatigue of friction stir welds, and comparison with MIG and TIG. *Int. J. Fatigue* **2003**, *25*, 1379–1387. [CrossRef]
31. Lukin, V.I.; Ioda, E.N.; Skupov, A.A.; Panteleev, M.D.; Ovchinnikov, V.V.; Malov, D.V. Effect of the surface roughness of friction stir welded joints on the fatigue characteristics of welded joints in V-1461 and V-1469 aluminium–lithium alloys. *Weld. Int.* **2017**, *31*, 974–978. [CrossRef]

Article

Effect of FSW Traverse Speed on Mechanical Properties of Copper Plate Joints

Tomasz Machniewicz [1], Przemysław Nosal [1], Adam Korbel [1] and Marek Hebda [2,*]

1 Faculty of Mechanical Engineering and Robotics, AGH University of Science and Technology, A. Mickiewicza Av. 30, 30-059 Krakow, Poland; machniew@agh.edu.pl (T.M.); pnosal@agh.edu.pl (P.N.); korbel@agh.edu.pl (A.K.)

2 Institute of Materials Engineering, Faculty of Materials Engineering and Physics, Cracow University of Technology, Warszawska 24, 31-155 Krakow, Poland

* Correspondence: mhebda@pk.edu.pl; Tel.: +48-126-283-423

Received: 22 January 2020; Accepted: 14 April 2020; Published: 20 April 2020

Abstract: The paper describes the influence of the friction stir welding travel speed on the mechanical properties of the butt joints of copper plates. The results of static and fatigue tests of the base material (Cu-ETP R220) and welded specimens produced at various travel speeds were compared, considering a loading applied both parallel and perpendicularly to the rolling direction of the plates. The mechanical properties of the FSW joints were evaluated with respect to parameters of plates' material in the delivery state and after recrystallisation annealing. The strength parameters of friction stir welding joints were compared with the data on tungsten inert gas welded joints of copper plates available in the literature. The results of microhardness tests and fractographic analysis of tested joints are also presented. Based on the above test results, it was shown that although in the whole range of considered traverse speeds (from 40 to 80 mm/min), comparable properties were obtained for FSW copper joints in terms of their visual and microstructural evaluation, their static and especially fatigue parameters were different, most apparent in the nine-fold greater observed average fatigue life. The fatigue tests turned out to be more sensitive criteria for evaluation of the FSW joints' qualities.

Keywords: FSW; copper; butt joint; mechanical properties; fatigue performance; traverse and rotation speed

1. Introduction

Friction stir welding (FSW) is a relatively novel technique for joining materials, and thanks to the wide possibilities of application, it has been gaining popularity rapidly recent years [1]. In particular, this technique can be used as one of few methods for joining metals that are hard to weld or have physicochemical properties significantly different from each other, i.e., in cases when the use of conventional fusion welding is strongly limited and braze welding does not yield sufficient joint strength [2]. For this reason, FSW is commonly used not only for welding aluminium alloys [3,4] but also for many other metallic materials such as magnesium alloys [1,5–7], titanium alloys [1,8], steel [9,10], copper [11–16], and different combinations of dissimilar metals [1].

Friction stir welding, as a solid-state process, is based on the plastic deformation of joined materials, elicited by using a special tool consisting in general of a shoulder and a mixing pin (Figure 1). The basic function of this tool is to generate heat, which enables the plasticisation of the materials, and then their mixing. Around the FSW joint, as shown in Figure 1, three characteristic zones within which the material properties have changed in relation to the base material (BM) are generally specified. These zones are: (i) the heat affected zone (HAZ), (ii) the thermo-mechanically affected zone (TMAZ) and (iii) the weld nugget (WN), where the material is fully recrystallized [1]. In addition, due to the complex motion of the working tool, there are two distinctive sides of the formed joint: the advancing side—where the tangential vector of the rotational speed of the tool is compatible with its travel speed vector, and the retreating side—where the senses of mentioned vectors are opposite (Figure 1).

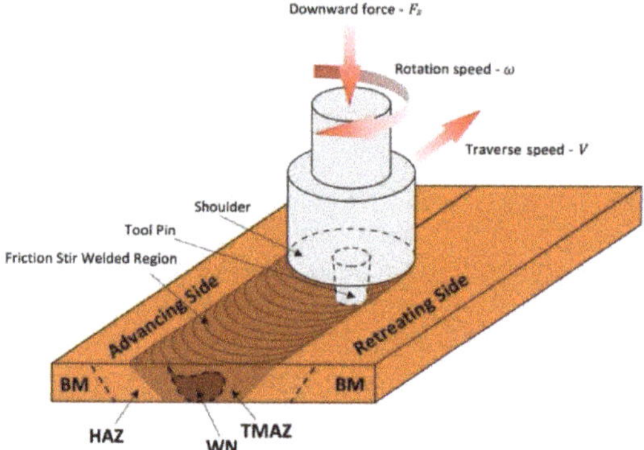

Figure 1. Schematic drawing of the friction stir welding (FSW) process.

The kinematics of the FSW process consist of two main motions, rotational and progressive, along the interface of joined materials. These motions correspond to the two basic parameters of the technological process which are the rotation speed (ω) and the traverse speed of the welding (V). These parameters affect the amount of heat generated during the process and thus the quality and properties of the wrought weld. Thus, for a given tool geometry, the mechanical properties of FSW joints are greatly influenced by the abovementioned technological parameters applied to the welding process as confirmed by numerous examples in literature [1,3,12,15,16]. A poor choice of these parameters may lead to various kinds of imperfections of the joints, significantly reducing their strength properties [17–22]. Although extensive research on the influence of FSW process parameters on the weld quality has already been conducted, existing publications most often concern aluminium alloys and static joint properties. The database on non-aluminium materials is more modest, and only rudimentary information can be found on the fatigue properties of produced joints. In particular, one of the materials that is still poorly known in this respect is copper, despite the fact that, due to the fact of its physical and chemical properties, copper combined with FSW technology is increasingly used in many areas of industry including the nuclear [22], energy [23] and automotive sectors [24]. Elements made in this way must have the required strength, not only in static but also in fatigue loading conditions, which can only be achieved through the appropriate choice of FSW process parameters.

A review of the literature provides divergent data on FSW parameters optimal for copper sheets. There are examples of significantly different combinations of rotation speed/traverse speed/plate thickness recommended for welding copper, e.g., 1000 rpm/30 mm·min^{-1}/2 mm [11], 400 rpm/100 mm·min^{-1}/3 mm [12], 250 rpm/61 mm·min^{-1}/4 mm [13] and 1300 rpm/170 mm·min^{-1}/6 mm [14]. Xue et al. [15], while welding copper plates with a thickness of 5 mm, applied various rotary speeds $\omega \geq 400$ rpm for the traverse speed of 50 mm/min and various traverse speeds $V \geq 50$ mm/min for the rotary speed of 800 rpm. In the entire range of the considered parameters, they observed the systematic effect of joint strength increasing with increasing traverse speed for constant $\omega = 800$ rpm, and with decreasing rotation speed at constant $V = 50$ mm/min. For the same sheet thickness, Khodaverdizadeh et al. [16] used both higher ($V = 75$ mm/min, $\omega = 600$–900 rpm) and lower ($V = 25$ mm/min, $\omega = 600$ rpm) process parameters relative to the research conducted by Xue et al. [15]. The obtained test results confirmed that the highest mechanical parameters of the joint were received for the combination of the lowest of the considered rotation speeds (600 rpm) and the highest traverse speed (75 mm/min). These results, however, are contrary to the generally expressed view that, due to the high thermal conductivity and high melting temperature of copper, a moderate welding speed is recommended, so that the right amount of heat energy can be generated [13]. On the other hand, excessive heat generation during welding of copper sheets hardened beforehand by cold rolling causes a more intensive reduction of strength parameters of the joint within the weld, due to the recrystallisation process. Because of the high degree of complexity of the processes occurring during FSW, it is important to determine the optimal technological parameters in order to obtain the highest possible strength properties of joints. For this purpose, the results of mechanical tests of butt joint specimens of electrolytic copper (99.998% Cu) with a thickness of t = 5 mm made by FSW with different traverse speeds are presented in this paper. Based on these, the influence of the applied process parameters on structural changes, microhardness and mechanical properties of the fabricated joints were analysed. When selecting optimal FSW traverse speed, static tests results are usually taken into account during research [11–15], but this work also considers the fatigue properties of FSW joints (i.e., fatigue lives). This is an approach rarely found in the literature but is considered by the authors as more sensitive criteria for evaluation of the FSW joints' qualities. Moreover, data on the impact of FSW welding parameters on the properties of copper joints are usually limited to one specimen orientation. The results in this paper demonstrate that the variations of FSW parameters may have a qualitatively different effect on the properties of friction stir welds oriented longitudinally and transversely to the rolling direction of the jointed plates. After considering these additional criteria, the optimal combination of transverse and rotation speed selected in respect to the best strength parameters of FSW copper joints proved to be different from the proposals recommended so far in the literature.

2. Materials and Methods

Base material samples and FSW butt joints were fabricated from 5 mm thick plates of high purity electrolytic copper (Cu-ETP R220) with a chemical composition containing more than 99.9% Cu, according to the EN 573-1 standard. The FSW welds were performed on a conventional Jafo CNC milling machine at a constant rotation speed ω = 580 rpm and at three different traverse speeds: V = 40, 60 and 80 mm/min. The other FSW parameters were zero tilt angle, plunged depth of 0.3 mm, tool plunging speed of 2.5 mm/min and dwelling time of 5 s. The dimensions of the plate welded in such a way were approximately 300 mm × 300 mm. A view of the tool geometry used for welding and a representative example of the weld are presented in Figure 2. Samples from base material and welded plates for static and fatigue tests, with geometry shown in Figure 3, were cut out using the water-jet technique. The good quality (defect-free) of the FSW joints produced in accordance with the above parameters was confirmed by visual observation and ultrasonic evaluation.

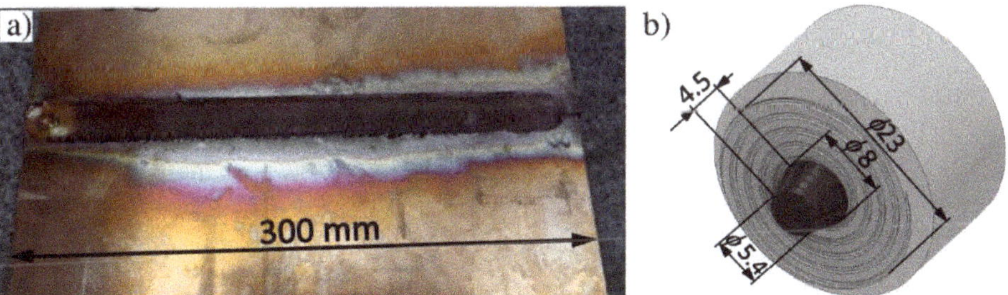

Figure 2. The FSW joint made at ω = 580 rpm, V = 60 mm/min (**a**) and geometry of the tool used for welding (**b**) (dimensions in mm).

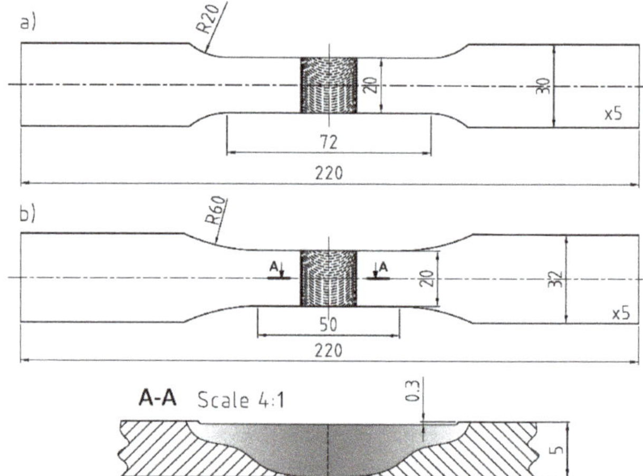

Figure 3. Specimen geometry used in static tests (**a**) and fatigue tests (**b**) (dimensions in mm).

For FSW joints, the welds were located in the centre of the samples as presented in Figure 3. Specimens, both from the base material and from welded plates, were made for two sheet configurations, i.e., longitudinally (orientation L) and transversally (orientation T) to the direction of their rolling process, as presented schematically in Figure 4. The test plan was additionally extended by tensile tests of specimens subjected to earlier annealing at 600 °C. These results were considered as a reference point for the assessment of weld strength properties in the heat affected zone (HAZ). The annealing temperature was adopted based on temperature maps, as presented in Figure 5, registered by the thermal imaging camera on the surface of sheets during the welding process in the tool working area. Although the temporary weld temperature (*T*) just behind the rotating shoulder was about 800 °C, it stabilised a short distance from the tool at 600 °C.

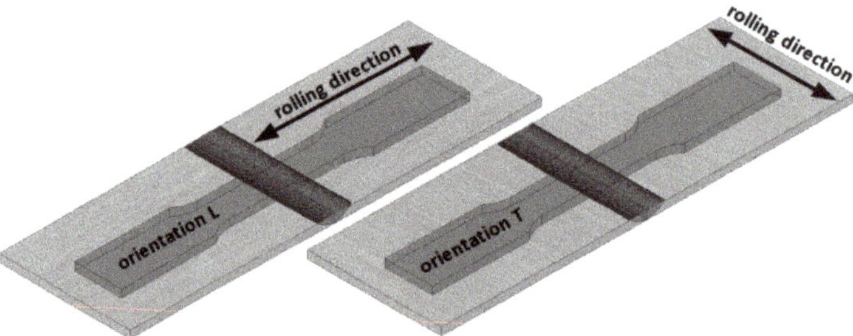

Figure 4. Schematic presentation of the longitudinal (L) and transversal (T) specimen orientation.

Static and fatigue tests were carried out using the MTS 810 universal testing machine, with a load limit of 100 kN. Tensile tests were conducted according to the EN ISO 6892-1:2016 standard under displacement control at a rate of 0.5 mm/min. The strains were measured using axial extensometer (Epsilon 3542-025M-025-ST) with a gauge base of 25 mm and a measuring range of ±6.25 mm. The extensometer was mounted in the centre of the samples, covering the width of the weld when FSW joints were tested.

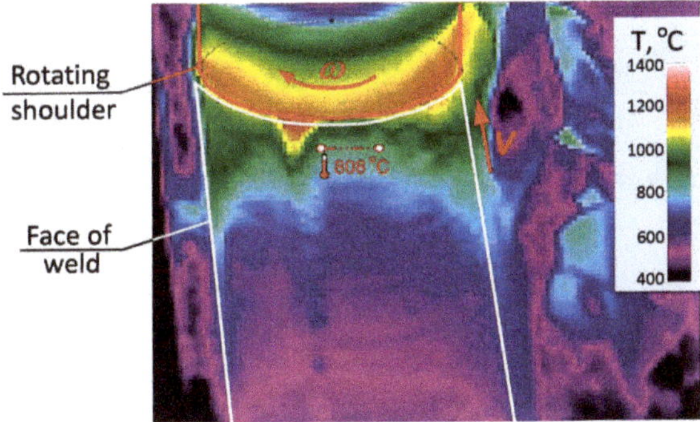

Figure 5. The temperature field (T) registered during FSW process by a thermal imaging camera on the copper plate surface behind the tool.

The stress-controlled constant-amplitude fatigue tests were performed with the stress ratio $R = 0$ using sinusoidal waveform loading at frequency $f = 20$ Hz. For the solid samples of the base material, four stress ranges were considered: $\Delta S = 210, 220, 230, 240$ MPa. In the case of samples with an FSW joint, comparative fatigue tests were carried out only in the stress range of $\Delta S = 160$ MPa due to the fact of their lower strength and large scatter of the results. The number of specimens subjected to monotonic tensile tests and fatigue tests is summarised in Table 1.

Table 1. Tensile and fatigue test matrix.

Kind of Test:		Tensile		Fatigue	
Specimen Orientation:		Longitudinal	Transversal	Longitudinal	Transversal
Specimen type	Base Cu-ETP R220	5	5	3	3
	FSW: $V = 80$ mm/min	3	3	3	3
	FSW: $V = 60$ mm/min	3	3	3	3
	FSW: $V = 40$ mm/min	3	3	3	3
	Annealed Cu-ETP	3	-	-	-

The microhardness measurements were performed for base material and three FSW joint specimens (i.e., one specimen per considered travel speed) in accordance with the ISO 6507-1 standard on Nexus 423A hardness equipment with a Vickers indenter, applying a loading force of 25 g and a measuring time of 10 s. The metallographic investigation was conducted using an Eclipse ME600 light optical microscope.

3. Results and Discussion

3.1. Microhardness Tests and Metallographic Analysis

The microhardness distribution determined for the various traverse speeds on the cross-section of the joints along the centre of the specimens' thickness is presented in Figure 6.

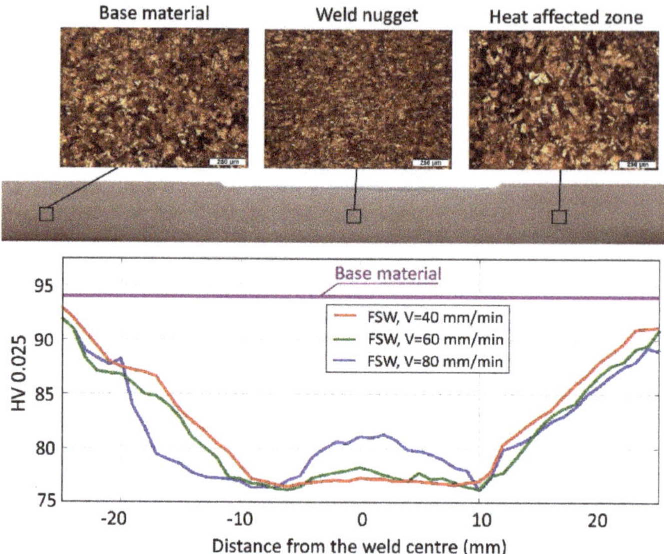

Figure 6. Microhardness profiles of the FSW joints corresponding to various traverse speeds and comparison of representative microstructure images in selected zones of the FSW copper joints produced with a traverse speed of 80 mm/min.

This figure also presents the representative microstructure images of the weld nugget, heat affected zone and base material for a joint welded at a speed of V = 80 mm/min. For this V-speed value, the profile of microhardness had a characteristic "W" shape (Figure 6) which is typical for most FSW joints [13,15–18,25]. For each of the traverse speeds used, a sudden decrease of microhardness in the HAZ was found. The minimum value of the microhardness (76.2 HV) was measured on the border of the HAZ and WN. The above trends corresponded well to the average grain sizes observed in the respective zones of the joint. Thus, the higher hardness in the WN zone than in HAZ was related to smaller grain size which was 10 μm for WN and 90 μm for HAZ. The base material, despite the larger grain size (approximately 70 μm), displayed a hardness and strength properties higher than those measured in the welded zone which is the effect of strain hardening arising from the sheet rolling process. A lower traverse speed value increases the amount of heat generated during the FSW process which has a direct impact on the lower hardness in the TMAZ and is manifested by a flatter microhardness profile for speeds of 60 and 40 mm/min (Figure 6). These trends, although widely shown in the literature [13,15,16,25], are contrary to other available data [26–28] showing that microhardness of the FSW copper joints increasing over the level related to the base material. For example, in the work of Berenji [26], for pure copper plates with the same thickness as those described in this paper (5 mm) and with very similar tool geometry and nearly the same rotation speed (600 rpm), a consistent increase of microhardness in the weld zone, over 75 HV for base material, was observed only when the traverse speed was above 25 mm/min. As in other works [27,28], this effect increased with the traverse speed. The highest reported ratios of such obtained microhardness of weld zone over the microhardness of the base material were (in HV): 90/75 [26], 114/77 [27] and 105/60 [28]. It may be noted that increased microhardness in the region of the FSW joints was observed in the case of soft copper conditions (microhardness below 80 HV) [26–28], while base material under higher strain with microhardness of over 80 HV results in a drop in hardness in the welded zone [13,15,16,25] which was also confirmed by the results of this paper.

3.2. Monotonic Tests Results

The tensile curves (σ—engineering stress; ε—engineering strain) for solid samples, including the base material specimens (specimens TBM), loaded parallel and perpendicularly to the rolling direction, and annealed specimens are presented in Figure 7. Determined from the static tensile tests, mechanical parameters of the solid and welded specimens, represented by yield stress (YS) ultimate stress (UTS) and reduction of area (AR), are summarised in Table 2.

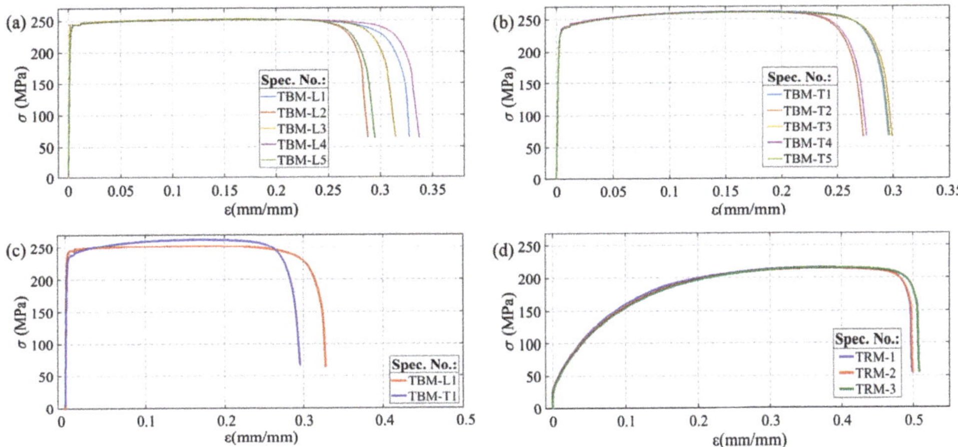

Figure 7. Tensile curves of solid samples: (**a**) base material—orientation L; (**b**) base material—orientation T, (**c**) base material—comparison of the representative curves for L and T orientation, (**d**) base material after annealing at 600 °C.

Table 2. Mechanical properties of tested samples.

Sample Type	Longitudinal Orientation (L)			Transversal Orientation (T)		
	YS (MPa)	UTS (MPa)	AR (%)	YS (MPa)	UTS (MPa)	AR (%)
Base Cu-ETP R220	242 ± 1.5	252.7 ± 1.5	67.3 ± 6.7	232.0 ± 0.7	261.2 ± 0.4	69.2 ± 1.0
FSW: V = 40 mm/min	87.7 ± 0.5	216.0 ± 0.8	69.4 ± 3.5	101.3 ± 0.5	220.7 ± 0.5	53.1 ± 7.2
FSW: V = 60 mm/min	87.3 ± 0.5	215.0 ± 0.8	70.5 ± 1.5	92.7 ± 0.5	218.7 ± 0.5	60.1 ± 4.2
FSW: V = 80 mm/min	82.0 ± 7.3	209.0 ± 8.5	40.3 ± 2.9	82.3 ± 0.9	216.3 ± 1.7	61.2 ± 2.2
Annealed Cu-ETP	33.0 ± 0.8	215.7 ± 0.5	71.3 ± 1.9	-	-	-
TIG welded joint [16]	53	168	5.7	-	-	-

As presented in Figure 7, the strength properties of solid sheets for the given orientation (L and T) were characterised by high repeatability, apart from a few percent spread of the strain at failure. Samples loaded along the rolling direction (TBM-L) above the elastic strain range showed almost perfectly plastic characteristics (Figure 7a). Comparison of the tensile curves for both orientations (Figure 7c) demonstrate that the strain at failure was slightly higher in the case of loading in accordance with the rolling direction of the sheet. In the transversal direction (specimen TBM-T, Figure 7c), where a slight strengthening effect was observed, the yield stress was lower by about 4% and the ultimate tensile strength was higher by about 3%, compared to longitudinal orientation (Table 2). The recrystalising annealing at 600 °C (specimens TRM, Figure 7d) caused a seven-fold decrease in yield stress and a decrease of about 17% in ultimate strength, with a simultaneous increase in strain at failure by close to 60%, and an increase in the value of reduction of area at failure by about 2%–4% (Table 2).

Monotonic tests of FSW joints comprised three tensile tests for each of the three traverse speeds considered. The cracking of the joints under static loading always occurred on the retreating side of the welds in the area of the TMAZ as shown in Figure 8. The exemplary tensile curves presented in

Figure 9, corresponding to the highest traverse speed (i.e., V = 80 mm/min—FSW80 specimens), were characterised by the largest observed scatter, which was particularly pronounced for the longitudinal orientation (specimens FSW80-L, Figure 9a). Two samples from this series fractured before reaching the extreme point on the static tensile curve (Figure 9a), which was also manifested by lower values of strain at failure, as well as lower ultimate strengths (Table 2). The tensile curves representative of the joints welded at different traverse speeds are compared in Figure 10.

Figure 8. Fracture location of the tensile FSW joint samples.

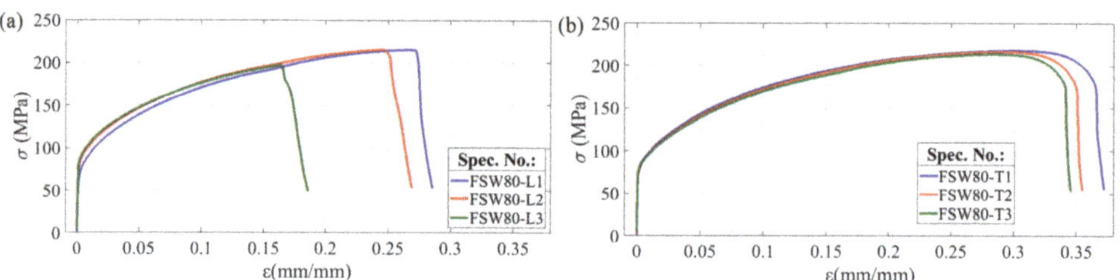

Figure 9. Tensile curves of FSW joint specimens produced with rotary speed ω = 580 rpm and traverse speed V = 80 mm/min: (**a**) orientation L; (**b**) orientation T.

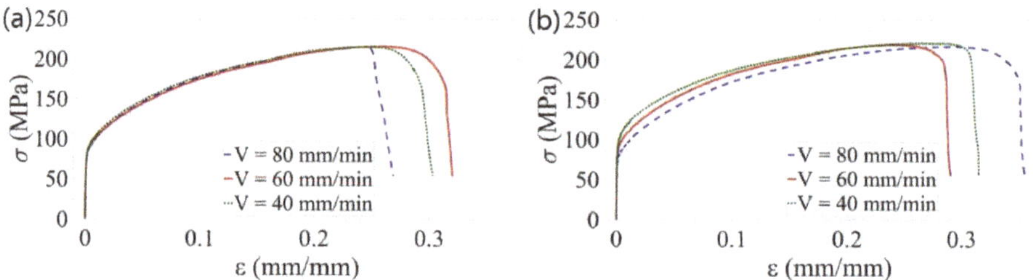

Figure 10. Comparison of the representative tensile curves of the specimens joined by FSW with various traverse speeds: (**a**) orientation L; (**b**) orientation T.

As may be observed, for longitudinal orientation the tensile curves regardless of V-speed are almost overlapped, except for the aforementioned effect of the lower ductility associated with the traverse speed of 80 mm/min (Figure 10a). In the transverse orientation (Figure 10b), a consistent trend of decrease of YS and UTS values with increasing traverse speed was found, which is qualitatively different to the available literature data [15,16,29]. Assuming that reduction of the joint strength parameters is induced by a copper recrystallisation process in the HAZ, this effect should be more intensive for a lower V-value, which corresponds to the higher amount of thermal energy generated during the FSW process. Besides the aforementioned experimental data, this mechanism is also confirmed by numerical simulations of the FSW welding process [30]. Macroscopic analysis of tensile fracture surfaces of FSW joints did not show distinct defects in the weld structure. The only dark streaks were observed on the fracture of the FSW specimen with the lowest strength parameters (sample FSW80-L1) as presented in Figure 11. Energy-dispersive spectroscopy (EDS) analysis of the areas marked on Figure 11 revealed the presence of iron, probably from the FSW tool, with content ranging from 1% to 4%.

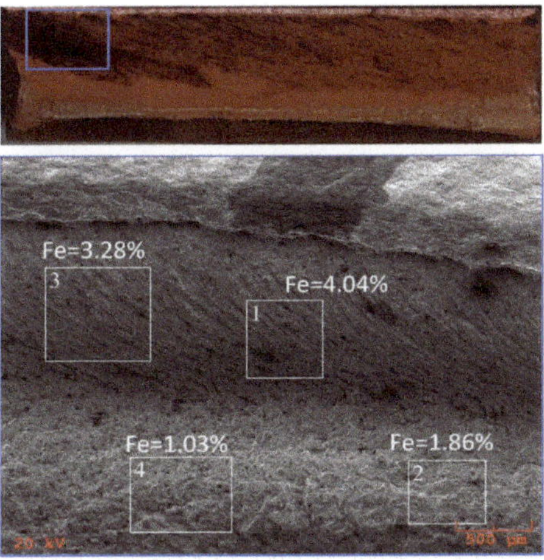

Figure 11. The fracture surface of the FSW80-L1 sample, showing areas of EDS analysis and iron content.

Reduction of the strength properties of FSW joints in relation to solid material samples, as the effect of the copper recrystallisation in the heat affected zone, is clearly visible in Figure 12.

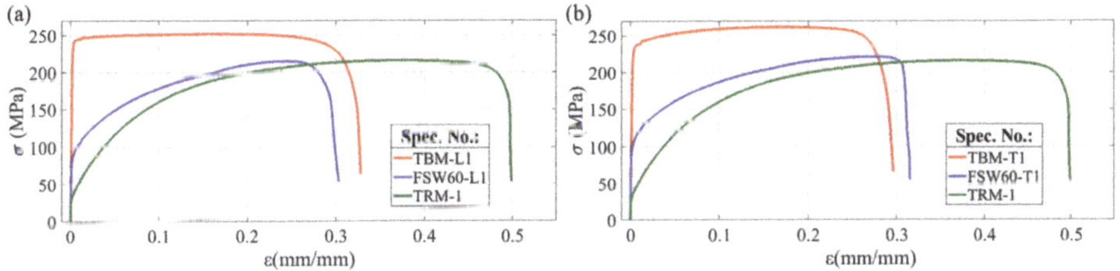

Figure 12. Comparison of representative tensile curves of base material (TBM), annealed material (TRM) and FSW specimens welded with V = 60 mm/min (FSW60): (**a**) orientation L; (**b**) orientation T.

The ratios of the mechanical parameters observed for FSW joints at different traverse speeds to the parameters specific to the solid material (YS/YS_{BM} and UTS/UTS_{BM}) for both sample orientations are shown in Figure 13. Resulting from the FSW process, the YS decreased on average by more than 60% (Figure 13), i.e., from 230–240 MPa to 80–100 MPa (Table 2). This effect was almost independent of the rolling direction of the sheet. A smaller but still significant decrease, amounting to 15%, was also observed for UTS. Figure 13 reveals a consistent trend of decreasing strength parameters of FSW joints with increasing traverse speed. Although (with the exception of the YS changes for transversal orientation (Figure 13b)) the variations presented in Figure 13 are very moderate, these trends are, in qualitative terms, opposite to those described in the literature [15,16,27]. In general, with increasing traverse speed, the amount of thermal energy generated during FSW decreases [4].

This contributes to maintaining higher strength parameters due to the lower intensity of the copper recrystallisation process. The excessive scatter of the strength parameters observed in Figure 13a (longitudinal orientation) for the speed V = 80 mm/min confirms the earlier mentioned objections regarding this series of joints.

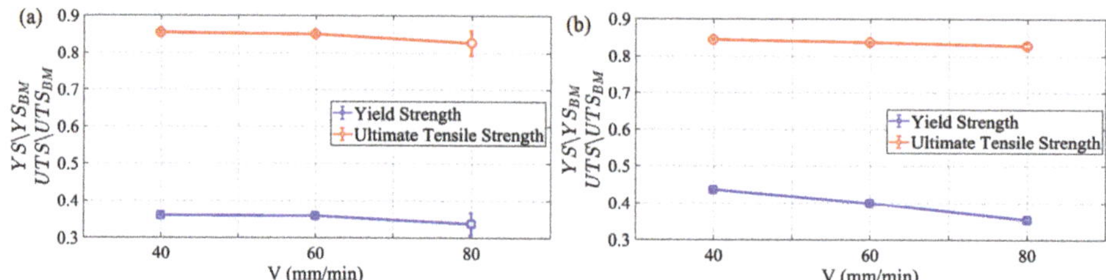

Figure 13. The mechanical parameters of FSW copper joints (YS and UTS) at different traverse speeds, normalised by base material properties (YS_{BM} and UTS_{BM}): (**a**) longitudinal orientation, (**b**) transversal orientation.

Nevertheless, the strength parameters of FSW welds, particularly the YS, remain higher in relation to the properties of the copper subjected to recrystalising annealing (Figure 12, Table 2). In addition, these FSW joint parameters were also significantly higher—UTS by close to 30%, and YS by about 67%—than the literature values for properties of butt joints welded from pure copper sheets using the tungsten inert gas (TIG) method [31]. This may be due to the strong plastic deformation of the material caused by the tool acting on the plates being joined by the FSW process, which does not occur in fusion welding (including TIG welding). The result is a refined microstructure formed in the region of the weld nugget and in the zone of thermo-mechanical impact (see Section 3.1.) which improves the mechanical properties of the joints. In addition, the temperature of the FSW process is lower than at the fusion welding, thus avoiding problems of porosity, cracking, sheet distortion and large size of heat affected zone. The higher strength of FSW joints compared to solid copper samples after annealing indicates that the reduction of mechanical properties of these joints is predominantly related to the loss of the cold-work hardening effect as a result of the recrystallisation process in the heat affected zone of the joint. Therefore, the improvement of strength parameters of the joints made using the FSW technique from strain-hardened copper sheets is possible by limiting the material's heating during welding. This, in addition to selecting the optimal FSW parameters, may be implemented using a cooling medium during the FSW process [29].

3.3. Fatigue Test Results

The fatigue test results for base material specimens for both orientations are compared in the form of *S–N* curves in Figure 14. As may be observed, the specimens loaded perpendicularly to the rolling direction of the sheet repeatedly showed slightly lower fatigue strength (of about 4%) than the samples of longitudinal orientation. This is opposite to the trend of ultimate strength changing for both orientations under monotonic loading (Table 2).

In all specimens with T-orientation, and in specimens with L-orientation welded at a speed of V = 60 mm/min, fatigue cracks occurred on the retreating side in the plane of the notch formed along the edge of the rotating tool plunge as illustrated in Figure 15a. In the samples with longitudinal orientation welded at speeds of V = 40 mm/min and 80 mm/min, fatigue cracks were also located on the retreating side but, similar to static tests, at some distance from the weld edge (Figure 15b). The plane of cracks developed at the welds' edges was nearly perpendicular to the face of the sample (Figure 15a), while in the TMAZ (Figure 15b) the cracks propagated on a plane inclined to the face of sample at an angle of about 45–60 degrees. In general, the cracks initiated in the corner of the sample on the face side of weld as illustrated in Figure 15c.

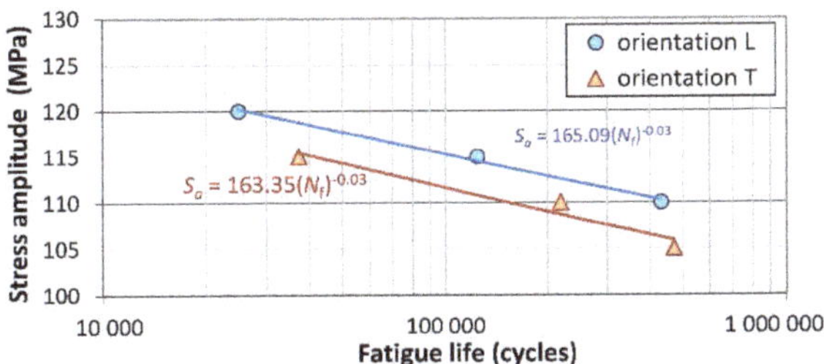

Figure 14. *S–N* curves of Cu-ETP R220 base material specimens.

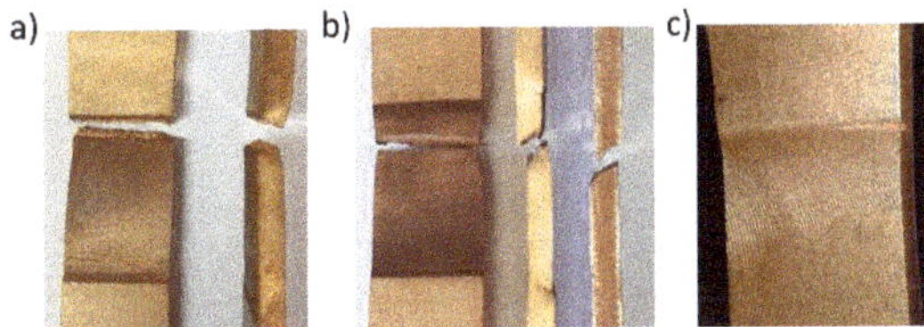

Figure 15. Location of the fatigue cracks observed in FSW joints: (**a**) crack at the edge of the weld, (**b**) crack at some distance from the edge of the weld, (**c**) typical mode of fatigue crack initiation.

Fatigue strength of the specimens containing FSW joints was significantly lower compared to base material specimens. In the tests of the FSW joint specimens with longitudinal orientation (Figure 16a), the highest fatigue lives (N_f), and simultaneously the most repetitive results were obtained for samples welded at a speed of V = 60 mm/min. For this configuration, other considered traverse speeds, i.e., the lower (V = 40 mm/min) and the higher (V = 80 mm/min), led to significant lower fatigue lives and much greater scatter of the test results. One of the specimens welded with a traverse speed of 80 mm/min cracked during the first loading sequence. For all samples from the two series, of the lowest fatigue life was characterised by the distinct location and morphology of the cracks. FSW joint specimens oriented transversally to the rolling direction (Figure 16b) were characterised by more repeatable results, regardless of the traverse speed used. For this orientation, the difference between the average fatigue lives for specimen series associated with all considered V-values did not exceed 22%. In contrast to samples with L orientation, the highest fatigue lives were observed for the joint specimens welded at a speed of V = 80 mm/min. Concurrently, the speed of V = 60 mm/min, which had given the highest durability for the samples with L-orientation, provided the lowest fatigue lives in the case of transversally oriented specimens.

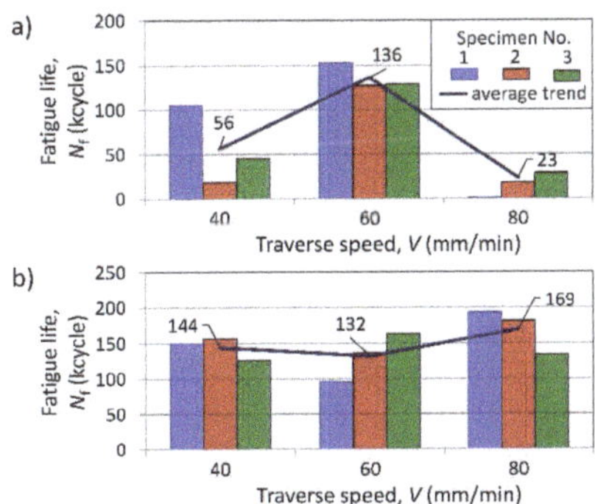

Figure 16. Influence of traverse welding speed on fatigue life of FSW joint specimens (R = 0, ΔS = 160 MPa): (**a**) orientation L; (**b**) orientation T.

Comparing the fatigue test results obtained for specimens cut parallel (L) and perpendicularly (T) to the rolling direction, it may be pointed out that the traverse speed $V = 60$ mm/min was the most advantageous of the parameters used, due to the highest level of repeatability of observed fatigue lives. For this V-value, the difference in average fatigue lives for both specimen orientations was only 3%, while for the speed of 40 mm/min the average fatigue life for the transversal orientation was more than three times higher, and for a speed of 80 mm/min it was more than ten times higher, compared to the longitudinal orientation of the joints. Even for the most favourable traverse speed, the fatigue strength of specimens containing FSW joints, at fatigue life of 130,000 cycles, was lower by about 30% compared to the base material specimens.

Available research papers concerning the impact of FSW parameters on the mechanical properties of copper FSW joints [11–16,25–29] usually focus on the microstructure, hardness profiles and static properties of joints. To the best knowledge of the authors, there are no publications on FSW copper joint properties which reference various joint orientation and/or fatigue properties. However, as the results of this paper demonstrate, accounting for the last factors may qualitatively change the conclusions about the strength properties of considered joints. As follows from Section 3.2. (Table 2 and Figure 13), the traverse speed had a rather moderate effect on the static strength parameters of the tested FSW joints. For tested FSW copper joints, contrary to available literature data [15,16,27], YS- and UTS-values decreased as traverse speed increased. However, this trend was significant only for YS-variation in the case of joints with T-orientations, for which the YS decreased by almost 19% with an increase of V-speed from 40 to 80 mm/min. For the other cases, the changes in YS and UTS values did not exceed 6%. Taking into consideration only the tensile and microhardness properties of joints, the traverse speed of 40 mm/min could be selected as the optimal one. Fatigue tests, however, proved to be more sensitive criteria for evaluation of the FSW joints' qualities. These tests clearly confirmed the weakness of the joints made at a speed of 80 mm/min with longitudinal orientation (which was earlier indicated only by the scatter of tensile test results (Figure 13)) and revealed the poor quality of the joints produced at a speed of 40 mm/min for the same orientation which had not been demonstrated earlier by other studies.

4. Conclusions

In this work, the properties of pure copper FSW joints were studied by means of microhardness, tensile and fatigue tests as well as microstructure and fractography analysis. As the main variables in conducted tests, three values of traverse speed and two different plate orientations in terms of rolling direction were considered. The constant rotation speed value (580 rpm) and different traverse speeds (40, 60 and 80 mm/min) were preliminarily selected for producing defect-free joints in terms of their visual and ultrasonic evaluations. The most relevant conclusions can be summarised as follows:

1. In the area of tested FSW joints, a significant decrease in microhardness was observed relative to the base material property. The microhardness profiles determined for the cross-section of the joints had a W-type shape with a sudden decrease in microhardness in the HAZ and a tendency towards a slight increase in WN which, in the qualitative sense, correlated well with the average grain sizes observed in these zones. However, the data available in the literature also indicate that opposite trends of microhardness changes are possible. Summarising the obtained results and the literature data, it may be generalised that increase of the microhardness in the region of FSW joints is observed for softer copper (microhardness below 80 HV), while the drop of hardness in the FSW zone is typical for copper with a higher degree of strain hardening.
2. Tensile strength parameters of tested FSW joints were significantly lower than those of base material properties which was predominantly related to the loss of the cold-work hardening effect as a result of the recrystallisation process in the HAZ. Resulting from the FSW process, the yield stress decreased on average by more than 60%, and ultimate stress fell by about 15%. For each of considered traverse speed, the FSW joints showed considerably higher static strength properties than similar joints made by TIG welding. These properties were also higher than those

of the base material after annealing. Regardless of traverse speed, the tensile strength parameters of FSW joints were higher for transversal than for longitudinal orientation. Tensile tests revealed a consistent trend of decreasing strength parameters of FSW joints (i.e., yield stress and ultimate stress) with increased traverse speed. Although this effect is moderate, it is qualitatively different from the data reported so far in the literature.

3. The fatigue tests turned out to be more sensitive criteria for evaluation of the FSW joints' qualities compared to other kind of examinations applied in this work. These tests clearly confirmed the weakness of the FSW joints produced at a speed of 80 mm/min in longitudinal orientation, which was earlier indicated only by the scatter of tensile test results, and revealed the poor quality of the joints made at a speed of 40 mm/min for the same orientation, which had not been predicted earlier by other studies. All samples from the two series of the lowest fatigue life were characterised by distinct location and morphology of the cracks. Considering all static and fatigue tests—for given plates, adopted tool geometry and specified rotary speed—the traverse speed of 60 mm/min proved to be the most advantageous. It should be noted that conclusions on the quality of the FSW joints resulting from fatigue tests would be different if they were referred separately to each of the considered joint orientations. This means that the variations of FSW parameters may have a qualitatively different effect on the properties of friction stir welds oriented longitudinally and transversely to the rolling direction of the joined plates.

Author Contributions: Conceptualization, T.M. and P.N.; methodology, T.M., M.H. and P.N.; static and fatigue tests, T.M., P.N. and A.K.; microhardness tests and metallographic analysis, M.H. and P.N.; visualization, P.N. and T.M.; data curation, P.N., T.M. and M.H.; writing—original draft preparation, T.M.; writing—editing and revision, T.M., P.N., A.K. and M.H.; correspondence, M.H.; funding acquisition, T.M., P.N. and M.H. All authors have read and agreed to the published version of the manuscript.

Funding: The financial support from subsidy granted by the Polish Ministry of Science and Higher Education within project no. 16.16.130.942 is acknowledged. This work has been supported within the framework of the project: The innovative system for coke oven wastewater treatment and water recovery with the use of clean technologies - INNOWATREAT that has received funding from the Research Fund for Coal and Steel under grant agreement no. 710078 and from The Ministry of Science and Higher Education Poland from financial resources on science in 2017–2019.

Conflicts of Interest: The authors declare no conflict of interest.

References

1. Mishra, R.S.; Sarathi, D.P.; Kumar, N. *Friction Stir Welding and Processing*; Springer: Cham, Switzerland; Heidelberg, Germany; New York, NY, USA; Dordrecht, The Netherlands; London, UK, 2014.
2. Jenney, C.L.; O'Brien, A. (Eds.) *Welding Handbook*, 9th ed.; American Welding Society: Miami, FL, USA, 2001.
3. Threadgill, P.L.; Leonard, A.J.; Shercliff, H.R.; Witherset, P.J. Friction stir welding of aluminium alloys. *Int. Mater. Rev.* **2009**, *54*, 49–93. [CrossRef]
4. Bevilacqua, M.; Ciarapica, F.E.; Forcellese, A.; Simoncini, M. Comparison among the environmental impact of solid state and fusion welding processes in joining an aluminum alloy. *Proc. Inst. Mech. Eng. Part B J. Eng. Manuf.* **2020**, *234*, 140–156. [CrossRef]
5. Johnson, R.; Threadgill, P. Friction Stir Welding of Magnesium Alloys. In Proceedings of the TMS Symposium on Magnesium Technology, San Diego, CA, USA, 2–6 March 2003.
6. Bruni, C.; Buffa, G.; Fratini, L.; Simoncini, M. Friction stir welding of magnesium alloys under different process parameters. *Mater. Sci. Forum* **2010**, *638–642*, 3954–3959. [CrossRef]
7. Forcellese, A.; Martarelli, M.; Simoncini, M. Effect of process parameters on vertical forces and temperatures developed during friction stir welding of magnesium alloys. *Int. J. Adv. Manuf. Technol.* **2016**, *85*, 595–604. [CrossRef]
8. Russell, M.J. Friction Stir Welding of Titanium Alloys—A Progress Update. In Proceedings of the 10th World Conference, Titanium, Hamburg, Germany, 13–18 July 2003.
9. Perrett, J.; Martin, J.; Peterson, J. Friction Stir Welding of Industrial Steels. In Proceedings of the TMS Annual Meeting, San Diego, CA, USA, 27 February–3 March 2011.
10. Fujii, H. Friction stir welding of steels. *Weld. Int.* **2011**, *25*, 731–744. [CrossRef]

11. Sakthivel, T.; Mukhopadhyay, J. Microstructure and mechanical properties of friction stir welded copper. *J. Mater. Sci.* **2007**, *42*, 8126–8129. [CrossRef]
12. Liu, H.J.; Shen, J.J.; Huang, X.Y.; Kuang, L.Y.; Liu, C.; Li, C. Effect of tool rotation rate on microstructure and mechanical properties of friction stir welded copper. *Sci. Technol. Weld. Join.* **2009**, *12*, 557–583. [CrossRef]
13. Lee, W.B.; Jung, S.B. The joint properties of copper by friction stir welding. *Mater. Lett.* **2004**, *58*, 1041–1046. [CrossRef]
14. Okamoto, K.; Doi, M.; Hirano, S.; Aota, K.; Okamura, H.; Aono, Y.; Ping, T.C. Fabrication of Backing Plate of Copper Alloy by Friction Stir Welding. In Proceedings of the 3rd International Symposium on Friction Stir Welding, Kobe, Japan, 27–28 September 2001.
15. Xue, P.; Xie, G.M.; Xiao, B.L.; Ma, Z.Y.; Geng, L. Effect of heat input conditions on micro-structure and mechanical properties of friction-stir-welded pure copper. *Metall. Mater. Trans. A* **2010**, *41*, 2010–2021. [CrossRef]
16. Khodaverdizadeh, H.; Mahmoudi, A.; Heidarzadeh, A.; Nazari, E. Effect of friction stir welding (FSW) parameters on strain hardening behaviour of pure copper joints. *Mater. Des.* **2012**, *35*, 330–334. [CrossRef]
17. Serio, L.M.; Palumbo, D.; de Filippis, L.A.C.; Galietti, U.; Ludovico, A.D. Effect of friction stir process parameters on the mechanical and thermal behavior of 754-H111 aluminum plates. *Materials* **2016**, *9*, 122. [CrossRef]
18. Costa, J.D.; Ferreira, J.A.M.; Borrego, L.P.; Abreu, L.P. Fatigue behaviour of AA6082 friction stir welds under variable loadings. *Int. J. Fatigue* **2012**, *37*, 8–16. [CrossRef]
19. Le, J.T.; Morgeneyer, T.F.; Denquin, A.; Gourgues-Lorenzon, A.F. Fatigue lifetime and tearing resistance of AA2198 Al–Cu–Li alloy friction stir welds: Effect of defects. *Int. J. Fatigue* **2015**, *70*, 463–472.
20. Chen, H.B.; Yan, K.; Lin, T.; Chen, S.B.; Jiang, C.Y.; Zhao, Y. The investigation of typical welding defects for 5456 aluminum alloy friction stir welds. *Mater. Sci. Eng. A* **2006**, *433*, 64–69. [CrossRef]
21. Cui, L.; Yang, X.; Zhou, G.; Xu, X.; Shen, Z. Characteristics of defects and tensile behaviors on friction stir welded AA6061-T4 T-joints. *Mater. Sci. Eng. A* **2012**, *543*, 58–68. [CrossRef]
22. Cederrqvist, L. Friction Stir Welding of Copper Canisters. Ph.D. Thesis, Lund University, Lound, Sweden, 2011.
23. Pikula, J.; Kwieciński, K.; Porembski, G.; Pietras, A. FEM Simulation of the FSW Process of Heat Exchanger Components. *Inst. Weld. Bull.* **2016**, *60*, 23–29. [CrossRef]
24. Lipowsky, H.; Arpaci, E. Copper in the Automotive Industry. *Wiley Intersci.* **2007**. [CrossRef]
25. Salahi, S.; Yapici, G.G. Fatigue behavior of friction stir welded joints of pure copper with ultra-fine grains. *Proc. Mater. Sci.* **2015**, *11*, 74–78. [CrossRef]
26. Barenji, R.V. Influence of heat input conditions on microstructure evolution and mechanical properties of friction stir welded pure copper joints. *Trans. Indian Inst. Met.* **2016**, *69*, 1077–1085. [CrossRef]
27. Nia, A.A.; Shirazi, A. Effects of different friction stir welding conditions on the microstructure and mechanical properties of copper plates. *Int. J. Miner. Metall. Mater.* **2016**, *23*, 799–809. [CrossRef]
28. Azizi, A.; Barenji, R.V.; Barenji, A.V.; Hashemipour, M. Microstructure and mechanical properties of friction stir welded thick pure copper plates. *Int. J. Adv. Manuf. Technol.* **2016**, *86*, 1985–1995. [CrossRef]
29. Farrokhi, H.; Heidarzadeh, A.; Saeid, T. Frictions stir welding of copper under different welding parameters and media. *Sci. Technol. Weld. Join.* **2013**, *18*, 697–702. [CrossRef]
30. Asadi, P.; Givi, M.K.B.; Akbari, M. Microstructural simulation of friction stir welding using a cellular automaton method: A microstructure prediction of AZ91 magnesium alloy. *Int. J. Mech. Mater. Eng.* **2015**, *10–20*, 1–14. [CrossRef]
31. Lin, J.W.; Chang, C.H.; Wu, M.H. Comparsion of mechanical properties of pure copper welded using friction stir welding and tungsten inert gas welding. *J. Manuf. Process.* **2014**, *16*, 296–304. [CrossRef]

Article

Effect of Tilt Angle and Pin Depth on Dissimilar Friction Stir Lap Welded Joints of Aluminum and Steel Alloys

Veerendra Chitturi *, Srinivasa Rao Pedapati and Mokhtar Awang

Department of Mechanical Engineering, Universiti Teknologi PETRONAS, Seri Iskandar 32610, Malaysia; srinivasa.pedapati@utp.edu.my (S.R.P.); mokhtar_awang@utp.edu.my (M.A.)
* Correspondence: Veerendra_16000644@utp.edu.my; Tel.: +60-182122064

Received: 31 October 2019; Accepted: 23 November 2019; Published: 26 November 2019

Abstract: Automobile, aerospace, and shipbuilding industries are looking for lightweight materials for cost effective manufacturing which demands the welding of dissimilar alloy materials. In this study, the effect of tool rotational speed, welding speed, tilt angle, and pin depth on the weld joint were investigated. Aluminum 5052 and 304 stainless-steel alloys were joined by friction stir welding in a lap configuration. The design of the experiments was based on Taguchi's orthogonal array for conducting the experiments with four factors and three levels for each factor. The microstructural analysis showed tunnel defects, micro voids, and cracks which formed with 0° and 1.5° tilt angles. The defects were eliminated when the tilt angle increased to 2.5° and a mixed stir zone was formed with intermetallic compounds. The presence of the intermetallic compounds increased with the increase in tilt angle and pin depth which further resulted in obtaining a defect-free weld. Hooks were formed on either side of the weld zone creating a mechanical link for the joint. A Vickers hardness value of HV 635.46 was achieved in the mixed stir zone with 1000 rpm, 20 mm/min, and 4.2 mm pin depth with a tilt angle of 2.5°, which increased by three times compared to the hardness of SS 304 steel. The maximum shear strength achieved with 800 rpm, 40 mm/min, and a 4.3 mm pin depth with a tilt angle of 2.5° was 3.18 kN.

Keywords: dissimilar materials; friction stir lap welding; pin depth; tilt angle; tool rotational speed; welding speed

1. Introduction

The joining of two different materials which are completely different in their mechanical, physical, and chemical properties is always a challenge. Aluminum and steel are very difficult to weld because of the large difference in their melting temperatures. Friction stir welding (FSW), which was introduced in 1991 at The Welding Institute, has the advantage over fusion welding techniques in terms of joining dissimilar materials. Friction stir welding depends entirely on how much heat is produced during the process, an outcome of the friction between the rotation tool and the fixed plates. If the heat generated is too high, the aluminum may reach its melting point which will not result in a sound joint; if the heat generated is not enough to stir the steel into aluminum, it will result in a joint with many defects. Thus, process parameters, such as tool rotational speed, welding speed, pin depth, and tilt angle, play a crucial role in deciding the weld quality [1,2].

Friction stir lap welding (FSLW) of aluminum and steel is usually carried with the aluminum on top and the steel on the bottom so that the heat generated will be enough for plastic deformation of the materials to form a strong joint. Elrefaey et al. [3] welded commercially pure aluminum and low carbon steel plates with different ranges in process parameters to investigate the joint strength and concluded that a minimal difference in the pin depth will have a large impact on the joint strength.

Lap joints between AA 5083 and SS 304 of 3 mm thickness were produced by Kimapong et al. [4] with a pin depth of 3–3.3 mm and found that increasing the rotational speed decreased the joint strength and vice versa with traverse speed and more intermetallic compounds forming with increased pin depth. Chen et al. [5] welded three different types of steel with AC4C alloy and found that zinc-coated steel had better fracture strength when compared with brushed and mirror-finished steel which was nearly 97.7% of the chosen steel. The effect of the cylindrical pin on 1.5 mm AA6181-T4 and HC340LA joints was studied by Coelho et al. [6] where 0.2 µm thick intermetallic compounds were formed and the weld had 73% joint strength of the aluminum base material. When Movahedi et al. [7] welded 3 mm thick AA 5083 and St-12 with different tool rotational speeds and welding speeds, they found that the joint strength increased with decreasing welding speed, and the microstructure revealed the presence of intermetallic compounds. The effect of tool rotational speed and traverse speed on 6063 aluminum and 1 mm thickness zinc-coated steel with a pin depth of 3.1 mm was investigated by Das et al. [8] and achieved 58.6% failure load of the base aluminum alloy. Studies on FSLW of 1060 aluminum alloy and 1Cr18Ni9Ti steel were investigated by Wei et al. [9] which revealed that the joint strength increased with the increase in pin depth when the tilt angle was maintained at 0°. The maximum shear load of 7500N was achieved by Wan et al. [10] when 6082-T6 aluminum and Q235A steel were friction stir lap welded with a tilt angle of 2.5° and resulted in defect-free weld. Various process parameters were used by Nguyen et al. [11] to weld DP800 and AA6351 sheets, and they noticed that the weld joints with high intermetallic thickness had a higher joint strength.

Not only do the process parameters and plunge depth effect the joint properties, but the pin geometry along with pin material will also have an impact on the joint strength. A new cutting-edge pin geometry was introduced by Xiong et al. [12]; the pin was made with YG8 cemented carbide rotary burrs and inserted into an H13 tool holder to weld 1100 aluminum alloy and 1Cr18Ni9Ti stainless steel sheets which yielded a shear strength of 89.7 MPa, which is stronger than that of the aluminum base alloy. Shamsujjoha et al [13] used W-Re-HfC pin tools to weld 1018 stainless steel and 6061 aluminum alloy and found that the longer pin tool achieved the highest joint strength which was 58% of the base aluminum alloy. Pseudo steady-heat transfer analysis conducted by Das et al. [14] proved that the presence of $Al_{13}Fe_4$ intermetallic compound (IMC) increases with the increase in rotational speed and decrease in welding speed which results in almost a 70% joint strength of steel. Further, the study on intermetallic compounds formed between the weld joints of 6061 aluminum and high-strength interstitial free steel proved that thermo-dynamically stable IMC's are formed, confirming that diffusion plays a major role in joint strength [15]. The microstructure of the lap welded joints of AA 6061-T6 and low carbon steels revealed that the thickness of the IMC layer decreases with the increase in welding speed which, in turn, will affect the joint strength [16]. Annealing was used to increase the joint strength of 3 mm 5083 aluminum alloy and 1 mm St-12 steel sheet by Movahedi et al. [17] when at 300 °C and 350 °C. The joint strength increased with the increase in annealing time, but the joint strength decreased as the temperature was increased to 400 °C. Haghshenas et al. [18] found that the joint strength achieved by diffusion bonding of 5754 aluminum alloy with coated high-strength steels increased with the decrease in welding speed and increase in pin depth. Fatigue crack propagation and fracture toughness of friction stir lap welded 3003 aluminum and SUS304 were examined by Hidehito et al. [19] who noticed that fracture toughness depends on crack propagation. A novel technique called self-riveting FSLW was developed by Huang et al. [20] to join a 6082-T6 aluminum alloy to QSTE340TM steel, and they proved that the riveting technique in FSLW will improve the fracture load of the joint.

It is evident from the abovementioned studies [1–20] that the selection of process parameters plays a significant role in determining the robustness of the weld. The objective of this study was to investigate the effect of tilt angle and pin depth on the welded joint. The tool rotational speed ranges used were from 800 rpm to 1200 rpm and the welding speed ranges used were from 20 mm/min to 40 mm/min. The pin depths chosen for the experimental procedure were between 4.1 mm and 4.3 mm with tilt angles ranging from 0° to 2.5°.

2. Experimental Methodology

The base metals chosen for this study were aluminum 5052 alloy with the dimensions 200 mm × 100 mm × 4 mm and 304 stainless steel alloy with the dimensions 200 mm × 100 mm × 2 mm. The chemical compositions of the base metals are shown in Table 1.

Table 1. Composition of 304 stainless steel and 5052 aluminum.

SS 304 Elements	Fe	Cr	Ni	Mn	N	S	C	Si	P
Percentage (wt.%)	Bal	18	8	2	0.1	0.03	0.08	0.75	0.045
AA 5052 Elements	**Al**	**Cr**	**Mg**	**Mn**	**Fe**	**Cu**	**Zn**	**Si**	**Others**
Percentage (wt.%)	Bal	0.15–0.35	2.2–2.8	0.1	0.4	0.1	0.1	0.25	0.15

The base metals are welded in lap joint configuration with CNC FSW machine with a constant inflow of argon gas to cool the tool from overheating. The schematic diagram of FSLW is shown in Figure 1. Tool material was made with Tungsten–Rhenium (W–Re), with a shoulder diameter of 20 mm, cylindrical shaped pin diameter of 5 mm with a pin length of 4.1 mm. Based on the preliminary results, the process parameters chosen for the design of experiments were tool rotational speed (TRS), welding speed (WS), pin depth (PD), and tilt angle (TA). The ranges of these parameters are shown in Table 2. The design of the experiments included four factors, each having three levels, and the optimized number of experiments were chosen according to the Taguchi L9 orthogonal array as shown in Table 3. The experiments were conducted using a load force of 7.5 kN.

Table 2. Process parameters and their levels.

Parameter	Level 1	Level 2	Level 3
Tool Rotational Speed (rpm)	800	1000	1200
Welding Speed (mm/min)	20	30	40
Pin Depth (mm)	4.1	4.2	4.3
Tilt Angle (Degree)	0	1.5	2.5

Table 3. Taguchi L9 orthogonal array.

Run	TRS (rpm)	WS (mm/min)	PD (mm)	TA (degree)
FSLW-1	800	20	4.1	0
FSLW-2	800	30	4.2	1.5
FSLW-3	800	40	4.3	2.5
FSLW-4	1000	20	4.2	2.5
FSLW-5	1000	30	4.3	0
FSLW-6	1000	40	4.1	1.5
FSLW-7	1200	20	4.3	1.5
FSLW-8	1200	30	4.1	2.5
FSLW-9	1200	40	4.2	0

The welded joints were cut according to ASTM E8 (sub-size) standards perpendicularly to the welding direction using wire-cut electrical discharge machining (EDM, Mitsubishi, Tokyo, Japan) for tensile shear testing. The dimensions of the tensile shear test specimen are shown in Figure 2. The tensile shear tests were carried out on a 100 kN universal testing machine at a speed rate of 3 mm/min. The hardness of the welded joints was determined with a Vickers hardness testing machine (Leco LM 247AT, St. Joseph, MI, USA) by applying a load of 500 gf with a dwell time of 15 s. The hardness values were taken in the mixed stirred zone along the cross-section. Samples for microstructural analysis were cut in 20 mm × 20 mm size with wirecut EDM. Keller's reagent was used to etch the 5052 aluminum side and Nital was used to etch the SS 304 side. Microstructure analysis was conducted using scanning electron microscopy (SEM, Phenom Pro X, Eindhoven, Netherlands).

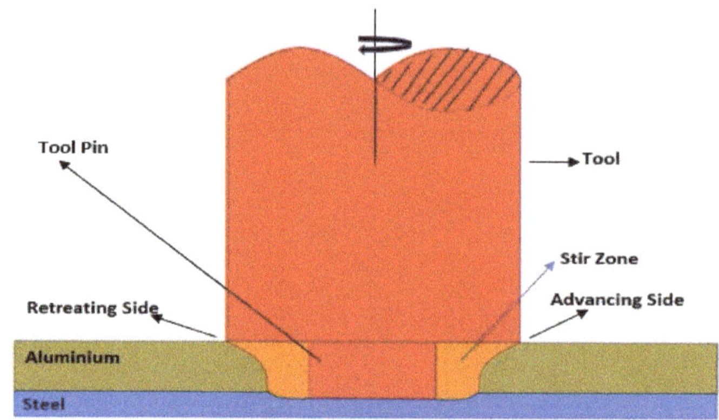

Figure 1. Schematic view of friction stir lap welding.

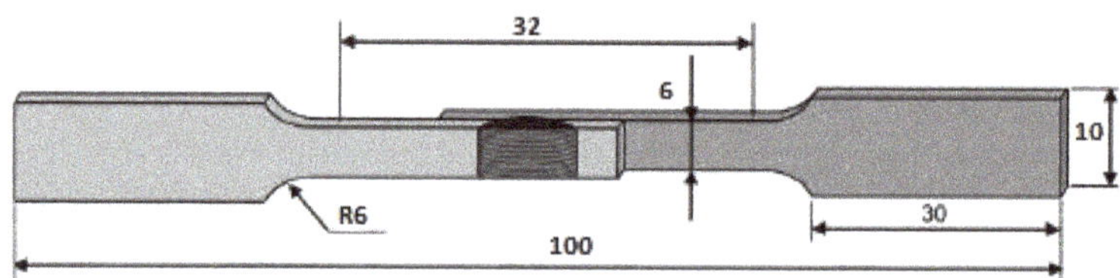

Figure 2. Dimensions of the tensile shear test specimen (mm).

3. Results and Discussion

3.1. Macrostructure of the Lap Joints

Macrostructures of the welded samples will give a good perception of the welded zone. The experiments were carried out with a 0° tilt angle, and the formation of a tunnel defect can be observed in Figure 3a. This was the result of not enough stirring of material and heat input, because the aluminum was thrown out of the mixed stir zone while stirring. The tunnel defect size increased with the increase in the depth and resulted in more flash formation during the welding. When the tilt angle was increased to 1.5 degrees with the same rotational and traverse speed and pin depth, the tunnel defect was reduced, but some micro voids were formed (Figure 3b) in FSLW-2 and FSLW-6 where the pin depth was 4.1 mm (0.1 penetration mm in steel) and 4.2 mm (0.2 penetration mm in steel). The micro voids were reduced to micro-cracks when the pin depth increased to 4.3 mm (0.3 penetration mm in steel) in FSLW-7. The macrostructure of the dissimilar FSLW joint is shown in Figure 3c, where a sound and defect-free joint was obtained. In the nugget zone, the material was extruded from the advancing side towards the retreating side of the weld, forming a mechanical link between the aluminum and steel plates.

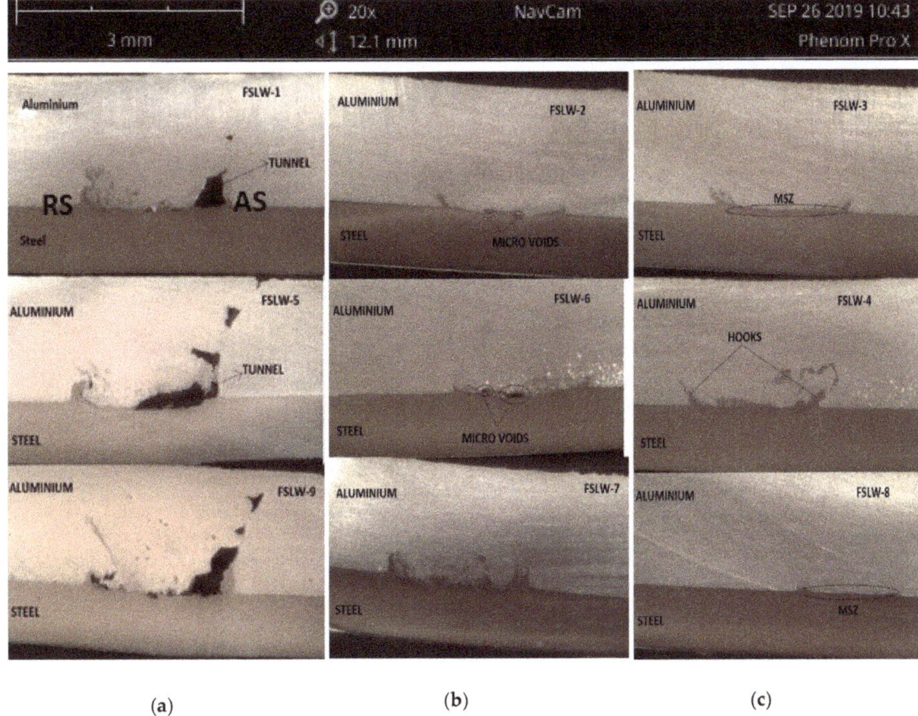

Figure 3. Macrostructure of the specimens with different tilt angles: (**a**) 0° Tilt Angle, (**b**) 1.5° Tilt Angle, (**c**) 2.5° Tilt Angle.

3.2. Microstructure of the Lap Joints

Microstructures of the welded zone were obtained using a scanning electron microscope. As a result of the tunnel defect shown in Figure 4, different sizes of scattered steel fragments were found in the mixed stir zone. The uneven distribution of steel fragments into aluminum increased with the increase in tool rotational speed. Due to the movement of the tool in the welding direction, the steel particles from the surface of the steel were dragged along the welding direction which is the reason for the uneven scattering of the steel fragments. The tunnel defect was observed on the advancing side whereas the stir zone on the retreating side was good and without any defects.

Increasing the tilt angle to 1.5° reduced the tunnel defect, but the formation of micro voids and root cracks can be observed. The uneven distribution of steel and aluminum fragments is observed in Figure 5 in the mixed stir zone (MSZ). The rotation of the tool was in the clockwise direction, and by increasing the tilt angle, the escaping of material from the stir zone was reduced, and because of the pin depth, a mixed stir zone was formed consisting of aluminum and steel fragments alongside the intermetallic compounds. A wave-like streamline was detected which is evidence that the mixture of both materials is occurring. However, the flash formed was less when compared to the 0° tilt angle, because the stirred material was pushed inside instead of escaping from the stirred zone.

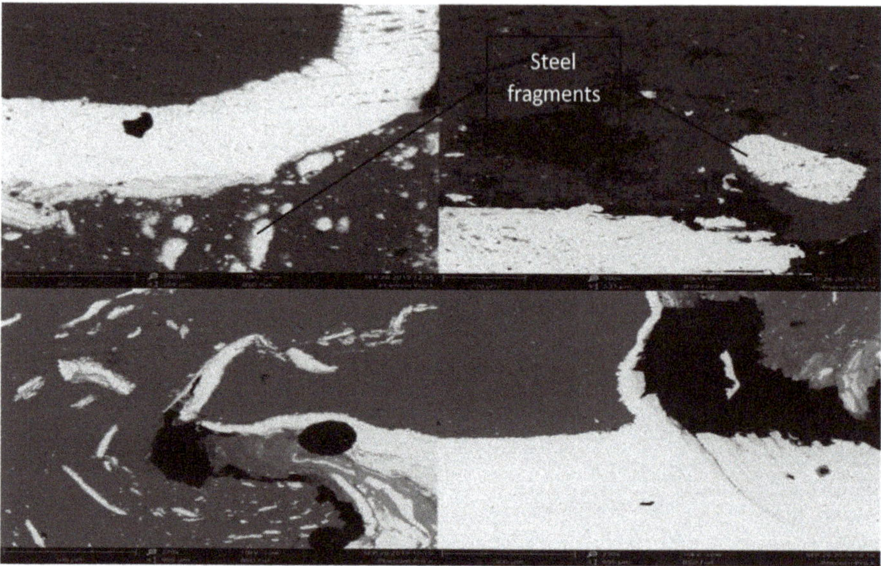

Figure 4. SEM images of the mixed stir zone with a 0° tilt angle.

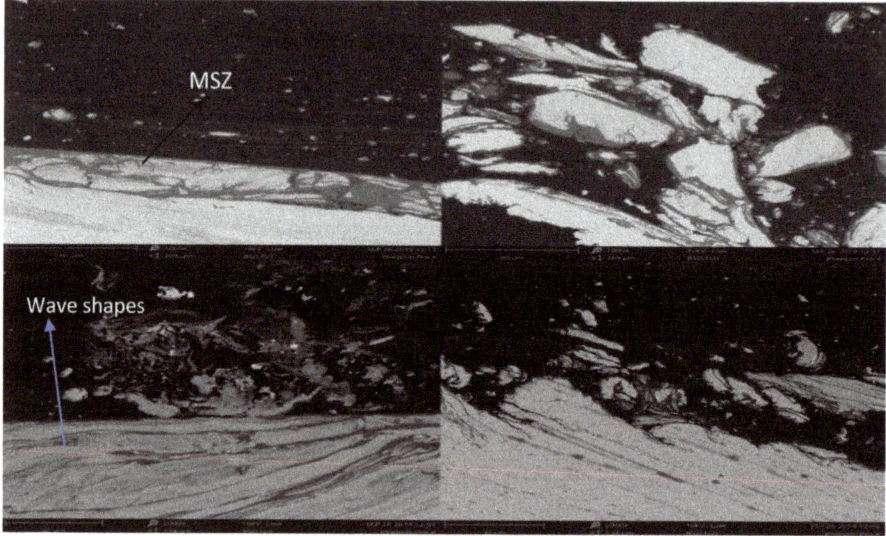

Figure 5. SEM images of the mixed stir zone with a 1.5° tilt angle.

Further increasing the tilt angle to 2.5°, a sound weld was obtained with proper mixing of aluminum and steel and without any defects as can be seen in Figure 6. The small voids which were seen with the 1.5° tilt angle were also eliminated. The increase in tilt angle and pin depth eliminated the uneven distribution of fragments in the stir zone. There was a hook-like mechanical link formed on either side of the stir zone as shown in Figure 6. The formation of the hook depends on the depth of the pin. Even here, the intermetallic compounds can be perceived in the mixed stir zone (MSZ) of the weld. The mechanical link which was formed on either side of the weld zone holds a crucial role in deciding the strength of the joint. The hook formed with FSLW-8 was less when compared to FSLW-3 and FSLW-4, and, as a result, the maximum tensile shear strength of 3.18 kN and 3.13 kN was achieved with FSLW-3 and FSLW-4, respectively.

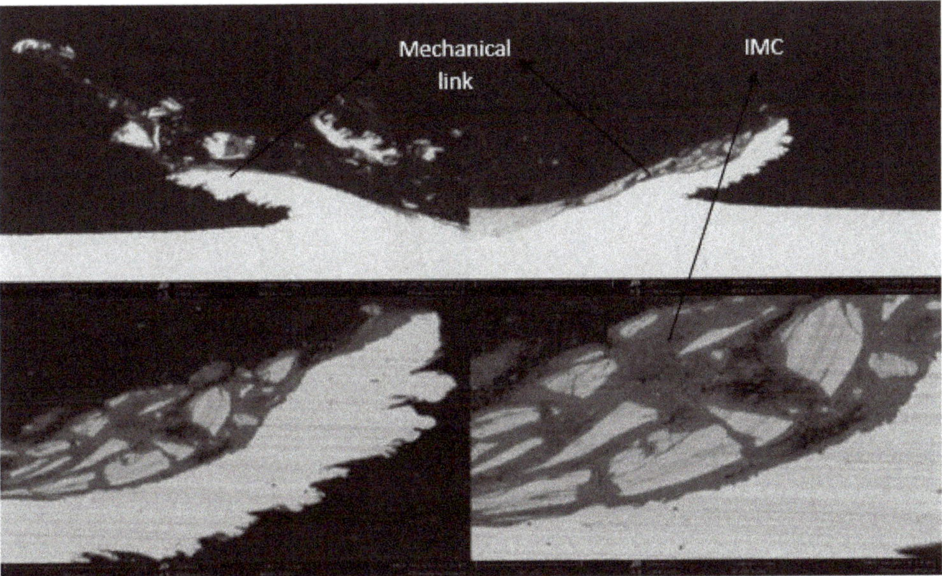

Figure 6. SEM images of the mixed stir zone with a 2.5° tilt angle.

3.3. Mechanical Properties of the Lap Joint

Tensile shear test results showed that a maximum of 3.18 kN shear strength can be achieved with a tool rotational speed of 800 rpm, welding speed of 40 mm/min with a pin depth of 4.2 mm when the angle was tilted to 2.5 degrees, and the least, 0.64 kN, was obtained with a TRS of 1000 rpm, WS of 30 mm/min, with a pin depth of 4.3 when the angle was not tilted. It is evident from the results that tensile shear strength mostly depends on the tilt angle and pin depth rather than tool rotational speed and welding speed. The top three tensile shear test results were obtained when the angle was tilted to 2.5 degrees, the least three were obtained when there was no tilt in the angle, and the medium results were obtained when the angle was tilted to 1.5 degrees.

The effect of pin depth and tilt angle on tensile shear strength is shown in Figure 7. At 0° tilt angle, the shear strength decreased with the increase in pin depth, and, at a 2.5° tilt angle, it increased with the increase in pin depth. This pattern shows that the increase in tilt angle can stir and hold the material together in the mixed stir zone such that the tunnel defects, which were formed at 0°, are eliminated. But, when at 1.5°, the pin depth did not show much variation in the joint strength.

The Vickers hardness results confirm that a maximum of 635.46 HV was obtained when the tool rotational speed was 1000 rpm, welding speed was 20 mm/min, the pin depth was 4.2, and the tilt angle 2.5°. The hardness results also followed the same pattern as the tensile shear test results, where the highest hardness was obtained when the tilt angle was 2.5° and the least when the tilt angle was 0°. The pin depth variation did not show much effect on the hardness like the tilt angle. As shown in Table 4, the hardness values with different pin depths at the same tilt angles were almost near to one another. The effect of the pin depth and tilt angle on hardness is shown in Figure 8. According to Basariya et al. [21], high hardness values can be obtained when intermetallic compounds are formed.

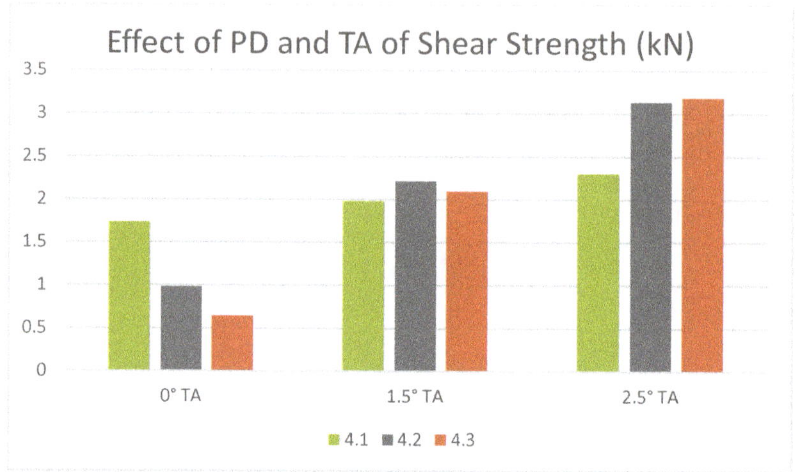

Figure 7. Effect of tilt angle on shear strength at various pin depths.

Table 4. Hardness and tensile shear strength values of the specimens.

Run	TRS (rpm)	WS (mm/min)	PD (mm)	Tilt Angle (degree)	Hardness (HV)	Shear Strength (kN)
FSLW-1	800	20	4.1	0	333.32	1.73
FSLW-2	800	30	4.2	1.5	524.02	2.21
FSLW-3	800	40	4.3	2.5	576.9	3.18
FSLW-4	1000	20	4.2	2.5	635.46	3.13
FSLW-5	1000	30	4.3	0	355.42	0.64
FSLW-6	1000	40	4.1	1.5	516.42	1.98
FSLW-7	1200	20	4.3	1.5	517.9	2.09
FSLW-8	1200	30	4.1	2.5	588.86	2.30
FSLW-9	1200	40	4.2	0	214.84	0.98

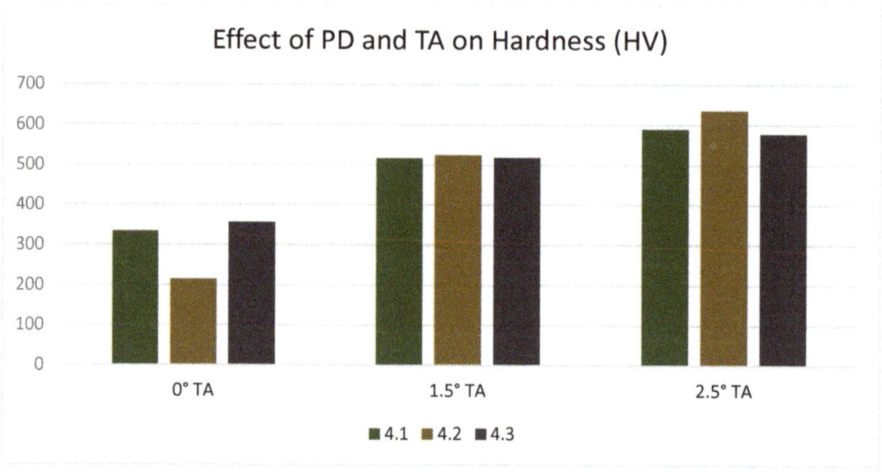

Figure 8. Effect of tilt angle on the hardness at various pin depths.

3.4. Signal-to-Noise Ratio Analysis

The obtained tensile shear strength and hardness values were further analyzed using Taguchi's signal-to-noise (S/N) ratios to obtain the influence of factors on the responses. The approach for the S/N ratio in this study was "the larger the better" in order to obtain the maximum response. The calculated S/N ratios for the criteria for "the larger the better" and mean responses are shown in Table 5.

Table 5. Signal-to-noise ratios of the experiments.

Level	Tool Rotational Speed	Welding Speed	Pin Depth	Tilt Angle
1	67.25	67.04	65.99	60.25
2	63.99	63.43	65.57	66.43
3	64.52	65.29	64.21	69.08
Delta	3.25	3.61	1.78	8.83
Rank	3	2	4	1

The output of S/N ratios are plotted in Figure 9. According to "the larger the better" criteria, it is revealed from the plot that the maximum tensile strength and hardness can be achieved with a tool rotational speed of 800 rpm, welding speed of 20mm/min, 4.1 mm pin depth, and a tilt angle of 2.5° where the tilt angle is the most influential factor ranked at 1 followed by welding speed, tool rotational speed, and pin depth.

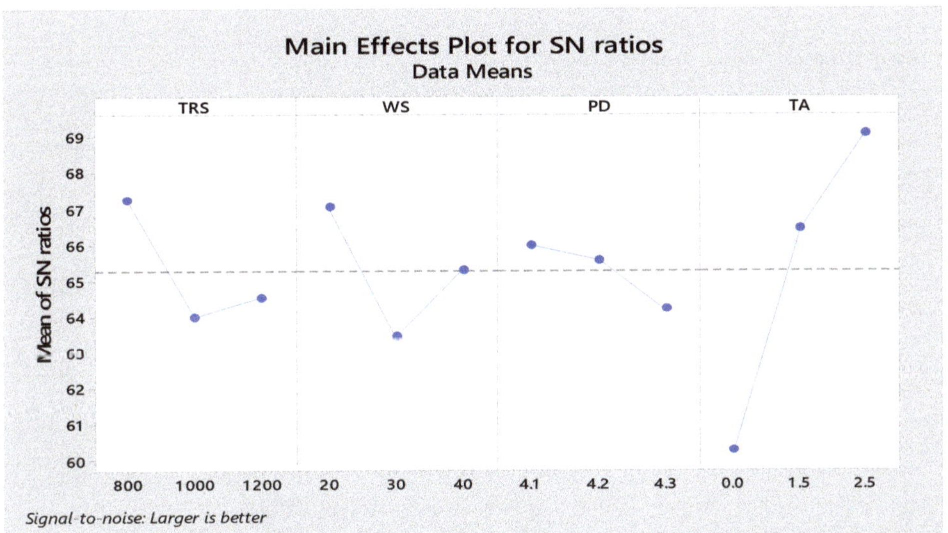

Figure 9. The signal-to-noise ratio plot for different process parameters.

4. Conclusions

In this study, friction stir lap welding of 5052 aluminum and 304 stainless-steel alloys were successfully achieved without any defects, and the following conclusions were drawn from the results.

1. Microstructural analysis revealed that a defect-free joint can be achieved with a tilt angle of 2.5° if the mixed stir zone is visible with respect to pin depth. Micro voids are formed if the tilt angle is reduced to 1.5° with steel fragments scattering in the mixed stir zone, and tunnel defects are clearly visible on the advancing side of the stir zone and uneven distribution of steel particles on the retreating side when the tilt angle is further reduced to 0°.

2. Tensile shear test results showed that a maximum strength of 3.18 N was achieved with 800 rpm, 40 mm/min, and a pin depth of 4.3 mm when the angle was tilted to 2.5° due to the fact of a good mixed stir zone with intermetallic compounds.
3. Eliminating the micro voids and tunnel defects achieved a maximum hardness of HV 635.36 with 1000 rpm, 20 mm/min, and a pin depth of 4.2 mm when the angle was tilted to 2.5° and when a defect-free weld was obtained.
4. From the microstructural analysis and tensile shear strength result, it is evident that a better joint strength is achieved with a tilt angle of 2.5° and a pin depth of 4.3 mm.

Author Contributions: The research work was carried out under the supervision of S.R.P. and M.A. The resources for the research were procured by S.R.P. The methodology, experiments, characterization and other tests were carried out by V.C. The original draft was written by V.C., S.R.P. and M.A. have contributed their suggestions in reviewing and editing the manuscript.

Funding: This research received no external funding.

Acknowledgments: The work was supported by the Institute of Transport Infrastructure, Universiti Teknologi PETRONAS, Malaysia.

Conflicts of Interest: The authors declare no conflict of interest. The funders had no role in the design of the study; in the collection, analyses, or interpretation of data; in the writing of the manuscript, or in the decision to publish the results.

References

1. Wan, L.; Huang, Y. Friction stir welding of dissimilar aluminium alloys and steels: A review. *Int. J. Adv. Manuf. Technol.* **2018**, *99*, 1781–1811. [CrossRef]
2. Mehta, K. A review on friction-based joining of dissimilar aluminum–steel joints. *J. Mater. Res.* **2019**, *34*, 78–96. [CrossRef]
3. Elrefaey, A.; Gouda, M.; Takahashi, M.; Ikeuchi, K. Characterization of aluminium/steel lap joint by friction stir welding. *J. Materi. Eng. Perform.* **2005**, *14*, 10–17. [CrossRef]
4. Kimapong, K.; Watanabe, T. Lap joint of A5083 aluminum alloy and SS400 steel by friction stir welding. *Mater. Trans.* **2005**, *46*, 835–841. [CrossRef]
5. Chen, Y.; Nakata, K. Effect of the Surface State of Steel on the Microstructure and Mechanical Properties of Dissimilar Metal Lap Joints of Aluminum and Steel by Friction Stir Welding. *Metall. Mat. Trans. A* **2008**, *39*, 1985–1992. [CrossRef]
6. Coelho, R.S.; Kostka, A.; Sheikhi, S.; Dos Santos, J.; Pyzalla, A.R. Microstructure and Mechanical Properties of an AA6181-T4 Aluminium Alloy to HC340LA High Strength Steel Friction Stir Overlap Weld. *Adv. Eng. Mater.* **2008**, *10*, 961–972. [CrossRef]
7. Movahedi, M.; Kokabi, A.H.; Reihani, S.S.; Najafi, H. Mechanical and Microstructural characterization of Al-5083/St-12 lap joints made by friction stir welding. *Procedia Eng.* **2011**, *10*, 3297–3303. [CrossRef]
8. Das, H.; Basak, S.; Das, G.; Pal, T.K. Influence of energy induced from process parameters on the mechanical properties of friction stir welded lap joint of aluminium to coated steel sheet. *Int. J. Adv. Manuf. Technol.* **2013**, *64*, 1653–1661. [CrossRef]
9. Wei, Y.; Li, J.; Xiong, J.; Zhang, F. Effect of tool pin insertion depth on friction stir lap welding of aluminium to stainless steel. *J. Materi. Eng. Perform.* **2013**, *22*, 3005–3013. [CrossRef]
10. Wan, L.; Huang, Y. Microstructure and Mechanical properties of Al/Steel Friction Stir Lap Weld. *Metals* **2017**, *7*, 542. [CrossRef]
11. Nguyen, V.N.; Nguyen, Q.M.; Thi HT, D.; Huang, S.C. Investigation on lap joint friction stir welding between AA6351 alloys and Dp800 steels. *Sādhanā* **2018**, *43*, 160–167. [CrossRef]
12. Xiong, J.T.; Li, J.L.; Qian, J.W.; Zhang, F.S.; Huang, W.D. High strength lap joint of aluminium and stainless steels fabricated by friction stir welding with cutting pin. *Sci. Technol. Weld. Join.* **2012**, *17*, 196–201. [CrossRef]
13. Shamsujjoha, M.; Jasthi, B.K.; West, M.; Widener, C. Friction Stir Lap Welding of Aluminum to Steel Using Refractory Metal Pin Tools. *J. Eng. Mater. Technol.* **2015**, *137*, 021009–021017. [CrossRef]

14. Das, H.; Jana, S.S.; Pal, T.K.; De, A. Numerical and experimental investigation on friction stir lap welding of aluminium to steel. *Sci. Technol. Weld. Join.* **2014**, *19*, 69–75. [CrossRef]
15. Das, H.; Ghosh, R.N.; Pal, T.K. Study on the formation and characterization of the Intermetallics in Friction stir welding of Aluminium alloy to coated steel sheet lap joint. *Metall. Mat. Trans. A* **2014**, *45*, 5098–5106. [CrossRef]
16. Boumerzoug, Z.; Helal, Y. Friction stir welding of dissimilar materials aluminum AL6061-T6 to low carbon steel. *Metals* **2017**, *7*, 42. [CrossRef]
17. Movahedi, M.; Kokabi, A.H.; Reihani, S.S.; Cheng, W.J.; Wang, C.J. Effect of annealing treatment on joint strength of aluminum/steel friction stir lap weld. *Mater. Design* **2013**, *44*, 487–492. [CrossRef]
18. Haghshenas, M.; Abdel-Gwad, A.; Omran, A.M.; Gökçe, B.; Sahraeinejad, S.; Gerlich, A.P. Friction stir weld assisted diffusion bonding of 5754 aluminum alloy to coated high strength steels. *Mater. Design* **2014**, *55*, 442–449. [CrossRef]
19. Nishida, H.; Ogura, T.; Hatano, R.; Kurashima, H.; Fujimoto, M.; Hirose, A. Fracture toughness and fatigue crack behaviour of A3003/SUS304 lap friction stir welded joints. *Weld. Int.* **2017**, *31*, 268–277. [CrossRef]
20. Huang, Y.; Wang, J.; Wan, L.; Meng, X.; Liu, H.; Li, H. Self-riveting friction stir lap welding of aluminium alloy to steel. *Mater. Lett.* **2016**, *185*, 181–184. [CrossRef]
21. Basariya, M.I.R.; Mukhopadhyay, N.K. Chapter 5, Structural and mechanical behaviour of Al-Fe intermetallics. In *Intermetallics Compounds*; IntechOpen: London, UK, 2018.

Review

Manufacturing Parameters, Materials, and Welds Properties of Butt Friction Stir Welded Joints–Overview

Aleksandra Laska * and Marek Szkodo

Department of Materials Engineering and Bonding, Faculty of Mechanical Engineering, Gdansk University of Technology, Narutowicza 11/12, 80-233 Gdansk, Poland; mszkodo@pg.edu.pl
* Correspondence: aleksandra.laska@pg.edu.pl; Tel.: +48-698-071-526

Received: 2 October 2020; Accepted: 31 October 2020; Published: 3 November 2020

Abstract: The modern and eco-friendly friction stir welding (FSW) method allows the combination of even such materials that are considered to be non-weldable. The development of FSW technology in recent years has allowed a rapid increase in the understanding of the mechanism of this process and made it possible to perform the first welding trials of modern polymeric and composite materials, the joining of which was previously a challenge. The following review work focuses on presenting the current state of the art on applying this method to particular groups of materials. The paper has been divided into subchapters focusing on the most frequently used construction materials, with particular emphasis on their properties, applications, and usage of the FSW method for these materials. Mechanisms of joint creation are discussed, and the microstructure of joints and the influence of material characteristics on the welding process are described. The biggest problems observed during FSW of these materials and potential causes of their occurrence are quoted. The influence of particular parameters on the properties of manufactured joints for each group of materials is discussed on the basis of a wide literature review.

Keywords: friction stir welding; FSW; solid type welding; mechanical properties; weld strength

1. Introduction

Friction stir welding (FSW) is a method invented at the Welding Institute of the United Kingdom and patented by Wayne Thomas in 1991 [1]. It is considered to be one of the most prospective material joining developments in the last 30 years. Primarily, this method was dedicated to joining aluminum and its alloys, but today it is widely used for titanium and its alloys, magnesium and its alloys, steel and ferrous alloys, and copper, but also polymers and composites. The FSW process is defined as a solid-state method. Materials to be joined do not melt during the process. Since the melting point is not reached, typical problems of fusion welding techniques are eliminated. These problems are usually related to a change of state, such as changes of volume and solubility of gases, and these effects are not observed during friction stir welding process [2–4].

During the process, a specially designed tool is put into linear movement along a joint line, rotating at the same time. The kinetic energy of the tool is transformed into thermal energy, generated by the friction on the interface between the tool and the components. The heated material is plasticized by a tool and extruded around the pin in a backward direction of a tool moving along the edge of a contact line. The FSW method is usually used to produce butt welds, but it also allows the fabrication of joints of other types, such as corner welds, T-welds, lap welds, and fillet welds [5–8]. A schematic illustration of a friction stir welded butt joint is shown in Figure 1. Nowadays, the FSW method is widely used in many industrial areas, such as aerospace (wings, fuel and cryogenic tanks, fuselages) [9–12], railways (underground carriages, wagons, container bodies) [13–15], marine and shipbuilding (deck panels, hulls, booms, masts, offshore accommodation) [14,16], construction industry (frames, bridges, pipelines) [17,18] and land transportation (wheel rims, mobile cranes, tail lifts) [19]. The FSW technology is also applied in sectors such as machinery equipment, electronics, metalworking, and the R&D sector [20].

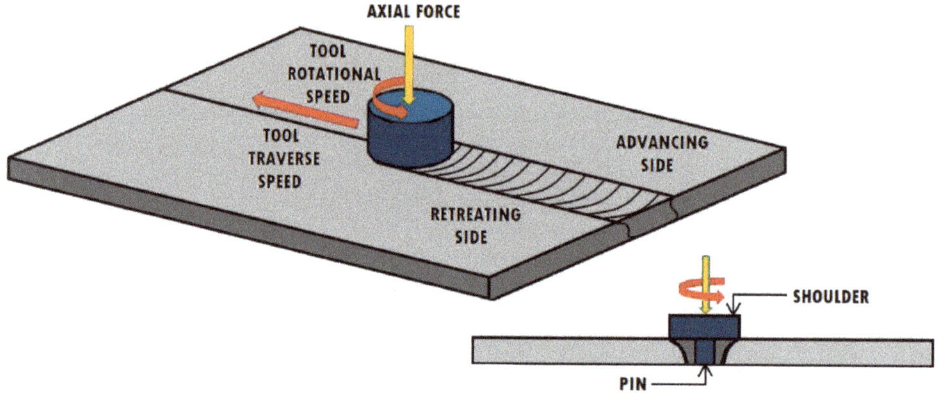

Figure 1. Schematic illustration of a friction stir welding process.

In the cross-section of the friction stir welded joint, a specific microstructure is observed. Due to the solid-state nature of the process, the zones that are not found in welds produced by conventional welding methods can be distinguished. Based on thermomechanical actions of the FSW tool, four distinct zones can be observed: weld nugget (stir zone, SZ), thermo-mechanically affected zone (TMAZ), heat affected zone (HAZ), and base material (unaffected zone, BM). The presence of the stir zone is a result of the recrystallization in the middle part of the thermo-mechanically affected zone. The nugget is formed by fine grain sized metal. The material of the SZ experiences plastic deformation due to the interactions with the tool. Rhodes et al. [21] and Liu et al. [22] claimed that in the recrystallized grains, a low density of dislocations is observed. However, other studies proved that the recrystallized grains of the SZ contain high density of sub-boundaries [23], dislocations [24] and subgrains [25]. Between the SZ and the HAZ, a unique transition zone, called the thermo-mechanically affected zone, can be observed. TMAZ is exposed to both temperature and deformations during the process. Because of insufficient deformations, strain recrystallization is not observed. Exposure to high temperatures during welding might cause the dissolution of precipitates in TMAZ. Beyond TMAZ, the heat-affected zone is observed. In that region, there are no plastic deformations, but it is still subjected to a thermal cycle. The alteration of properties in HAZ, compared to the base material, includes changes in ductility, toughness, corrosion susceptibility, and strength. The changes in grain size or chemical makeup are not observed [26]. The typical cross-section of the FSW joint and the microstructures of different zones are shown in Figure 2.

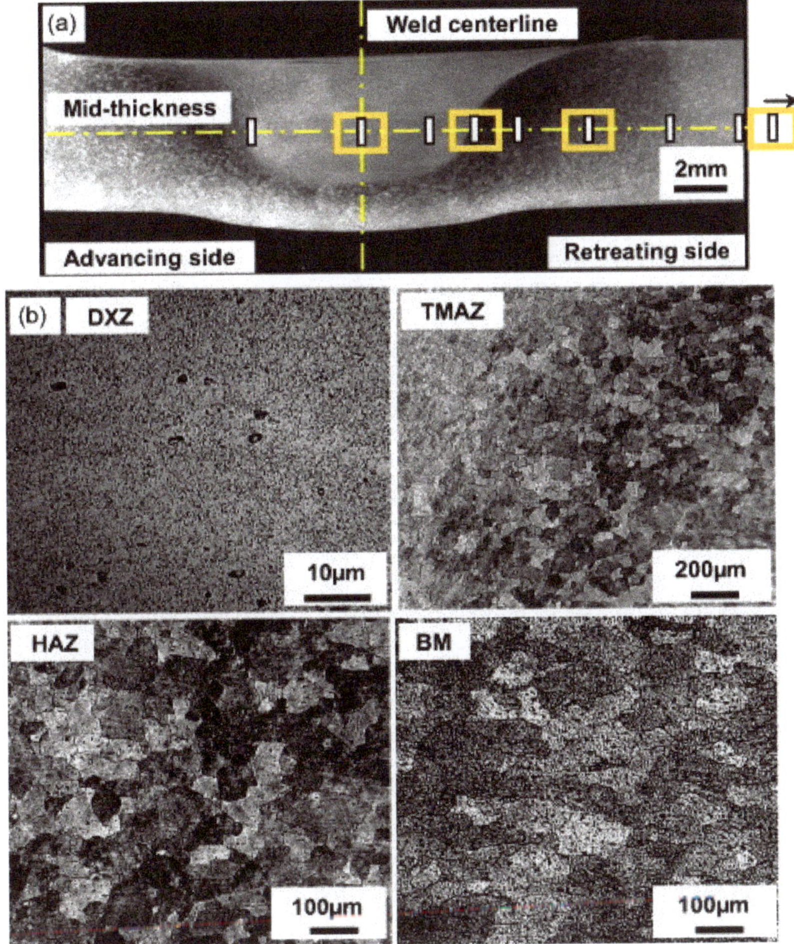

Figure 2. (**a**) Typical weld macrostructure of 6061–T6 Al alloy in FSW, (**b**) microstructure of the joint-stir zone (DXZ, dynamic recrystallized zone), TMAZ, HAZ, and BM taken at yellow square marks shown in (**a**) [27].

The FSW technology is classified as a green technology. As a melting point is not reached during the process, less energy is used in comparison to fusion welding techniques. Moreover, CO_2 emission to the atmosphere can be significantly reduced [28]. It is relatively easy to control the process and setting optimal parameters allows for a subsequent reduction of necessary non-destructive testing. The pollutions generated by sprays for visual and magnetic inspection and exposure to radiation in the case of X-ray tests are reduced. Additionally, if the optimal parameters are set, post-weld heat treatment is not required [28,29]. It results in the reduction of CO_2, energy consumption, and other pollutions emitted to the atmosphere. The FSW process uses a non-consumable tool; the use of shielding gas is not necessary.

One of the most important process parameters is the geometry of the tool. The tool consists of a specially designed pin and shoulder. The movement of the plasticized material depends on its geometrical features [30]. Its geometry also conditions the workpiece thickness, possible materials to be welded, and type of the joints [7]. The pin is an element of the tool that is directly plunged between the surfaces of the workpieces. It is plunged into the material until the shoulder reaches the contact with the surface of the components [31]. The tool design governs the process loads and microstructure of the weld, when from the heating aspects, the most important parameter is a ratio of a shoulder diameter to a pin diameter [32]. The FSW method is mostly characterized by two parameters related to its kinematics—rotational speed (ω) and traverse speed (v) along the joint line [33]. Selecting the optimum tool traverse and rotational speed is a crucial concern in the design of the FSW process.

The FSW technique allows the joining of many types of materials, even hard materials such as steel and engineering alloys. Recent studies have also been investigating the joining of metals to polymers through the FSW technique [34–36]. According to the analysis of Magalhaes et al. [20], most of the research and patents on FSW welding have concerned aluminum and its alloys, followed by ferrous alloys, magnesium, titanium, and their alloys. There are also studies regarding the welding of composites, copper, and polymers. The published papers analyze not only the FSW industrialization [37–40], but also the process mechanism, welding parameters, weld properties, and material microstructure. This paper presents the latest results of research on various materials welded using the FSW method and the influence of technological parameters of the process on the mechanical properties of welds and their microstructure, with particular reference to butt joints.

2. Methodology

The purpose of the overview is to establish the influence of the process parameters on the mechanical properties of the joints of different groups of materials. The method underpinning this paper is a systematic literature review. The number of papers published on FSW technology grows quickly, and the studies have become very diverse, making important the understanding and assessment of its impact on research and development level. In order to achieve this objective, an analysis of the literature published since 1991 to the current date was performed. The bibliographical references analysed were selected mostly from Scopus and ScienceDirect databases, using keywords such as "FSW", "friction stir welding", "solid state welding", and "solid type welding". Several hundred papers from the top journals publishing on FSW were analysed. Almost 70% of the articles presented in the review have been published in the last ten years; therefore, the paper is an analysis of the latest developments in the field of FSW process. A meticulous literature review on the applied FSW welding parameters allowed the presentation of the properties of the produced welds, if such information was presented in the quoted articles.

3. Aluminum and Its Alloys

Aluminum and its alloys are materials widely processed by FSW. Due to the difficulty of welding aluminum using traditional methods, FSW offers an excellent solution for joining these materials, ensuring reliability, ease of control of process parameters, and minimized risk of defects contributing to a reduction in the mechanical properties of the welds. During the FSW process of aluminum and its alloys, the temperature usually stays below 500 °C [41–43]. The experimental validation of the temperature on the tool surface is difficult to identify due to large deformations at the interface between the material and the tool, but Colegrove et al. [44] suggested that it can be near the solidus temperature.

Aluminum alloys can be divided into precipitation-hardened and solid solution-hardened alloys [32]. Although the precipitation-hardened aluminum alloys are easily welded by FSW, the heat-affected zone might be severely softened, essentially characterized by the dissolution or coarsening of the existent primary precipitates of the original thermal cycle [23]. It is reported that the hardness profile depends mostly on the precipitate distribution, and the grain size is of minor importance [23,45]. The most relevant to the hardness profile of the FSW joints of precipitation-hardened alloys is frictional heating during the process. The thermal hysteresis has an influence on distribution, size, and volume fraction of the strengthening precipitates [45].

The analysis of the state of the art has highlighted that there is no general dependence of mechanical properties of welds as a function of particular process parameters. In the studies of

Krasnowski et al. [46], it was reported that the ultimate tensile strength (UTS) of the AA6082-T6 joints initially decreases as a function of the tool traverse speed and then increases, reaching the maximum value of the UTS for the relatively highest tool traverse speed at the constant rotational speed for three different tool geometries. Opposite results has been observed by Rao et al. [47] during the FSW of IS:65032 aluminum alloy at the tool rotational speed of 1300 rpm and triangular pin shape, but the square pin shape for the same tool rotational speed confirmed the relationship observed before by Krasnowski et al. [46]. The above examples clearly indicate that the shape of the tool plays a key role in the FSW process, and changing only this parameter may cause the opposite effect of other process parameters on the properties of welds. Considering the influence of the tool rotational speed on the UTS of FSW joints, it is worth quoting studies on AA6061 alloy by Emamian et al. [48]. In these studies, it was observed that for linear speed v = 40 mm/min, the initial increase in the tool rotational speed causes an increase in UTS, but when it reaches its maximum, the UTS decreases with the increase in the speed. In the same study, for the tool traverse speed of 100 mm/min, UTS initially decreases with the increase in the tool rotational speed, but when it reaches its minimum, UTS increases with the increase in the tool rotational speed. Rajendran et al. [49] investigated the influence of the tilt angle on the hardness of the nugget zone of AA2014-T6 FSW lap joints. The tilt angle of 2° resulted in the maximum value of the hardness. In the studies on Al 5754 alloy, conducted by Barlas et al. [50], it was reported that the tilt angle equal to 2° provides better mechanical properties of the joints compared to the zero tilt angle. The investigations of Peel at al. [51] showed that the ratio of the shoulder diameter to the pin diameter (D/d) equal to 3.6 resulted in the highest ultimate tensile strength and yield strength of the AA5083 FSW joints, while the studies of Khan et al. [52] proved that in the range from 2.6 to 3.2 of the D/d ratio, the lowest value resulted in the best UTS and elongation of the AA6063-T6 joints. The reason for the different results may be the shape of the tool. The above studies used pins with different geometry—Peel et al. [51] used a threaded pin, while a smooth cylindrical pin was used by Khan et al. [52]. What is more, it is worth noting that the alloys with different chemical compositions were used in the studies, especially in the magnesium content, which could also affect the results obtained.

In the literature, there are few reports about FSW welding of metallic foams. However, the FSW technique is not an ideal solution to join the foams due to their compressibility. The pressure necessary to create frictional forces between the tool and the material is not sufficient after inserting the tool between the components to be joined, or the resulting forces destroy the porous structure of the material. A more popular solution is to implement the FSW method to produce sandwich structures, where a porous structure is placed between two sheets of solid material. In the research of Peng et al. [53], FSW was adopted to prepare aluminum foam sandwich. For this purpose, aluminum foam and solid aluminum AA6061-T6 plates. The aluminum foam panel was inserted between two solid plates and welded on both sides. It was concluded that the FSW technology offers better mechanical properties of the foam sandwiches compared to traditional adhesion and brazing. Busic et al. [54] investigated the influence of tool traverse speed and tilt angle on the mechanical properties of FSW of aluminum foam sandwich panels. Butt welds were produced by double side welding applying insertion of extruded aluminum profile. The studies proved that both tool traverse speed and tilt angle have significant influence on the UTS and flexural strength of the welds. In general, the current state of the art is poor in this type of research. The joining of foams still needs more studies, especially when permanent metallurgical bonding has to be obtained.

It is widely reported that the process parameters play a crucial role in the mechanical properties of the welds. However, the above examples prove there are no generally defined relationships. Table 1 summarizes analyzed studies on friction stir welding of different aluminum alloys. Table 1 presents the selected parameters that provided the highest mechanical properties of the welds, and in parentheses, the properties of the parent material are given for each example.

Table 1. FSW of aluminum and its alloys—process parameters and mechanical properties of the joints.

Material	Plate Thickness [mm]	Process Parameters					Weld Properties					Reference
		v [mm/min]	ω [rpm]	Tool Shape	Pin D/d Ratio	Tilt Angle [°]	UTS [MPa]	Yield Strength [MPa]	Hardness of the Stir Zone [HV]	Elongation [%]	Defects	
AA2195-T8	7.4	300	400	Cone shape threaded pin, threaded surface of the shoulder	-	-	445.0 (607.9)	-	-	12.50 (12.49)	No defects	[55]
AA356.0-T6 (double side welded)	8	200	1200	Threaded conical pin	3.25	-	200 (244)	123 (140)	-	16.3 (13.4)	-	[56]
AA5083	3	100	-	M6 threaded pin	3.6	2	304 (457)	154 (392)	-	-	-	[51]
AA5086-O	5	150	900	Tapered pin with 3 threads and concave shoulder surface	3.3	-	250 (253)	123 (112)	-	-	-	[31]
AA6013-T4	2.5	450	1400	-	-	-	300 (320)	-	-	-	-	[25]
AA6013-T6	2.5	400	1400	-	-	-	295 (394)	-	-	-	-	[25]
AA6061	5	100	1300	Cylindrical smooth	-	-	227 (308.5)	~153 (266.6)	-	~7.3 (16.28)	Defect free	[47]
AA6061	10	100	1600	Threaded cylindrical pin	3	-	214.4 (305)	-	-	-	No defects	[48]
AA6061-T6	5	150	900	Tapered pin with 3 threads and concave shoulder surface	3.3	-	285 (315)	241 (278)	57	-	-	[31]
AA6061-T651 (double side welded)	8	200	1200	Threaded conical pin	3.25	-	218 (299)	142 (264)	-	23.3 (27.2)	-	[56]
AA6063	5	100	2800	Threaded cylindrical pin	3	3	~150 (220)	~70 (170)	-	4.5 (13)	-	[57]
AA6063-T5	4	600	-	-	-	-	~155 (~220)	~105 (~185)	-	10 (~19)	No defects	[58]
AA6063-T6	4.75	40	900	Cylindrical smooth pin	2.6	1.5	145.34 (220)	-	-	20.85 (14.00)	-	[52]
AA6082-T6	8	900	710	Cylindrical pin with threads and three flutes	3.3	1.5	243.4	-	-	-	-	[46]

Table 1. *Cont.*

Material	Plate Thickness [mm]	Process Parameters					Weld Properties					Reference
		v [mm/min]	ω [rpm]	Tool Shape	Pin D/d Ratio	Tilt Angle [°]	UTS [MPa]	Yield Strength [MPa]	Hardness of the Stir Zone [HV]	Elongation [%]	Defects	
AA6352 (double side welded)	6	115	1350	Tapered pin	-	2	172 (250)	-	90.2 (93.5)	-	No significant defects	[33]
AA7050-T7451	6.35	103	396	-	-	-	429 (555)	304 (489)	-	6 (16.7)	-	[24]
AA7075-T651	6.35	127	-	-	-	-	525 (622)	365 (571)	-	15 (14.5)	-	[42]
SSM 356	4	160	1750	Cylindrical pin	4	3	173.5 (168.7)	138.8 (134.9)	40.9 (36.4)	3.1 (5.3)	-	[59]
SSM 356-T6	4	160	1750	Cylindrical pin	4	3	172.9 (295.6)	138.3 (236.5)	68.4 (61.2)	4.5 (4.8)	-	[59]

4. Magnesium and Its Alloys

Among commonly used structural materials, magnesium has the lowest density. Because of its hexagonal close-packed (hcp) structure at room temperature, the formability of magnesium is very constrained; however, it increases significantly at temperatures of 230–310 °C [60]. Most of the commercially used magnesium alloys are ternary alloys containing aluminum, zinc, silicon, and rare earth metals [61]. In Mg-Al series, the most common alloys are AZ (Mg-Al-Zn) and AS (Ag-Al-Si) [62,63]. The successful method to join magnesium alloys is arc welding, but some difficulties might occur in joining, especially the cast grades alloys [60]. During fusion welding of aluminum alloys, the shielding gases are necessary due to oxidation at welding temperatures. The most significant problems occurring during fusion welding of magnesium alloys are the porosity of the welds [32], distortions due to high thermal conductivity and thermal expansion of magnesium alloys [10], evaporation, and solute atoms segregation, which leads to softening of the joint area. In addition to the application of the FSW technique mentioned above, friction stir welded magnesium alloys find their applications in industrial equipment of nuclear energy, due to their low neutron absorption, excellent thermal conductivity, and good resistance to carbon dioxide [64].

There are plenty of studies on the microstructure of the FSW magnesium joints. Xin et al. [65] reported that the primary texture does not significantly affect the final microstructure and texture of the nugget zone. However, texture distribution in the thermo-mechanically affected zone influences the mechanical properties of the joints [66]. In the studies of Yang et al. [67], it was reported that the shoulder size does not have an impact of texture modification in the nugget zone of friction stir welded Mg-3Al-1Zn alloy, but it weakens the (0002) texture in the thermo-mechanically affected zone. Commin et al. [68] observed that during FSW of AZ31 hot-rolled base material, the structure is not significantly changed when the shoulder diameter is equal to 13 mm, but the shoulder diameter of 10 mm resulted in the strong texture modification. In their studies, it was also reported that the highest tensile residual stress was observed in the thermo-mechanically affected zone. It was observed that a larger diameter of the shoulder reduced the residual stress due to the higher heat delivered to the welded material.

The analysis of the research conducted so far does not allow the drawing of general conclusions concerning the optimization of process parameters. On the basis of studies carried out by Lim et al. [69], it was concluded that the tensile properties of AZ31B-H24 welds are not significantly affected by FSW process parameters, whereas Lee et al. [70] reported that with an increase of the tool rotational speed, the strength of the joints of the same alloy increased. Wang et al. [11] and Kumar et al. [71] reported that for AZ31 butt welds, the UTS, yield strength, elongation, and hardness primarily increase with an increase of the tool traverse speed and after reaching the maximum value, decrease with a further increase of the welding speed. In the studies of Han et al. [72], the ultimate tensile strength of Mg-Gd alloy increases with an increase of the tool traverse speed. The opposite dependence was presented by Sahu et al. [73] for AM20 butt welds. It should be noted that both tests were performed at a different tool rotational speed, so the amount of heat generated was different. Moreover, both tests were different in the geometry of the tools used. In the study of Sahu et al. [73], the influence of D/d ratio on the mechanical properties was also investigated. In the range from 2 to 4, the highest D/d ratio provided the highest UTS. Sevvel et al. [74] and Pareek et al. [75] investigated the influence of the tool rotational and traverse speed on the mechanical properties on AZ31 magnesium alloy. The results of the tests do not allow the drawing of a general conclusion. Sevvel et al. [74] proposed the lowest tool traverse speed and the highest rotational speed to obtain the highest ultimate tensile strength and the highest yield strength of the welds, while in the studies of Pareek et al. [75], the highest tool traverse speed and the highest rotational speed resulted in the best mechanical properties of the welds. Sevvel et al. proposed the tool rotational speed equal to 1000 rpm, while in the studies of Pareek et al., it was set as 2000 rpm. In this case, a higher tool traverse speed could provide enough heat to the weld, which could be insufficient if the speed was lower, as in the studies of Sevvel et al. Studies on the hardness of friction stir welded magnesium alloys show contradictory conclusions. Esparza et al. [76] and

Park et al. [77] reported that the welds exhibit almost the same hardness in the various zones. On the contrary, Xie et al. [78] and Zhang et al. [79] noted that the nugget of the welds has significantly higher hardness than the other zones. It can be explained by breaking up large intermetallic compounds Al_2Ca in Mg-Al-Ca alloy studied by Zhang et al. [79] and Mg-Zn-Y phases in Mg-Zn-Y-Zr studied by Xie et al. [78] and their dispersion in the stir zone, which resulted in the increase of the hardness. As mentioned earlier, aluminum alloys are divided in two types: precipitation-hardened and solid solution-hardened alloys. Thus, in ternary magnesium alloys containing aluminum as the main alloying element, the hardness of magnesium alloy varies according to the percentage of aluminum present in the structure.

Table 2 presents the properties of FSW joints of magnesium and its alloys and the mechanical properties of the parent material if presented by the authors.

Table 2. FSW of magnesium and its alloys–process parameters and mechanical properties of the joints.

Material	Plate Thickness [mm]	Process Parameters					Weld Properties					Reference
		v [mm/min]	ω [rpm]	Tool Shape	Pin D/d Ratio	Tilt Angle [°]	UTS [MPa]	Yield Strength [MPa]	Hardness of the Stir Zone [HV]	Elongation [%]	Defects	
AM20	4	63	600	Cylindrical pin	4	-	132.17 (202)	115.56 (160)	61 (46)	2.17 (7)	-	[73]
AZ31	4	90	1500	-	3	-	255 (275)	-	-	-	-	[80]
AZ31	8	120	1200	Conical pin	2	2.5	225.1 (249.5)	130.5 (156.3)	-	5.4 (14)	-	[11]
AZ31-O	2	200	1000	-	2.6	-	~170 (~250)	~90 (~150)	-	-	-	[68]
AZ31B	5	0.5	1000	Tapered cylindrical pin	3	-	183 (262)	101 (179)	-	-	No defects	[74]
AZ31B	5	40	1400	Threaded conical pin	3	2.5	186.76 (215)	139.1 (171)	71 (69)	5.00 (14.7)	-	[81]
AZ31B	5	40	1120	Taper threaded pin	3	2	188 (215)	148 (171)	121 (69.3)	7.3 (14.3)	-	[19]
AZ31B	6	40.2	1600	Threaded cylindrical pin	3.0	0	205 (215)	166 (171)	75 (69.3)	7.3 (14.7)	Defect free	[82]
AZ31B	6	50.8	1200	-	-	-	248	-	67.95 HB	-	-	[71]
AZ31B-O	5	60	1200	Left handed threaded pin	3	-	187.8 (206)	-	64.77 (50)	16.73 (20)	-	[13]
AZ31-H24	3.175	204	2000	-	-	-	225.6 (307.7)	115.3 (227.6)	-	-	No defects	[75]
AZ31B-H24	4.95	4	1000	-	-	-	208 (315)	115 (202)	-	-	-	[83]
AZ61	4	25	1400	Left-handed threaded pin with three flutes	3	-	220 (270)	175 (219)	81 (70)	7.2 (~8.2)	-	[16]
AZ91	6	28	710	Threaded straight cylindrical pin	3	-	76.17	-	-	-	No defects	[84]
AZ91	6	60	600	-	2.8	2.5	262 (106)	132 (55)	-	18.9 (15.2)	No defects on the top surface	[85]

Table 2. *Cont.*

Material	Plate Thickness [mm]	Process Parameters					Weld Properties						Reference
		v [mm/min]	ω [rpm]	Tool Shape	Pin D/d Ratio	Tilt Angle [°]	UTS [MPa]	Yield Strength [MPa]	Hardness of the Stir Zone [HV]	Elongation [%]	Defects	
AZ91D	3	75	500	Left-handed tapered cylindrical pin	2.6	2.5	107 (107)	-	-	-	Defect free	[86]
AZ91D	3	90	1200	-	2	-	200 (220)	140 (150)	-	2 (3.6)	-	[87]
MB3	3	120	1500	-	-	-	240 (245)	-	-	-	No macro defects	[88]
Mg-Y-Nd alloy (double side welded)	20	240	700	Threaded conical pin with three flutes	2	3	277.6 (336.1)	204.1 (245.9)	-	7.27 (10.43)	Defect free	[89]

5. Steel and Ferrous Alloys

The FSW method was initially dedicated to aluminum and its alloys, but with the development of this technology, other materials are successfully joined. Steel and ferrous alloys are still a challenge due to their high hardness. The biggest problem when welding steel and ferrous alloys is choosing the right tool for this process. The tool material must have high resistance to frictional wear, resistance to cracking, high strength, and resistance to chemical degradation at high temperatures achieved during the process [90,91]. Finding the proper material is a major engineering challenge. There are also studies on various ceramic options [92]. Nevertheless, composite tools made of polycrystalline boron nitride/tungsten rhenium (pcBN/W-Re) are also commonly used in the FSW of steel [93].

The friction stir welding method offers a reduction of the metallurgical changes in the heat-affected zone due to lower heat input compared to fusion welding techniques. The FSW method is a good alternative for joining difficult to fusion weld steel grades. Furthermore, during fusion welding, the normal source of hydrogen might lead to hydrogen cracking, while this problem is eliminated during FSW [32]. Early studies on FSW of steels proved that the peak temperature during the process of 1000–1200 °C is much lower than that observed during conventional welding [10,94,95]. Hence, the region of the heat-affected zone with pearlitic steels, which becomes fully austenitic, is supposed to be narrower. Moreover, the size of the grains of austenite is expected to be finer than in the case of arc welding. The unfavorable transformations, such as untampered martensite, can be avoided in the FSW method [10].

The thermo-mechanical nature of the FSW process induces phase transformations controlled by the selection of appropriate process parameters, such as tool rotational speed and tool traverse speed. Changes in the microstructure of carbon steel depending on the process temperature were presented in the paper of Fujii et al. [96]. In the study of Cui et al. [97], different microstructures of high-carbon steel were observed by controlling both the tool rotational speed and the tool traverse speed. Saeid et al. [98] obtained defect-free welds of duplex stainless steel by the FSW method in relatively low temperatures, which led to the avoidance of phase transformation and the ratio between phases was not changed. In the research of Ghosh et al. [99], the dependency of temperature and rate of deformation on microstructure for high-strength M190 steel was examined. Miura et al. [100] reported that the FSW method on Cr-Mo steel results in the increase of the volume fraction of retained austenite, and the joints perform high ultimate tensile strength and elongation. A similar observation for ferritic stainless steel was noted by Fujii et al. [101]. However, there is no general explanation for this mechanism.

The influence of welding parameters of steel on the mechanical properties of welds is not fully determined. In the research of Mahoney et al. [102], the HSLA-65 alloy was welded and the mechanical properties of the welds in dependence on the rotational speed and traverse speed of the tool were examined. The results showed that the tensile strength of the welds increases with both rotational and linear speed, while the elongation of the welds decreases. The studies of Miura et al. [103] show an inverse relationship: for iron alloys with nickel and carbon, the ultimate tensile strength of the welds and their yield strength decreases with an increase in tool speed. It should be noted that these incompatibilities result from differences in the range of selected parameters. Mahoney et al. applied a tool traverse speed up to 152.4 mm/min, while Miura et al. applied one with maximum 400 mm/min. However, similarly to the studies on HSLA-65 steel alloy of Mahoney et al. [102], the elongation of the welds decreases as the rotational speed increases. In the research of Fujii et al. [104], three carbon steels with different carbon contents were subjected to the FSW method. Their tensile strength was tested depending on the established traverse speed. The results showed that for IF (interstitial-free) steels, this parameter does not significantly affect the UTS, while for S12C and S35C steels, this effect was more visible but not uniform. For S12C, UTS increased with the tool traverse speed, while for S35C, the tensile strength first increased and then decreased with an increase of the tool traverse speed. For both steels, higher UTS than the value for the parent material was achieved. The same dependence as for S35C steel was observed by Reynolds et al. [90] for DH36 steel. Tensile strength and yield strength first increased and then decreased as both traverse and rotational speed of the tool increased. The mechanical properties of the welds were higher than those of the parent material. In the research of Meshram et al. [105] on ASIS 316 steel, the highest tool rotational speed and the lowest traverse speed of the tool influenced the mechanical properties of the joints. The UTS and hardness exceeded the ones of the base metal. The studies of Maltin et al. [106] on DH36 steel showed that the highest both tool traverse and rotational speed resulted in the highest UTS and yield stress, exceeding the values for the parent material.

Some researchers claimed that FSW of steel is an attractive alternative in comparison to the fusion welding, and the feasibility of the method was proved by many studies, although more scientific research in this field is needed, especially with improving tools' geometry and the proper selection of the tool materials [94]. In contrast, Bhadeshia and DebRoy [107] suggested that the FSW technology is not expected to be widely applied, because fusion welding techniques already allow producing reliable, cost-effective joints.

Table 3 presents the results of selected studies on friction stir welded steels. The process parameters that provided the best mechanical properties of the welds were presented, and the properties of the parent material were given in brackets.

Table 3. FSW of steel and ferrous alloys—process parameters and mechanical properties of the joints.

Material	Plate Thickness [mm]	Process Parameters					Weld Properties					Reference
		v [mm/min]	ω [rpm]	Tool Shape	Pin D/d Ratio	Tilt Angle [°]	UTS [MPa]	Yield Strength [MPa]	Hardness of the Stir Zone [HV]	Elongation [%]	Defects	
HSLA-65	6	154.2	600	Convex step-spiral scroll shoulder	-	-	852 (538-690)	662 (448)	-	22.3 (min 18)	-	[102]
HSLA DMR-249A	5	30	600	Tapered pin with no threads	5	0	664 (610)	-	410 (270)	19 (29)	Free from macro-level defects	[108]
IF	1.6	400	400	Cylindrical pin with no threads	3	3	~310 (284)	-	-	-	-	[104]
S12C	1.6	400	400	Cylindrical pin with no threads	3	3	~480 (317)	-	-	-	-	[104]
S35C	1.6	200	400	Cylindrical pin with no threads	3	3	~780 (574)	-	-	-	-	[104]
DH36	6	450	600	-	-	0	832.51 (531.62)	656.68 (376.71)	-	5.57	-	[106]
DH36	6.4	306	526	Slightly tapered pin with no threads	-	2.5	~940 (~580)	~650 (~350)	-	-	No volumetric defects	[90]
Ultrafine grained AISI 304L	2	80	630	Conical pin	~3	3	~760 (920)	~500 (720)	285 (330)	~42 (47)	-	[109]
AISI 316	4	8	1100	-	-	-	610 (608)	-	230 (190)	35 (49)	-	[105]
AISI 316	4	8	1000	-	-	-	630 (608)	-	-	37 (49)	Defect free	[110]
AISI 1018	5	50	1000	Tapered pin with no threads	2.2	-	457 (421)	424 (361)	-	20 (27)	-	[111]
HNAS (High nitrogen nickel-free austenitic stainless steel)	2.4	100	400	Tapered pin	3.3	0	~1100 (~1060)	~760 (~680)	400 (370)	~37 (~43)	No groove-like defects	[112]
Fe-18,4Cr-15,8Mn-2,1Mo-0,66N-0,04C	2	100	800	-	3	-	980 (967)	580 (604)	-	30 (53)	-	[113]
Fe-18Cr-16Mn-2Mo-0,85N	3	50	800	-	-	2	1375 (1234)	908 (782)	-	25.13 (39.8)	-	[114]
Fe-24Ni-0,1C	1.6	400	200	-	3	3	1283 (793)	390 (336)	-	29.0 (5.6)	-	[103]

6. Titanium and Its Alloys

Titanium and its alloys are characterized by very good mechanical properties such as high strength, high corrosion resistance, and a very good strength-to-weight ratio, but their processing at temperatures higher than 550 °C is difficult due to their low resistance to oxidation. In addition, in single-phase titanium alloys, a tendency to grain growth is observed, which results in a decrease in the mechanical properties of the material [115]. When using the FSW method for soft alloys such as aluminum or magnesium, the problem of tool wear and its material selection is not a challenge, but in the case of titanium joining, these problems may arise. The most commonly used tools during the FSW process are those with a developed pin geometry, such as threads, flats, or flutes. When friction stir welding titanium and its alloys, conventional tools can be significantly damaged. It is therefore necessary to modify the geometry in such a way that the material is properly mixed while minimising the problem of wear. These solutions consist of uncomplicated tools with a columnar or conical pin, and the mixing of the material is assisted by appropriate shoulder surface modification, such as scrolls, ridges, knurling, grooves, or concentric circles. Usually, the ratio of shoulder diameter (D) to pin diameter (d) is chosen to be equal to 3 to provide the best mechanical properties of the joint [116,117]. In the case of FSW of titanium and its alloys, a heat generated by the shoulder cannot flow to the join root, and a relatively small pin is not able to properly stir the plasticized material. Therefore, usually tools with smaller shoulder diameter and larger pin are used, and the D/d ratio is smaller, which can be observed in the examples summarized in Table 4. Another important aspect of the tool design process is the selection of the right material. The significant strength of titanium and its alloys at hot working temperatures makes it necessary to select a material for the tool that will be resistant to high forces during the process and be inert for reactive titanium at temperatures reaching 0.8 of its melting point. The most popular materials for FSW of titanium and its alloys are W-, Re-, Mo-based alloys, and TiC [118–123]. Titanium and its alloys have a relatively low thermal conductivity and a high melting point, so a temperature gradient between the advancing and retreating side of the components may appear when friction stir welding. Applying the FSW method to titanium alloys might be challenging due to the thickness of the components and the tool geometry limitation, mostly for alpha and near-alpha alloys. In case of such alloys, the lower thermal conductivity of alpha phase, its higher low stress, and higher heat capacity of titanium makes it difficult to select the proper tool material for titanium alloy with high β trans temperature. β or α + β alloys are susceptible to the temperature of β transus during friction welding in dependence on welding parameters and thermal distribution during the process. Examples of tools used during the process of FSW of titanium and its alloys are tungsten carbide (WC) and titanium carbide (TiC) tools produced by sintering. Specially designed water cooling systems are also successfully used to better dissipate heat from the tool [124].

The mechanical properties of welds are directly influenced by the evolved micro- and macrostructure of the joints. The macrostructure observed in titanium alloys is clearly different from the banded elliptical macrostructure observed in aluminum and its alloys [32] and a parabolic shape of the weld nugget was observed in the research of Gangwar et al. [125] on titanium alloys. Fonda et al. [126], in their research on aluminum alloys, observed that the banding may be attributed to the fluctuations in the second phase particles density or the crystallographic texture changes, while in the titanium alloys, the absence of hard second phases or inclusions suggests the formation of banding formation in the nugget due to texture difference [127]. Gangwar et al. [127] in the review suggested that the elongation of the FSW titanium components is lower than the base metal due to microstructural gradients observed in the gauge length of the transverse tensile specimen, and the strains are mostly carried by the narrow areas of the thermo-mechanically affected zone. However, the examples presented in Table 4 show the opposite conclusions. In the studies of Kulkarni et al. [128] on Ti-54M plates, the increase in specimen elongation was observed with the increase in the tool traverse speed in the FSW process. On the contrary, in the studies of the same authors on Ti-6242 plates of the same thickness and the same process parameters, the decrease of the elongation with the increase of the tool traverse speed was observed. This confirms the assumption that the chemical composition of the material is very important for FSW welding. These studies also confirm the suggestion that α and near α alloys, among them Ti-6242, are more challenging. Similarly, the studies of Su et al. [129] on Ti-6Al-4V a decrease in elongation with an increase in the tool traverse speed was observed, while Mashinini et al. [130] studies on the same alloy showed the opposite relation. This effect is caused by differences in other applied parameters, including tool geometry, tool rotational speed, and tilt angle. In the same studies of Su et al. [129] and Mashinini et al. [130], opposite correlations between UTS and the tool traverse speed were also presented. In the studies of Fujii et al. [131] on pure titanium plates, the UTS firstly increases with the increase of the tool traverse speed and then decreases with the further increase of the tool traverse speed. The opposite relation was observed by Kulkarni et al. [128] in the studies on Ti-6Al-4V titanium alloy. Kulkarni et al. [128] presented a tendency of increasing UTS as a function of the tool rotational speed, while Zhang et al. [120] reported an opposite relation. An in-depth review of the literature did not allow the explanation of the reason for contradictory conclusions in the cited examples, but it should be noted that in all cited studies, tools with different geometry and made of different material were used, and this could have been the reason for obtaining different results.

Table 4 presents the results of friction stir welding on titanium and its alloys and the mechanical properties of the parent material, if presented by the authors.

Table 4. FSW of titanium and its alloys—process parameters and mechanical properties of the joints.

Material	Plate Thickness [mm]	Process Parameters					Weld Properties					Reference
		v [mm/min]	ω [rpm]	Tool Shape	Pin D/d Ratio	Tilt Angle [°]	UTS [MPa]	Yield Strength [MPa]	Hardness of the Stir Zone [HV]	Elongation [%]	Defects	
Pure Ti	2	200	200	-	2.5	-	~430 (420)	-	180 (146)	-	-	[131]
Pure Ti	5.6	50	110	-	-	-	430 (440)	-	-	20 (25)	Defect free	[119]
Ti-54M fine grain	0.1	100	275	-	-	-	~950 (972)	~780 (889)	-	~5.9 (16.5)	Defect free	[128]
Ti-6Al-4V fine grain	0.1	125	325	-	-	-	~950	~760	-	~4.3	Defect free	[128]
Ti-6Al-4V standard grain	0.1	100	275	-	-	-	~930 (950)	~720 (880)	-	~7.7 (14)	Defect free	[128]
Ti-6Al-4V	2	101.6	900	Smooth cylindrical pin	1.6	2.5	1156.2 (1014.7)	1067.4(941.8)	-	21.7 (23.1)	Processing defects	[129]
Ti-6Al-4V	2	50	250	Tapered pin	2	-	813 (1013)	-	-	3.2 (8.5)	-	[128]
Ti-6Al-4V	3	75	300	Small shoulder and large tapered pin	-	-	1025.0 (higher than BM)	973.6 (higher than BM)	-	9.7	-	[132]
Ti-6Al-4V	3	60	300	Convex shoulder and tapered pin	~3	-	~1050 (~920)	~950 (~830)	~315	~33 (~21)	Cavity defects	[120]
Ti-6Al-4V	3	45	550	Tapered pin	1.75	0.5	1059 (1000)	-	-	4 (18)	Small root flaws	[133]
Ti-6Al-4V	3.17	40	500	Flat shoulder surface and tapered smooth pin	2	1.5	1040 (1017)	-	-	9 (20)	-	[130]
Ti-6Al-4V	6	100	280	-	-	-	1016 (1045)	971 (978)	335.6 (315.4)	9 (16)	-	[134]
Ti-6242 fine grain	0.1	125	325	-	-	-	~950	~730	-	~4.1	Defect free	[128]
Ti-6242 standard grain	0.1	125	325	-	-	-	~880 (1000)	~730 (895)	-	~8.4 (12)	Defect free	[128]
TC4	2	50	400	Smooth tapered pin	-	2.5	953 (1036)	-	~345 (~325)	-	-	[135]

7. Copper and Its Alloys

Copper and its alloys are widely used in many engineering applications due to their properties such as good electrical and thermal conductivity, relatively good mechanical strength, high corrosion resistance, and high formability. The most popular copper alloying elements are zinc, aluminum, nickel, and tin [136]. After steel and ferrous alloys and aluminum and its alloys, copper and its alloys are the most commonly used materials in industries, especially in the marine, aerospace, electronics, and military sectors. However, pure copper strength is not high enough for load bearing components, but it increases by alloying. The most popular copper alloys are solid solution hardened (single-phase). Copper is characterised by low galvanic reactivity, so the risk of reaction or corrosion is low. It is also characterized by high plasticity and resistance to oxidation. The most important applications for copper and its alloys include heat sinks, electrodes for resistance welding, and rotating target neutron sources. The FSW method was also successfully used for joining components for nuclear waste canisters [137,138]. It is also used in processes of soldering and brazing [139].

Joining copper and its alloys using traditional welding methods is difficult due to the high thermal conductivity and high melting point, which makes it necessary to generate a large amount of heat, thus increasing the cost of the process, and the resulting welds may exhibit porosity, distortion, and solidification cracks. Conventional fusion welding techniques require very fast heat delivery due to 10–100 times higher heat conductivity than that of steels [10,32]. Copper and its alloys are ranked as hard-to-weld materials [140]. The most significant problems occurring during the conventional welding of Cu and its alloys are high distortion, irregularities of the weld surface, decrease of strength at the weld surface connected with the formation of ZnO (for high Zn-content alloys), insufficient penetration because of the high thermal conductivity of Cu, and colour change because of the oxidation process. The FSW method is an excellent solution, because the melting temperature is not reached during the process.

Due to the high thermal conductivity of copper and its alloys, the required heat input should be higher than for FSW of other materials. This means that the process is usually conducted at lower tool traverse speed and/or higher tool rotational speed. It is required mostly for FSW of pure copper, which has higher thermal conductivity than its alloys. As for all friction stir welded metal joints, four specific zones can be observed in the cross-section–stir zone (SZ), thermo-mechanically affected zone (TMAZ), heat-affected zone (HAZ), and base metal (BM). For the FSW joints of copper alloys, the HAZ is not highly distinguishable [141,142]. The recrystallization process occurs relatively easily in copper and its alloys, especially single-phase, so the SZ extends almost the TMAZ and the boundaries between those two zones are hard to determine.

In the research of Machniewicz et al. [143], 5 mm pure copper plates were friction stir welded in both the longitudinal direction and perpendicular to the rolling direction. For both examples, the ultimate tensile strength decreased with an increase in the tool traverse speed. Moreover, the microhardness of the welds was measured. The microhardness profile presented a "W" shape, which is characteristic for most friction stir welded joints [144–149]. In the profile of such welds of any metal, a sharp decrease in hardness can be observed in the heat-affected zone, and then the hardness slightly increases in the direction of the weld nugget. This phenomenon is related to the difference in grain size in various zones—in the weld zone, the microstructure is more fine-grained than in the heat-affected zone, and therefore higher hardness is observed there. In the studies of Khodavardizadeh et al. [146], Xue et al. [145], and Surekha et al. [150] on pure copper plates, the ultimate tensile strength of the welds as a function of the tool traverse speed was also determined. The results of those studies are not consistent with those of Machniewicz et al. [143], and the UTS of the welds increased with an increase of the tool traverse speed. In the studies of Khodavardizadeh et al. [146] and Surekha et al. [150], it was observed that the elongation of welded samples increases with an increase of the tool traverse speed, while Xue et al. [145] noted the opposite relationship for the same material. The above quoted studies were carried out with different process parameters and the values of applied tool rotational speed were more than twice as high for tests of Xue et al. [145] as for Surekha et al. [150]. In the studies of Liu et al. [151] on pure copper plates, the ultimate tensile strength as a function of the tool rotational speed was measured. It was observed that the UTS firstly increased and then slowly decreased with the increase of the tool rotational speed. The same relation was observed while measuring the elongation. The maximum UTS was equal to the value for the base material. In the studies of Xue et al. [145] and Khodavardizadeh et al. [147], the UTS of the welds decreases with an increase of the tool rotational speed, while in the studies of Sahlot et al. [152], Xie et al. [153], and Cartigueyen et al. [154] on the pure copper plates, the opposite relationship was observed. Cartigueyen et al. [154], Xie et al. [153], and Xue et al. [145] reported that the elongation of the copper FSW joints increased with the increase of the tool rotational speed, while the opposite relationship was observed by Khodavardizadeh et al. [147]. In this study, it was also noted that the hardness of the copper weld nuggets increases with the increase of the tool traverse speed, while Surekha et al. [150] observed almost no changes of the hardness, and the measured values were similar to the one of the base material. Khodavardizadeh et al. [146] noted that the hardness of the copper weld nuggets decreases with an increase of the tool rotational speed, while Xie et al. [153] and Cartgueyen et al. [154] noted an opposite relationship. In another study of Cartigueyen et al. [155] on 6 mm thick copper plates, the influence of the pin geometry was investigated. The threaded cylindrical pin provided better mechanical properties of the friction stir welded joints than square, triflute, and hexagonal pins. Different tool geometries were used in all the quoted research or these geometries are not presented in the papers. It should be noted that tool geometry is a key process factor, and it is necessary to define it in published studies.

Table 5 shows an overview of the studies on copper and its alloys conducted so far. Each example contains a set of parameters (if given) that provided the best mechanical properties of the welds and the values of these properties including the properties of the parent material (values in brackets).

Table 5. FSW of copper and its alloys—process parameters and mechanical properties of the joints.

Material	Plate Thickness [mm]	Process Parameters					Weld Properties					Reference
		v [mm/min]	ω [rpm]	Tool Shape	Pin D/d Ratio	Tilt Angle [°]	UTS [MPa]	Yield Strength [MPa]	Hardness of the Stir Zone [HV]	Elongation [%]	Defects	
Pure copper	2	50	1200	Conical pin	2	0	217.56 (237.81)	-	-	2.03 (39)	Defect free	[156]
Pure copper	2	30	1000	-	-	-	231 (273)	1.2 (3.1)	136	-	Defect free	[157]
Pure copper	3	30	2000	Tapered pin	2.9	-	~195 (217)	-	-	-	Defect free	[152]
Pure copper	3	100	400	Right hand threaded cylindrical pin and concave shoulder	4	-	282 (282)	-	-	16.4	Defect free	[151]
Pure copper	3	25	1100	-	-	2.5	194 (212)	70 (68)	-	22.8 (28.1)	-	[158]
Pure copper	3	40	900	Flat shoulder and cylindrical pin	3	3	168 (260)	109 (231)	85 (110)	13.5 (31)	Defect free	[159]
Pure copper	3	250	300	Non-threaded cylindrical pin	2.4	-	328 (270)	261 (209)	113.6 (84.6)	23 (22)	Defect free	[150]
Pure copper	4	61	1250	-	-	3	~225 (~260)	-	~90 (105-110)	-	Defect free	[144]
Pure copper	5	30	910	Straight cylindrical pin	3	-	216.9	-	77.37 HB	9.2	No defects	[160]
Pure copper	5	40	580	Taper pin	~3	0	220.7 (261.2)	101.3 (232.0)	-	-	Defect free	[143]
Pure copper	5	75	600	-	-	-	221 (234)	127 (178)	88 (107)	43 (47)	-	[146]
Pure copper	5	50	400	-	-	2.5	235.9 (236.7)	207.7 (222.9)	-	15.1 (27.7)	Defect free	[145]
Pure copper	5	50	800	Cylindrical threaded pin	~3.3	2.5	~240 (~240)	~140 (~225)	63.1 (82.2)	~45 (~28)	Defect free	[153]
Pure copper	5	112	500	Threaded cylindrical pin	2.5	2.5	326 (331)	134 (137)	~105 (~78)	31 (29)	Defect free	[161]
Pure copper	6	50	350	Threaded cylindrical pin	3	0	228 (279)	209 (271)	92 (82)	-	-	[155]
Pure copper	6	50	350	Square pin	-	0	207 (279)	203 (271)	88 (82)	-	-	[155]
Pure copper	6	50	350	Triflute pin	-	0	196 (279)	196 (271)	90 (82)	-	-	[155]
Pure copper	6	50	350	Hexagonal pin	-	0	165 (279)	163 (271)	97 (82)	-	-	[155]

Table 5. *Cont.*

Material	Plate Thickness [mm]	Process Parameters					Weld Properties					Reference
		v [mm/min]	ω [rpm]	Tool Shape	Pin D/d Ratio	Tilt Angle [°]	UTS [MPa]	Yield Strength [MPa]	Hardness of the Stir Zone [HV]	Elongation [%]	Defects	
Pure copper	6	50	500	Concave shoulder and threaded cylindrical pin	3	0	229 (279)	229 (271)	88 (82)	49.9 (34.4)	Defect free	[154]
Pure copper	6	315	630	Square pin	2.4	2	215 (150)	190 (145)	-	33 (14)	-	[162]
2200 Copper Alloy	5	31.25	900	Taper pin with no threads	3.3	1	-	-	-	26.98	-	[163]
2200 Copper Alloy	5	31.25	900	Cylindrical with no threads	3.3	1	-	-	-	10.56	-	[163]
Brass 60%-Cu, 40%-Zn	2	500	1000	-	3	3	~390 (381)	~180 (192)	~132 (97)	~52 (61)	Defect free	[141]

8. Polymers

Polymeric materials are generally characterised by different properties than metallic materials. Although the FSW method was initially dedicated to metal bonding, with the increasing use of polymers in various industries, it has been successfully used to join also this group of materials. The research on the friction stir welding of polymeric materials conducted so far focuses mainly on joining of polyethylene (PE) [164–169], high-density polyethylene (HDPE) [166,170–172], polyamide Nylon 6 [173–175], acrylonitrile butadiene styrene (ABS) [176–179], polyvinyl chloride (PVC) [171], polypropylene (PP) [180,181], and polyethylene terephthalate glycol (PETG) [182].

Using conventional tools for friction stir welding of polymers is the easiest but not the most effective approach, which will be concluded later. Panneerselvam et al. [183] successfully butt-joined 10 mm polypropylene plates using the FSW method, but for some pin geometries, the insufficient material soften; the tool damage or the blowhole defects were observed in the joint line. Design of experiments analysis (DOE) was used for modelling and analysing the influence of the process parameters [184]. The analysis of the obtained results allowed the authors to conclude that insufficient heat on the retreating side did not allow for proper mixing of the material in this zone and increased the probability of weld defects. The same conclusions were presented by Simoes et al. [185]. The morphology of the polymethyl methacrylate joints welded by the FSW method was analysed, and it was claimed that the advancing side of the welds performs almost the same transparency as the base material, while in the retreating side, the insufficient stirring and voids were observed. The studies conducted so far allow the conclusion that low heat conductivity of polymeric materials is not favourable for an efficient FSW process. Sufficient softening and plasticization of the material that is not in direct contact with the tool is difficult to achieve. In the literature, approaches can be found related to the application of an additional heating system that would compensate for heat deficiencies related to low thermal conductivity and friction coefficient of polymer materials. Squeo et al. [186] friction stir welded 3 mm polyethylene sheets using a pin previously heated with a hot air gun. This solution was not considered reliable due to the rapid cooling of the tool. Another approach was to use a hot plate between the CNC table and the components. The plate was heated up to 150 °C and ensured high quality of joints, while the biggest disadvantage of this method was low repeatability during the process. Arbegast [187] proposed a model of the FSW joint that describes the conditions of the process and a mechanism of creating weld defects. The theory also confirms that volumetric defects are likely to be observed on the retreating side of the weld.

The FSW method using a conventional tool to perform the welding process usually does not bring the expected results, and the properties of the welds are relatively low. In order to minimize the risk of weld defects and increase the efficiency of the process, a modification of the FSW method—stationary shoulder friction stir welding (SSFSW)—is used. The mechanism of the SSFSW process consists of a rotating pin that runs in a non-rotating shoulder element sliding on the material surface during welding. The stationary shoulder, usually called a shoe, minimizes the risk of the plasticized weld material being expelled from the weld seam [181]. The literature review on the SSFSW method can lead to the conclusion that this modification should be used for joining polymers in order to obtain non-defects and high mechanical properties welds [168,176,182,188]. Rezgui et al. [188] applied the SSFSW method with a wooden stationary shoulder to weld 15 mm thick HDPE. The temperature detected during the process was in a range from 120 to 180 °C, which means that the material reached its melting point, and the process did not have a solid-state nature. Other studies also confirm that friction stir welding of polymers is not a solid-state process, unlike the FSW of metals [185,189]. In order to obtain better results from the SSFSW method, a tool called "hot shoe" was developed and patented by Nelson et al. [190]. The aluminum static shoulder with the polytetrafluoroethylene (PTFE) coating with a heater component inside of the shoe allows the attainment of the tensile strength of 75% of the base material of ABS. The concept of SSFSW with a hot shoe as a shoulder was also successfully used in the studies of Bagheri et al. [176], Banjare et al. [191], and Laieghi et al. [192].

Sahu et al. [180] successfully welded 6 mm polypropylene sheets using three pin geometries—cylindrical, square, and conical. Only cylindrical and square pins enabled the production of defect-free welds. The influence of the tool traverse speed and the tool rotational speed on the ultimate tensile strength was determined. It was observed that the UTS of the welds firstly increased with an increase in the tool rotational speed and then slowly decreased. The same dependence was observed for the tensile strength function of welds on the tool traverse speed. The same conclusions for both tool traverse and tool rotational speed were confirmed by Pirizadeh et al. [178] on the friction stir welded 5 mm thick ABS plates and Arici et al. [165] on double pass friction stir welded Nylon 6 plates, where the influence of the tool traverse speed on the UTS of the welds was measured. In the studies, it was concluded that the tilt angle of the tool equal to 1° results in better mechanical properties of the welds than the tilt angle of 0°. The studies of Youssif et al. [175] on 13 mm thick Nylon 6 plates proved that the UTS of the welds decreases with an increase of the tool traverse speed. The same dependence was observed for the UTS as a function of the tool rotational speed. The same conclusions were presented in the studies of Zafar et al. [174] on 16 mm Nylon 6 plates. In the studies of Bagheri et al. [176] on 5 mm thick ABS plates, the UTS always increased with the increase of the tool rotational speed for all of the values of applied tool traverse speed. The maximum value of UTS reached almost 90% of the value for the base material.

Table 6 summarizes the best process parameters applied during the friction stir welding of polymeric materials and presents the mechanical properties of welds and parent material if given by the authors.

Table 6. FSW of polymers—process parameters and mechanical properties of the joints.

Material	Plate Thickness [mm]	Process Parameters					Weld Properties					Reference
		v [mm/min]	ω [rpm]	Tool Shape	Pin D/d Ratio	Tilt Angle [°]	UTS [MPa]	Yield Strength [MPa]	Hardness of the Stir Zone [HV]	Elongation [%]	Defects	
Polyethylene (PE)	5	25	1000	Cylindrical pin with no threads	3.2	1	19.30 (20.00)	-	-	-	-	[165]
Polypropylene (PP)	6	15	750	Square pin with no threads	2	1	19.74 (33)	-	-	-	Peeling defects	[180]
Polyvinyl chloride (PVC)	5	10	1800	Right-hand screw pin	~4.2	0	23.5 (66.5)	-	Around 78% of base material	-	-	[171]
Nylon 6	10	10	1000	Left handed threaded cylindrical pin	4	-	34.8 (73.4)	-	64 SD (shore-D hardness) (70 SD)	-	Defect free	[173]
Nylon 6	13	10	1250	Right handed threaded cylindrical pin	-	-	25.75 (54)	-	-	8.7 (43)	-	[175]
Nylon 6	16	25	300	Right-hand threaded pin	2.4	0	27.22 (85)	-	-	-	Lack of bonding, minor weld defects at the bottom	[174]
Polyamide 6 (PA6)	5	40	440	Right-hand screw pin	~4.2	0	30 (67.1)	-	60% of base material	-	-	[171]
Polyamide (nylon 66)	8	42	1570	Smooth cylindrical pin	4	-	8.51 (15.57)	-	-	-	-	[193]
Acrylonitrile Butadiene Styrene (ABS)	5	20	1600	Right-hand threaded cylindrical pin	1.7	-	32.62 (36.76)	-	-	-	-	[176]
Acrylonitrile Butadiene Styrene (ABS) (double side welded)	5	40	400	Smooth cylindrical pin, flat shoulders surfaces	-	-	15.58 (34.14)	-	-	-	-	[178]
Acrylonitrile Butadiene Styrene (ABS) (double side welded)	5	40	400	Smooth convex pin, flat shoulders surfaces	-	-	20.70 (34.14)	-	-	-	-	[178]
Acrylonitrile Butadiene Styrene (ABS)	6	200	1500	Conical threaded pin	-	-	30.6 (40.5)	-	-	-	Defect free	[179]

Table 6. *Cont.*

Material	Plate Thickness [mm]	Process Parameters					Weld Properties						Reference
		v [mm/min]	ω [rpm]	Tool Shape	Pin D/d Ratio	Tilt Angle [°]	UTS [MPa]	Yield Strength [MPa]	Hardness of the Stir Zone [HV]	Elongation [%]	Defects		
Acrylonitrile Butadiene Styrene (ABS)	8	16	1400	Cylindrical with no threads	3.3	1	41.42 (41.80)	-	-	-	-		[177]
Acrylonitrile Butadiene Styrene (ABS)	8	25	900	Conical with no threads	3.3	2	41.95 (41.80)	-	-	-	-		[177]
High-density polyethylene (HDPE)	4	115	3000	-	3	2	19.4 (22.5)	-	-	-	-		[166]
High-density polyethylene (HDPE)	5	15	1240	Right-hand screw pin	~4.2	0	22.3 (31.9)	-	Above 90% of base material	-	No significant defects		[171]

9. Composites

Polymer matrix composites (PMCs) and metal matrix composites (MMCs), especially aluminum matrix composites (AMCs), replace metals and their alloys and polymers in many industries, as the mechanical properties of such materials can be controlled by the proper selections of the filler properties. High temperature during conventional welding methods of metal matrix composites can lead to degradation of the microstructure of the composite, which leads to deterioration of mechanical properties of the joint. As an example, the process of formation of a brittle Al_4C_3 phase during conventional welding of a SiC-reinforced aluminum matrix composite can be presented [60]. Phase changes might be avoided by using shorter thermal cycles or lower heat input. During the FSW process, the temperature of the components to be joined is also increased, but it is relatively lower than that during conventional welding processes, and this problem is reduced. The problems that might occur during the FSW of composites include, beyond in the case of MMCs, the possibility of forming the brittle phases, which results in deterioration of strength, the formation of clusters of reinforcement particles, or change of particles' configuration due to the plastic deformation and applied forces and temperature [194,195].

In the FSW process of composites, in addition to the standard process parameters such as the rotational and traverse speed of the tool and the tilt angle of the tool, the shape of the tool itself also has a significant impact on the properties of the joints. Vijay and Murugan [196] used three pin shapes to friction stir weld Al/TiB_2/10 composite—square, hexagonal, and octagon. Using the untampered square pin resulted in obtaining the maximum tensile strength of the joint equal to 99.47% of the base material. The authors explained that by obtaining the highest ratio of static volume to dynamic volume of the plasticized material equal to 1.56 for the square pin, the best mechanical properties of the joints were achieved, while for hexagonal and octagon pins, the values were equal to 1.21 and 1.11, respectively. The same phenomenon was confirmed by Hassan et al. [197] in the study on aluminum matrix composite containing Mg, SiC, and graphite particles. The square head pin provided better mechanical properties of the joints than hexagonal and octagonal pins. Mahmoud et al. [198] used four different pin shapes—circular with and without threads, triangular, and square—to fabricate composite surface layers with SiC particles dispersed in A1050-H24 aluminum plates. It was reported that the square pin enabled the formation of the most homogeneous microstructure of the nugget zone.

Another aspect often discussed in the case of FSW of composites is the change of shape and size of the particle size of the composite reinforcement. The first reports in this area indicated the identical number of particles before and after the process, which implies that there is no particle breakage during the process [199–201]. However, other tests conducted prove that the breakdown of the reinforcing particles takes place in the nugget zone during the FSW process [202–206]. Baxter and Reynolds [202] reported that for the composite with 7079 aluminum matrix and SiC reinforcement, the number of SiC particles the number of particles doubled without changing their volume percentage, which means that the particle breakage takes place during the process. In the studies of Acharya et al. [207], the particle size of the SiC reinforcement in AA6092 in friction stir welded material zones was measured. In the nugget zone, thermo-mechanically affected zone, and the base materials the size of the particles was equal to 4.03, 4.99, and 7.92 μm, respectively. Feng et al. [204,205], in the studies on Al2009-15vol% SiC composite, reported that the particle breakage takes place in the stir zone, and the particles are uniformly distributed. Moreover, it is generally observed that the matrix phase experienced a grain refinement due to the dynamic recrystallization resulting from the frictional heating [204–206].

Kumar et al. [208], in the studies on glass-filled Nylon 6 friction stir welded 5 mm thick plates, used different tool traverse and tool rotational speed, and the values of the tilt angle were equal to 0, 1, and 2°. It was observed that the highest ultimate tensile strength and the elongation were obtained for the joints with the highest tilt angle. The UTS of the joints increased with an increase of the tool rotational speed and decreased as a function of the tool traverse speed. In the studies of Bhushan et al. [209] on the AA6082/SiC/10p 6 mm thick plates, the lowest tilt angle of 1° and the highest tool rotational speed resulted in the highest UTS and elongation. Jafrey et al. [210] reported that for 5 mm thick plates of PP/C30B/EA nanocomposite, the highest tool traverse speed resulted in the highest UTS of the joints. Liu et al. [211] and Wang et al. [212], in their studies on AC4A/SiC/30p and 2009Al-T4/SiC/17p, respectively, also reported that the elongation and the UTS of the joints increase when the tool traverse speed increases. Mozammil et al. [213], in the studies on Al-4.5% Cu/TiB_2/2.5 p plates with thickness of 6 mm, noted that the highest tool traverse speed of 26 mm/min and the highest tool rotational speed of 931 rpm resulted in the highest UTS for all the joints fabricated with three tool shoulder geometries—full flat, 1 mm flat shoulder and 7° concave, and 2 mm flat and 7° concave. Among them, the tool with 1 mm flat shoulder and 7° concave provided the best mechanical properties of the joints. Vijayavel et al. [117] investigated the influence of the D/d ratio on the mechanical properties of the FSW welds of LM25AA-5% SiC. Among the values from 2 to 4 in steps of 0.5, the D/d ratio equal to 3 resulted in the highest UTS, hardness, and elongation of the welds. The UTS value of such a joint reached 123% of that for the base material.

Table 7 summarizes the process parameters and weld properties for composites, considering in brackets the properties of the parent material if given by the authors.

Table 7. FSW of composites—process parameters and mechanical properties of the joints.

Material	Plate Thickness [mm]	Process Parameters					Weld Properties					Reference
		v [mm/min]	ω [rpm]	Tool Shape	Pin D/d Ratio	Tilt Angle [°]	UTS [MPa]	Yield Strength [MPa]	Hardness of the Stir Zone [HV]	Elongation [%]	Defects	
Al-4.5%Cu/TiB2/2.5p	6	26	931	Full flat shoulder surface, cylindrical pin with no threads	~2.9	2	190.39 (no inf about parent material)	-	69.86	16.875	-	[213]
Al-4.5%Cu/TiB2/2.5p	6	26	931	1 mm flat shoulder and 7° concave, cylindrical pin with no threads	~2.9	2	198.62	-	57.63	18.825	-	[213]
Al-4.5%Cu/TiB2/2.5p	6	26	931	2 mm flat shoulder and 7° concave, cylindrical pin with no threads	~2.9	2	191.39	-	60.76	17.050	-	[213]
AA6061/SiC (10wt%) /fly ash (7.5 wt%)	6	60	1200	Square profile pin	-	2	-	-	~130 (102)	-	No major defects	[214]
AA2124/SiC/25p-T4	3	40	1120	Cylindrical left-handed screwed pin	3.3	2	366 (454)	-	170 (185)	1.4 (2.4)	-	[215]
AA6092/SiC/17.5p	6	120	1500	Taper cylindrical pin	3	2	347 (415)	290 (360)	140 (157)	5.46 (7.76)	No major defects	[208,216]
AC4A + 30vol%SiCp	5	150	2000	Columnar pin with right-handed threads	2.3	3	140 (163)	-	-	0.33	Defect free	[211]
17vol%SiCp/2009Al-T4	3	800	1000	Threaded conical pin	2.8	-	501 (514)	341 (344)	-	3.5 (4.0)	No defects	[212]
17vol%SiCp/AA2009	3	50	1000	Cylindrical pin	2.8	-	443 (581)	278 (508)	-	4.7 (4.3)	No defects	[217]
AA2124/SiC/25p-T4	3	45	900	-	3	2	355.15 (454)	-	-	-	Flash defects on the surface	[218]
25%SiC/2124Al	8	15	400	Tapered conical pin	3.5	1.5	359 (372)	-	-	-	Small voids	[219]
AA6061-10%SiCp	6	45	1100	Threaded cylindrical pin	3	-	206 (278)	126 (200)	95 (105)	6.5 (8.0)	No defects	[220]
15vol%SiCp/2009Al	6	100	800	Conical pin	2.5	-	441 (537)	306 (343)	-	5.4 (10.1)	No defects	[221]
AA6092/SiC/17.5p-T6	3.1	100	1500	Flat edge featureless concave shoulder and M6 threaded cylindrical pin with a domed end	-	2	314 (420)	220 (370)	-	5 (3.5)	-	[222]

Table 7. *Cont.*

Material	Plate Thickness [mm]	Process Parameters					Weld Properties						Reference
		v [mm/min]	ω [rpm]	Tool Shape	Pin D/d Ratio	Tilt Angle [°]	UTS [MPa]	Yield Strength [MPa]	Hardness of the Stir Zone [HV]	Elongation [%]	Defects	
AA6082/SiC/10p	6	100	1800	Cylindrical	2.5	1	359	-	-	-	-	[209]
LM25AA-5% SiCp	12	40	1000	Plain taper pin	3	-	192 (155)	-	~105 (68)	7.2 (2)	-	[116]
PP/C30B/EA nanocomposite	5	18	-	Square pin with no threads	1	-	13.533 (25.08)	-	-	-	-	[210]
PP/C30B/EA nanocomposite	5	18	-	Cylindrical pin with no threads	1	-	16.300 (25.08)	-	-	-	-	[210]
PP/C30B/EA nanocomposite	5	18	-	Triangle pin with no threads	1	-	13.500 (25.08)	-	-	-	-	[210]
Glass-filled Nylon 6	5	12	600	Cylindrical pin with cylindrical shank	3	2	36.51 (86.01)	-	-	7.35 (13.68)	Defect free	[208]

10. Dissimilar Materials

Joining of dissimilar materials is essential in applications that require different material properties in the same component, i.e., joining different aluminum alloys to other metals allows reducing weight [223]. However, FSW of different metals is very challenging due to the huge differences in mechanical and metallurgical properties of dissimilar materials [224,225]. Welding of dissimilar materials in the context of FSW may also refer to joining the same material family but different alloys or grades, or the same material but with different thickness of components. Recently, the FSW method has been successfully and progressively applied to weld dissimilar alloys because of its technical benefits and cost-effectiveness. Comprehensive analysis of material flow during the FSW process of dissimilar materials, studies on mechanical properties of joints, and selection of process parameters are necessary before applying the method in constructional applications [226,227]. One of the key parameters, besides the geometry of the tool and its rotational and traverse speed, is the offset of the tool toward one side. It should be noted that it is difficult to predict the amount of heat generated, material flow, and mechanical properties of dissimilar joints using theoretical analysis [228].

The FSW method allows for avoiding problems that occur during fusion welding. For instance, arc welding, laser beams, and electron beams might cause the formation of coarse grains or brittle intermetallic compounds in the dissimilar aluminum/magnesium joints [229,230]. The proper selection of welding parameters using the FSW method is extremely difficult in the case of aluminum–magnesium welding due to the possibility of liquid formation resulting from the relatively low eutectic temperature (437 and 450 °C) in the binary Al-Mg phase diagram [231]. This phenomenon may, as in the case of fusion welding techniques, result in the formation of intermetallic compounds. Abdollahzadeh et al. [232] improved the microstructural characteristics by using a zinc interlayer in the magnesium and aluminum butt joints. Intermetallic formation of Al-Mg was avoided. The most common phases in the stirred zone were both Mg-Zn and Mg-Al-Zn intermetallic compounds, as well as Al solid solution and residual Zn. It is also difficult to obtain aluminum to copper welds using FSW. The brittle intermetallic compounds such as AlCu, Al_2Cu, and Al_9Cu_4 are easy to create in such joints [224,233,234]. Zhang et al. [235] proposed the underwater process environment to minimize this problem.

Movement of the material around the pin during FSW of dissimilar materials affects the creation of the bond formation and material interlocking mechanism. It improves the mechanical properties of the weld and increases its strength. This phenomenon was reported for Al-Mg [236], Al-Cu [237], Mg-steel, and Al-steel [238] welds. The phenomenon of material interlocking depends on the characteristic of the complex material flow and depends mostly on the geometry of tool and the positioning of components to be welded. The mechanical interlocking occurs only in a stir zone, and it is an effect of the rotational movement of the tool. The phenomenon of interlocking is also observed in explosive welding where a periodic structure can be distinguished, but this organised structure is not observed in the case of friction stir welding [239].

As established by Fu et al. [240], the heat generated during the FSW process mostly derives from the friction (E_f), the viscous dissipation (E_v), and the plastic deformations at the interface between tool and components (E_d). Zettler et al. [241], in the studies on dissimilar FSW of AZ31 and AA6040 alloys, reported that at the boundary between magnesium alloy and the tool, the frictional coefficient is lower than that on the aluminum alloy and the tool boundary. It was recommended to give tool offset toward aluminum to provide greater contact between the tool and the aluminum to increase the contributions of E_f and E_v during the friction stir welding process. This phenomenon can be explained by referring to the crystal structures of these materials. In the case of aluminum with its face-centred cubic structure (FCC) and twelve slip systems, better deformability is observed in comparison to the hexagonal close-packed structure (HCP) with three slip systems of magnesium. It promotes higher heat input through E_v and E_d during the FSW of aluminum than of magnesium [223,240]. These types of phenomena have a significant impact during the FSW of dissimilar materials; therefore, an important process parameter in such cases is also the tool offset.

In the studies of Jamshidi et al. [31] two aluminum alloys were used to fabricate friction stir welded dissimilar joints—AA6061-T6 and AA5086-O. For the case with AA6061-T6 plate on the advancing side and AA5086-O plate on the retreating side, better UTS and yield strength of the join were observed, but the values of the reverse configuration were only slightly worse. Moreover, for both cases, it was observed that the highest tool traverse speed of 150 mm/min and the lowest tool rotational speed of 840 rpm resulted in the best mechanical properties of the joints. Kasai et al. [242] fabricated three dissimilar material joints—on the advancing side, low carbon steel plates were used, while on the retreating side, pure magnesium, AZ31, and AZ61 plates, respectively, were used. The highest UTS was noted for the joint with low carbon and AZ61, and it was observed that the UTS of steel/magnesium joints increases with the increase of Al content in the magnesium alloy. The reason for this phenomenon is simple—for all the used magnesium components, the UTS of the base material also increases with the increase in aluminum content in the material. In the studies of Zettler et al. [241], the dissimilar joints of Al6040-T61 and AZ31 were fabricated using the FSW method. The UTS of the joint of the configuration with aluminum alloy on the retreating side was about 50% higher than for the joint with reverse configuration. Peel et al. [243] used AA5083 and AA6082 aluminum alloys to fabricate dissimilar joints. It was observed that for both material configurations, the highest UTS was achieved when the tool traverse speed and the tool rotational speed were the highest and equal to 300 mm/min and 840 rpm, respectively. Furthermore, the joints with AA5083 plate on the advancing side exhibit higher UTS than for the reverse configuration, and for all the joints, no defects were observed. Similar studies were conducted by Khodir et al. [244]. The dissimilar welds of AA2024-T3 and AA7075-T6 were fabricated with both plates configurations. The welds with AA2024-T3 plate on the advancing side exhibit better mechanical properties, such as the UTS, the yield strength, and the elongation. Avinash et al. [245], in studies on AA2024-T3 on the advancing side and AA7075-T6 on the retreating side joints, reported that applying the lowest tool traverse speed of 80 mm/min and the highest tool rotational speed of 1400 rpm resulted in the highest UTS of the joint. Palanivel et al. [246] investigated the influence of the tool geometry on the UTS of dissimilar AA6351-T6 on the advancing side and AA5083-H111 on retreating side joints. Five different pins were used: Straight square, straight octagon, straight hexagon, tapered octagon, and tapered square. Among them, the straight square pin enabled the production of defect-free joints with the highest UTS, while all the welds fabricated with tapered octagon and tapered square pins exhibit tunnel defects along the joint line. Malarvizhi et al. [247] studied the influence of the D/d ratio of the tool on dissimilar AZ31B-O/AA6061-T6 joints. For the tapered smooth pin, the D/d ratios from 2 to 4 in steps of 0.5 were investigated. It was reported that the value equal to 3.5 resulted in the highest UTS, yield strength, and elongation of the produced joints.

Table 8 summarizes the studied results of research on welds from dissimilar materials. Table 8 includes the materials used for advancing and retreating side of the welds, process parameters, welds properties, and the properties of base materials if given by the authors.

Table 8. FSW of dissimilar materials—process parameters and mechanical properties of the joints.

Material		Process Parameters						Weld Properties					
Advancing Side	Retreating Side	Plate Thickness [mm]	v [mm/min]	ω [rpm]	Tool Shape	Pin D/d Ratio	Tilt Angle [°]	UTS [MPa]	Yield Strength [MPa]	Hardness of the Stir Zone [HV]	Elongation [%]	Defects	Reference
AA2024-T3	AA7075-T6	3	102	1200	Cylindrical threaded pin	3	-	423 (416 for AA2024-T3 and 593 for AA7075-T6)	290.0 (327 for AA2024-T3 and 498 for AA7075-T6)	-	14.9 (29.5 for AA2024-T3 and 17.7 for AA7075-T6)	-	[244]
AA2024-T6	AA7075-T6	5	12	1200	Flat shoulder and smooth cylindrical pin	3	-	356 (416 for AA2024-T6 and 485 for AA7075-T6)	-	-	-	Defect free	[116]
AA5052	AA6061	-	28	710	Cylindrical pin with two threads	3	-	180	-	82	10.6	-	[248]
AA5083	AA6082	3	300	840	Conical surface of the shoulder, threaded cylindrical pin	3	2	241	-	-	3.05	-	[243]
AA5086-O	AA6061-T6	5	150	840	Tapered pin with 3 threads and concave shoulder surface	3.3	-	221 (253 for AA5086-O; 315 for AA6061-T6)	136 (112 for AA5086-O; 278 for AA6061-T6)	-	-	-	[31]
AA6061-T6	AA5086-O	5	150	840	Tapered pin with 3 threads and concave shoulder surface	3.3	-	224 (315 for AA6061-T6; 253 for AA5086-O)	139 (278 for AA6061-T6; 112 for AA5086-O)	-	-	-	[31]
AA6082	AA5083	3	200	840	Conical surface of the shoulder, threaded cylindrical pin	3	2	227	-	-	2.67	-	[243]
AA6351-T6	AA5083-H111	6	60	950	Flat shoulder surface, straight square pin with no threads	3	0	273 (AA6351-T6: 310 AA5083H111: 308)	-	-	-	No defects	[246]
AA7075-T6	AA2024-T3	6.5 (AA7075-T6); 5 (AA2024-T3)	80	1400	Square pin	-	-	261	-	-	-	Defect-free	[245]
AA7075-T6	AA2024-T3	3	102	1200	Cylindrical threaded pin	3	-	381 (593 for AA7075-T6 and 416 for AA2024-T3)	280.0 (498 for AA7075-T6 and 327 for AA2024-T3)	-	9.0 (17.7 for AA7075-T6 and 29.5 for AA2024-T3)	-	[244]

Table 8. *Cont.*

Material		Plate Thickness [mm]	Process Parameters					Weld Properties					Reference
Advancing Side	Retreating Side		v [mm/min]	ω [rpm]	Tool Shape	Pin D/d Ratio	Tilt Angle [°]	UTS [MPa]	Yield Strength [MPa]	Hardness of the Stir Zone [HV]	Elongation [%]	Defects	
St37	304 austenitic stainless steel	3	50	600	-	2.9	3	494 (446 for St37 and 679 for 304 steel)	290 (305 for St37 and 287 for 304 steel)	240 (120 for St37 and 180 for 304 steel)	28 (42 for St37 and 111 for 304 steel)	No macro defects	[249]
St52 mild steel	AA5186	3	56	355	M3 threaded pin	6	3	246 (520 for St52 and 275 for AA5186)	-	-	-	No defects	[250]
Low carbon steel	Pure Mg	2	100	1000	Cylindrical with no threads	3	3	~70 (316 for low carbon steel and 170 for pure Mg)	-	-	-	-	[242]
Low carbon steel	AZ31	2	100	500	Cylindrical with no threads	3	3	~165 (316 for low carbo steel and 260 for AZ31)	-	-	-	-	[242]
Low carbon steel	AZ61	2	100	750	Cylindrical with no threads	3	3	~220 (316 for low carbon steel and 280 for AZ61)	-	-	-	-	[242]
AZ31	AA6040-T61	2	225	1400	Threaded tapered pin	2.6	2.5	189 (228–238 for AZ31 and 175-205 for Al6040-T61)	-	-	-	-	[241]
AA6040-T61	AZ31	2	200	1400	Threaded tapered pin	2.6	2.5	127 (175–205 for Al6040-T61 and 228–238 for AZ31)	-	-	-	-	[241]
AA5052-H	AZ31B	3	200	1000	-	3	3	147 (244 for A5052-H and 241 for AZ31B)	64.0 (181 for A5052-H and 200 for AZ31B)	-	3.4 (18.0 for A5052-H and 21.6 for AZ31B)	No defects	[251]
AA6061	AZ31	3	40	1000	Concave shoulder and cylindrical pin	5	2.5	178 (295 for 6061 Al and 235 for AZ31)	170 (235 for 6061 Al and 130 for AZ31)	-	2.4 (12.5 for 6061 Al and 18.7 for AZ31)	Defect free	[252]
AZ31B-0	AA6061-T6	6	20	400	Tapered smooth pin	3.5	-	192 (216 for AZ31B-0 and 311 for AA6061-T6)	153 (175 for AZ31B-0 and 280 for AA6061-T6)	-	10 (15 for AZ31B-0 and 20 for AA6061-T6)	Defect free	[247]

Table 8. *Cont.*

Material		Plate Thickness [mm]	Process Parameters					Weld Properties					Reference
Advancing Side	Retreating Side		v [mm/min]	ω [rpm]	Tool Shape	Pin D/d Ratio	Tilt Angle [°]	UTS [MPa]	Yield Strength [MPa]	Hardness of the Stir Zone [HV]	Elongation [%]	Defects	
AZ31B	AA1100	3	20	570	Cylindrical threaded pin	3.6	-	122 (228 for AZ31B; 175 AA1100)	101 (145 AZ31B; 105 AA1100)	-	9 (17 AZ31B; 11 AA1100)	-	[253]
SS400 mild steel	AZ31B-O	2	100	1250	Unthreaded cylindrical pin	5	-	178.5 (455 for SS400; 257 AZ31B-O)	-	-	~2.7 (39 for SS400 and 23.3 for AZ31B-O)	-	[254]
304L stainless steel	Pure Cu	3	31.5	1500	Taper pin	-	2	173.05 (574 for 304L; 227 for copper)	-	-	8.22 (51 for 304L; 23 for copper)	-	[255]
Pure Cu	AA1050	6	63	1400	-	-	-	88.466 (91% of AA1050)	-	-	-	-	[256]
Pure Cu	AA1060	5	100	600	Cylindrical pin	~3.3	-	110	-	-	-	Defect free	[257]

11. Conclusions and Future Challenges

Over the past two decades, FSW technology has developed significantly, and it is widely used to combine not only aluminum and other light metals but also titanium, steel, composites, and polymers. The above literature review has extensively discussed the current state of knowledge and addressed the most common problems arising in the process of welding different groups of materials. The analysis of material characteristics and the correct selection of parameters for a particular material is crucial for effective welding.

Despite numerous studies on the selection of suitable process parameters such as tool traverse speed and tool rotational speed, tool geometry, tilt angle, and tool offset, there is still a need for process optimization. It should not be forgotten that in addition to the main process parameters, consideration should be given to factors such as plunged depth, axial force, and tool material. In addition, the material flow mechanism needs to be standardized, especially when welding dissimilar materials with different mechanical properties. Optimization of process parameters should be based on the characteristics of the materials being processed. Appropriate tool geometry and its material selection should also be preceded by a material analysis. In the case of aluminum or copper welding, the problem of material wear is not as important as in the case of titanium or steel. For these materials, specially designed materials are used, most often composite materials, which allow minimizing the process of material wear. Titanium welding requires a tool with an uncomplicated geometry, while for light alloys, a more developed tool geometry is required. SSFSW technology is becoming increasingly important for polymer welding, while traditional FSW technique might be problematic with a conventional tool, it usually does not bring the expected results, and the properties of the welds are relatively low. Despite extensive knowledge of the welding of aluminum and its alloys, the problem of the welding of dissimilar materials still remains. The most recent reports indicate the use of an interlayer or water environment to minimize the risk of unwanted intermetallic compounds.

The environmental friendliness and cost-effectiveness of this method make its use increasingly widespread in many industries, but further work is needed to optimise this process to unify the conclusions about the impact of individual parameters on the properties of the resulting welds. It is also necessary to work on further improving the geometry of the tool and the proper selection of its material in order to minimize the wear process. For the development of friction stir welding technique, it is important to control tool wear state in the real time of the process in order to ensure the highest possible process repeatability. The tool life can be effectively predicted by implementing the numerical simulation of the interaction between the component's material and the tool. The material flow is the crucial phenomenon during FSW, and it still requires more understanding. There is a need for concerted research efforts towards the computer simulations of the process, which will help develop understanding of the mechanism of material flow and heat generation. Due to the wide application of FSW technology in the marine industry, special attention should be paid to the electrochemical properties of the produced welds. The current state of the art is poor in this type of research. Future studies should focus on the influence of particular process parameters not only on the mechanical properties of welds intended for the marine industry, but also on their corrosion properties.

Author Contributions: Conceptualization: A.L. and M.S.; formal analysis: M.S.; investigation: A.L.; writing—original draft preparation: A.L. All authors have read and agreed to the published version of the manuscript.

Funding: This research received no external funding.

Conflicts of Interest: The authors declare no conflict of interest.

References

1. Arunprasad, R.V.; Surendhiran, G.; Ragul, M.; Soundarrajan, T.; Moutheepan, S.; Boopathi, S. Review on Friction Stir Welding Process. *Int. J. Appl. Eng. Res. ISSN* **2018**, *13*, 5750–5758.

2. Kossakowski, P.; Wciślik, W.; Bakalarz, M. Macrostructural Analysis of Friction Stir Welding (FSW) Joints. *J. Chem. Inf. Model.* **2018**, *1*, 1689–1699. [CrossRef]
3. Kah, P.; Rajan, R.; Martikainen, J.; Suoranta, R. Investigation of weld defects in friction-stir welding and fusion welding of aluminum alloys. *Int. J. Mech. Mater. Eng.* **2015**, *10*, 26. [CrossRef]
4. Safeen, M.W.; Spena, P.R. Main issues in quality of friction stir welding joints of aluminum alloy and steel sheets. *Metals* **2019**, *9*, 610. [CrossRef]
5. Pasha, A.; Reddy, R.P.; Ahmad Khan, I. Influence of Process and Tool Parameters on Friction Stir Welding–Over View. *Int. J. Appl. Eng. Technol.* **2014**, *4*, 54–69.
6. Sato, Y.S.; Urata, M.; Kokawa, H. Parameters controlling microstructure and hardness during friction-stir welding of precipitation-hardenable aluminum alloy 6063. *Metall. Mater. Trans. A* **2002**, *33*, 625–635. [CrossRef]
7. Colligan, K.J. *The Friction Stir Welding Process: An Overview*; Woodhead Publishing Limited: Cambridge, UK, 2009.
8. Raweni, A.; Majstorović, V.; Sedmak, A.; Tadić, S.; Kirin, S. Optimization of AA5083 friction stir welding parameters using taguchi method. *Teh. Vjesn.* **2018**, *25*, 861–866. [CrossRef]
9. Micari, F.; Buffa, G.; Pellegrino, S.; Fratini, L. Friction Stir Welding as an effective alternative technique for light structural alloys mixed joints. *Procedia Eng.* **2014**, *81*, 74–83. [CrossRef]
10. Nandan, R.; DebRoy, T.; Bhadeshia, H.K.D.H. Recent advances in friction-stir welding-Process, weldment structure and properties. *Prog. Mater. Sci.* **2008**, *53*, 980–1023. [CrossRef]
11. Wang, W.; Deng, D.; Mao, Z.; Tong, Y.; Ran, Y. Influence of tool rotation rates on temperature profiles and mechanical properties of friction stir welded AZ31 magnesium alloy. *Int. J. Adv. Manuf. Technol.* **2017**, *88*, 2191–2200. [CrossRef]
12. Mishra, A. Friction Stir Welding of Aerospace Alloys. *Int. J. Res. Appl. Sci. Eng. Technol.* **2019**, *7*, 863–870. [CrossRef]
13. Singh, I.; Cheema, G.S.; Kang, A.S. An experimental approach to study the effect of welding parameters on similar friction stir welded joints of AZ31B-O Mg alloy. *Procedia Eng.* **2014**, *97*, 837–846. [CrossRef]
14. Gesella, G.; Czechowski, M. The application of friction stir welding (FSW) of aluminum alloys in shipbuilding and railway industry. *J. KONES* **2017**, *24*, 85–90. [CrossRef]
15. Kawasaki, T.; Makino, T.; Masai, K.; Ohba, H.; Ina, Y.; Ezumi, M. Application of friction stir welding to construction of railway vehicles. *JSME Int. J. Ser. A Solid Mech. Mater. Eng.* **2004**, *47*, 502–511. [CrossRef]
16. Singh, K.; Singh, G.; Singh, H. Investigation of microstructure and mechanical properties of friction stir welded AZ61 magnesium alloy joint. *J. Magnes. Alloy.* **2018**, *6*, 292–298. [CrossRef]
17. Sevvel, P.; Jaiganesh, V. An detailed examination on the future prospects of friction stir welding–a green technology. In Proceedings of the Second International Conference on Advances in Industrial Engineering Applications (ICAIEA 2014), Chennai, India, 6–8 January 2014; pp. 275–280.
18. Grimm, A.; Schulze, S.; Silva, A.; Göbel, G.; Standfuss, J.; Brenner, B.; Beyer, E.; Füssel, U. Friction Stir welding of Light Metals for Industrial Applications. *Mater. Today Proc.* **2015**, *2*, S169–S178. [CrossRef]
19. Singarapu, U.; Adepu, K.; Arumalle, S.R. Influence of tool material and rotational speed on mechanical properties of friction stir welded AZ31B magnesium alloy. *J. Magnes. Alloy.* **2015**, *3*, 335–344. [CrossRef]
20. Magalhães, V.M.; Leitão, C.; Rodrigues, D.M. Friction stir welding industrialisation and research status. *Sci. Technol. Weld. Join.* **2018**, *23*, 400–409. [CrossRef]
21. Rhodes, C.G.; Mahoney, M.W.; Bingel, W.H.; Spurling, R.A.; Bampton, C.C. Effects of friction stir welding on microstructure of 7075 aluminum. *Scr. Mater.* **1997**, *36*, 69–75. [CrossRef]
22. Liu, G.; Murr, L.E.; Niou, C.S.; McClure, J.C.; Vega, F.R. Microstructural aspects of the friction-stir welding of 6061-T6 aluminum. *Scr. Mater.* **1997**, *37*, 355–361. [CrossRef]
23. Sato, Y.; Kokawa, H.; Enomoto, M.; Jogan, S. Microstructural evolution of 6063 aluminum during friction-stir welding. *Metall. Mater. Trans. A Phys. Metall. Mater. Sci.* **1999**, *30*, 2429–2437. [CrossRef]
24. Jata, K.V.; Sankaran, K.K.; Ruschau, J.J. Friction-stir welding effects on microstructure and fatigue of aluminum alloy 7050-T7451. *Metall. Mater. Trans. A Phys. Metall. Mater. Sci.* **2000**, *31*, 2181–2192. [CrossRef]
25. Heinz, B.; Skrotzki, B. Characterization of a friction-stir-welded aluminum alloy 6013. *Metall. Mater. Trans. B Process Metall. Mater. Process. Sci.* **2002**, *33*, 489–498. [CrossRef]
26. Colligan, K.J. *Failure Mechanisms of Advanced Welding Processes*; Woodhead Publishing Limited: Cambridge, UK, 2010; pp. 137–163.

27. Woo, W.; Balogh, L.; Ungár, T.; Choo, H.; Feng, Z. Grain structure and dislocation density measurements in a friction-stir welded aluminum alloy using X-ray peak profile analysis. *Mater. Sci. Eng. A* **2008**, *498*, 308–313. [CrossRef]
28. Tasi, P.; Hajro, I.; Hodži, D.; Dobraš, D. Energy Efficient Welding Technology: Fsw. In Proceedings of the 11th International Conference on Accomplishments in Electrical and Mechanical Engieering and Information Technology, 30 May–1 June 2013; pp. 429–442.
29. Sivaraj, P.; Kanagarajan, D.; Balasubramanian, V. Effect of post weld heat treatment on tensile properties and microstructure characteristics of friction stir welded armour grade AA7075-T651 aluminum alloy. *Def. Technol.* **2014**, *10*, 1–8. [CrossRef]
30. Surendrababu, P.; Gopala Krishna, A.; Srinivasa Rao, C. Material Flow Behaviour in Friction Stir Welding Process-A Critical Review on Process Parameters and Modeling Methodologies. *Int. J. Emerg. Technol. Adv. Eng.* **2013**, *3*, 219–225.
31. Jamshidi Aval, H.; Serajzadeh, S.; Kokabi, A.H. Evolution of microstructures and mechanical properties in similar and dissimilar friction stir welding of AA5086 and AA6061. *Mater. Sci. Eng. A* **2011**, *528*, 8071–8083. [CrossRef]
32. Mishra, R.S.; Ma, Z.Y. Friction Stir Welding and Processing. *Mater. Sci. Eng.: R: Rep.* **2005**, *50*, 1–78. [CrossRef]
33. Hussain, A.K.; Azam, S.; Quadri, P. Evaluation of Parameters of Friction Stir Welding for Aluminum Aa6351 Alloy. *Int. J. Eng. Sci. Technol.* **2010**, *2*, 5977–5984.
34. Patel, A.R.; Kotadiya, D.J.; Kapopara, J.M.; Dalwadi, C.G.; Patel, N.P.; Rana, H.G. Investigation of Mechanical Properties for Hybrid Joint of Aluminum to Polymer using Friction Stir Welding (FSW). *Mater. Today Proc.* **2018**, *5*, 4242–4249. [CrossRef]
35. Liu, F.C.; Liao, J.; Nakata, K. Joining of metal to plastic using friction lap welding. *Mater. Des.* **2014**, *54*, 236–244. [CrossRef]
36. Patel, A.R.; Dalwadi, C.G.; Rana, H.G. A Review: Dissimilar Material Joining of Metal to Polymer using Friction Stir Welding (FSW). *IJSTE-Int. J. Sci. Technol. Eng.* **2016**, *2*, 702–706.
37. Murr, L.E.; Li, Y.; Trillo, E.; McClure, J.C. Fundamental Issues and Industrial Applications of Friction-Stir Welding. *Mater. Technol.* **2000**, *15*, 37–48. [CrossRef]
38. Threadgill, P.L.; Johnson, R. The Potential for Friction Stir Welding in Oil and Gas Applications. In Proceedings of the Fourteenth International Offshore Polar Engineering Conference, Toulon, France, 23–28 May 2004.
39. Shtrikman, M.M. Current state and development of friction stir welding Part 3. Industrial application of friction stir welding. *Weld. Int.* **2008**, *22*, 806–815. [CrossRef]
40. Thomas, W.M.; Nicholas, E.D. Friction stir welding for the transportation industries. *Mater. Des.* **1997**, *18*, 269–273. [CrossRef]
41. Tang, W.; Guo, X.; McClure, J.C.; Murr, L.E.; Nunes, A. Heat input and temperature distribution in friction stir welding. *J. Mater. Process. Manuf. Sci.* **1998**, *7*, 163–172. [CrossRef]
42. Mahoney, M.W.; Rhodes, C.G.; Flintoff, J.G.; Spurling, R.A.; Bingel, W.H. Properties of friction-stir-welded 7075 T651 aluminum. *Metall. Mater. Trans. A Phys. Metall. Mater. Sci.* **1998**, *29*, 1955–1964. [CrossRef]
43. Reynolds, A.P.; Lockwood, W.D.; Seidel, T.U. Processing-property correlation in friction stir welds. *Mater. Sci. Forum* **2000**, *331*, 1719–1724. [CrossRef]
44. Colegrove, P.A.; Shercliff, H.R. Experimental and numerical analysis of aluminum alloy 7075-T7351 friction stir welds. *Sci. Technol. Weld. Join.* **2003**, *8*, 360–368. [CrossRef]
45. Sato, Y.S.; Kokawa, H.; Ikeda, K.; Enomoto, M.; Jogan, S.; Hashimoto, T. Microtexture in the friction-stir weld of an aluminum alloy. *Metall. Mater. Trans. A Phys. Metall. Mater. Sci.* **2001**, *32*, 941–948. [CrossRef]
46. Krasnowski, K.; Sedek, P.; Łomozik, M.; Pietras, A. Impact of selected FSW process parameters on mechanical properties of 6082-T6 aluminum alloy butt joints. *Arch. Metall. Mater.* **2011**, *56*, 965–973. [CrossRef]
47. Srinivasa Rao, M.S.; Ravi Kumar, B.V.R.; Manzoor Hussain, M. Experimental study on the effect of welding parameters and tool pin profiles on the IS:65032 aluminum alloy FSW joints. *Mater. Today Proc.* **2017**, *4*, 1394–1404. [CrossRef]
48. Emamian, S.; Awang, M.; Hussai, P.; Meyghani, B.; Zafar, A. Influences of tool pin profile on the friction stir welding of AA6061. *ARPN J. Eng. Appl. Sci.* **2016**, *11*, 12258–12261.
49. Rajendran, C.; Srinivasan, K.; Balasubramanian, V.; Balaji, H.; Selvaraj, P. Effect of tool tilt angle on strength and microstructural characteristics of friction stir welded lap joints of AA2014-T6 aluminum alloy. *Trans. Nonferrous Met. Soc. China* **2019**, *29*, 1824–1835. [CrossRef]

50. Barlas, Z.; Ozsarac, U. Effects of FSW parameters on joint properties of AlMg3 Alloy. *Weld. J.* **2012**, *91*, 16S–22S.
51. Peel, M.; Steuwer, A.; Preuss, M.; Withers, P.J. Microstructure, mechanical properties and residual stresses as a function of welding speed in aluminum AA5083 friction stir welds. *Acta Mater.* **2003**, *51*, 4791–4801. [CrossRef]
52. Khan, N.Z.; Khan, Z.A.; Siddiquee, A.N. Effect of Shoulder Diameter to Pin Diameter (D/d) Ratio on Tensile Strength of Friction Stir Welded 6063 Aluminum Alloy. *Mater. Today Proc.* **2015**, *2*, 1450–1457. [CrossRef]
53. Peng, P.; Wang, K.; Wang, W.; Huang, L.; Qiao, K.; Che, Q.; Xi, X.; Zhang, B.; Cai, J. High-performance aluminum foam sandwich prepared through friction stir welding. *Mater. Lett.* **2019**, *236*, 295–298. [CrossRef]
54. Busic, M.; Kozuh, Z.; Klobcar, D.; Samardzic, I. Friction Stir Welding (FSW) of Aluminum Foam Sandwich Panels. *Metalurgija* **2016**, *55*, 473–476.
55. Yoo, J.T.; Yoon, J.H.; Min, K.J.; Lee, H.S. Effect of Friction Stir Welding Process Parameters on Mechanical Properties and Macro Structure of Al-Li Alloy. *Procedia Manuf.* **2015**, *2*, 325–330. [CrossRef]
56. Lim, S.; Kim, S.; Lee, C.G.; Kim, S. Tensile behavior of friction-stir-welded A356-T6/Al 6061-T651 bi-alloy plate. *Metall. Mater. Trans. A Phys. Metall. Mater. Sci.* **2004**, *35*, 2837–2843. [CrossRef]
57. Sayer, S.; Ceyhun, V.; Tezcan, Ö. The influence of friction stir welding parameters on the mechanical properties and low cycle fatigue in AA 6063 (AlMgSi0.5) alloy. *Kov. Mater.* **2008**, *46*, 157–164.
58. Sato, Y.S.; Kokawa, H. Distribution of tensile property and microstructure in friction stir weld of 6063 aluminum. *Metall. Mater. Trans. A Phys. Metall. Mater. Sci.* **2001**, *32*, 3023–3031. [CrossRef]
59. Boonchouytan, W.; Ratanawilai, T.; Muangjunburee, P. Effect of pre/post heat treatment on the friction stir welded SSM 356 aluminum alloys. *Adv. Mater. Res.* **2012**, *32*, 1139–1146. [CrossRef]
60. Çam, G. Friction stir welded structural materials: Beyond Al-alloys. *Int. Mater. Rev.* **2011**, *56*, 1–48. [CrossRef]
61. Singh, K.; Singh, G.; Singh, H. Review on friction stir welding of magnesium alloys. *J. Magnes. Alloy.* **2018**, *6*, 399–416. [CrossRef]
62. Dargusch, M.S.; Bowles, A.L.; Pettersen, K.; Bakke, P.; Dunlop, G.L. The effect of silicon content on the microstruture and creep behavior in die-cast magnesium AS alloys. *Metall. Mater. Trans. A Phys. Metall. Mater. Sci.* **2004**, *35*, 1905–1909. [CrossRef]
63. You, S.; Huang, Y.; Kainer, K.U.; Hort, N. Recent research and developments on wrought magnesium alloys. *J. Magnes. Alloy.* **2017**, *5*, 239–253. [CrossRef]
64. Froes, F.H.; Eliezer, D.; Aghion, E. The Science, Technology, and Applications of Magnesium. *J. Miner. Met. Mater. Soc.* **1998**, *50*, 30–34. [CrossRef]
65. Xin, R.; Liu, D.; Li, B.; Sun, L.; Zhou, Z.; Liu, Q. Mechanisms of fracture and inhomogeneous deformation on transverse tensile test of friction-stir-processed AZ31 Mg alloy. *Mater. Sci. Eng. A* **2013**, *565*, 333–341. [CrossRef]
66. Shang, Q.; Ni, D.R.; Xue, P.; Xiao, B.L.; Ma, Z.Y. Evolution of local texture and its effect on mechanical properties and fracture behavior of friction stir welded joint of extruded Mg-3Al-1Zn alloy. *Mater. Charact.* **2017**, *128*, 14–22. [CrossRef]
67. Yang, J.; Xiao, B.L.; Wang, D.; Ma, Z.Y. Effects of heat input on tensile properties and fracture behavior of friction stir welded Mg-3Al-1Zn alloy. *Mater. Sci. Eng. A* **2010**, *527*, 708–714. [CrossRef]
68. Commin, L.; Dumont, M.; Rotinat, R.; Pierron, F.; Masse, J.E.; Barrallier, L. Influence of the microstructural changes and induced residual stresses on tensile properties of wrought magnesium alloy friction stir welds. *Mater. Sci. Eng. A* **2012**, *551*, 288–292. [CrossRef]
69. Lim, S.; Kim, S.; Lee, C.G.; Yim, C.D.; Kim, S.J. Tensile behavior of friction-stir-welded AZ31-H24 Mg alloy. *Metall. Mater. Trans. A Phys. Metall. Mater. Sci.* **2005**, *36*, 1609–1612. [CrossRef]
70. Lee, W.B.; Yeon, Y.M.; Jung, S.B. Joint properties of friction stir welded AZ31B-H24 magnesium alloy. *Mater. Sci. Technol.* **2003**, *19*, 785–790. [CrossRef]
71. Kumar, R.; Pragash, M.S.; Varghese, S. Optimizing the process parameters of FSW on AZ31B Mg alloy by Taguchi-grey method. *Middle East J. Sci. Res.* **2013**, *15*, 161–167. [CrossRef]
72. Han, J.; Chen, J.; Peng, L.; Tan, S.; Wu, Y.; Zheng, F.; Yi, H. Microstructure, texture and mechanical properties of friction stir processed Mg-14Gd alloys. *Mater. Des.* **2017**, *130*, 90–102. [CrossRef]
73. Sahu, P.K.; Pal, S. Effect of FSW parameters on microstructure and mechanical properties of AM20 welds. *Mater. Manuf. Process.* **2018**, *33*, 288–298. [CrossRef]
74. Sevvel, P.; Jaiganesh, V. Characterization of mechanical properties and microstructural analysis of friction stir welded AZ31B Mg alloy thorough optimized process parameters. *Procedia Eng.* **2014**, *97*, 741–751. [CrossRef]

75. Pareek, M.; Polar, A.; Rumiche, F.; Indacochea, J.E. Metallurgical evaluation of AZ31B-H24 magnesium alloy friction stir welds. *J. Mater. Eng. Perform.* **2007**, *16*, 655–662. [CrossRef]
76. Esparza, J.A.; Davis, W.C.; Murr, L.E. Microstructure-property studies in friction-stir welded, thixomolded magnesium alloy AM60. *J. Mater. Sci.* **2003**, *38*, 941–952. [CrossRef]
77. Park, S.H.C.; Sato, Y.S.; Kokawa, H. Effect of micro-texture on fracture location in friction stir weld of Mg alloy AZ61 during tensile test. *Scr. Mater.* **2003**, *49*, 161–166. [CrossRef]
78. Xie, G.M.; Ma, Z.Y.; Geng, L.; Chen, R.S. Microstructural evolution and mechanical properties of friction stir welded Mg-Zn-Y-Zr alloy. *Mater. Sci. Eng. A* **2007**, *471*, 63–68. [CrossRef]
79. Zhang, D.; Suzuki, M.; Maruyama, K. Microstructural evolution of a heat-resistant magnesium alloy due to friction stir welding. *Scr. Mater.* **2005**, *52*, 899–903. [CrossRef]
80. Xunhong, W.; Kuaishe, W. Microstructure and properties of friction stir butt-welded AZ31 magnesium alloy. *Mater. Sci. Eng. A* **2006**, *431*, 114–117. [CrossRef]
81. Ugender, S.; Kumar, A.; Reddy, A.S. Microstructure and Mechanical Properties of AZ31B Magnesium Alloy by Friction Stir Welding. *Procedia Mater. Sci.* **2014**, *6*, 1600–1609. [CrossRef]
82. Padmanaban, G.; Balasubramanian, V. Selection of FSW tool pin profile, shoulder diameter and material for joining AZ31B magnesium alloy-An experimental approach. *Mater. Des.* **2009**, *30*, 2647–2656. [CrossRef]
83. Afrin, N.; Chen, D.L.; Cao, X.; Jahazi, M. Strain hardening behavior of a friction stir welded magnesium alloy. *Scr. Mater.* **2007**, *57*, 1004–1007. [CrossRef]
84. Patel, N.; Bhatt, K.D.; Mehta, V. Influence of Tool Pin Profile and Welding Parameter on Tensile Strength of Magnesium Alloy AZ91 During FSW. *Procedia Technol.* **2016**, *23*, 558–565. [CrossRef]
85. Chai, F.; Zhang, D.; Li, Y. Microstructures and tensile properties of submerged friction stir processed AZ91 magnesium alloy. *J. Magnes. Alloy.* **2015**, *3*, 203–209. [CrossRef]
86. Kadigithala, N.K.; Vanitha, C. Effects of welding speeds on the microstructural and mechanical properties of AZ91D Mg alloy by friction stir welding. *Int. J. Struct. Integr.* **2020**, *11*, 769–782. [CrossRef]
87. Kouadri-Henni, A.; Barrallier, L. Mechanical properties, microstructure and crystallographic texture of magnesium AZ91-D alloy welded by friction stir welding (FSW). *Metall. Mater. Trans. A Phys. Metall. Mater. Sci.* **2014**, *45*, 4983–4996. [CrossRef]
88. Wang, K.S.; Shen, Y.; Yang, X.R.; Wang, X.H.; Xu, K.W. Evaluation of Microstructure and Mechanical Property of FSW Welded MB3 Magnesium Alloy. *J. Iron Steel Res. Int.* **2006**, *13*, 75–78. [CrossRef]
89. Weng, F.; Liu, Y.; Chew, Y.; Lee, B.Y.; Ng, F.L.; Bi, G. Double-side friction stir welding of thick magnesium alloy: Microstructure and mechanical properties. *Sci. Technol. Weld. Join.* **2020**, *25*, 359–368. [CrossRef]
90. Reynolds, A.P.; Tang, W.; Posada, M.; DeLoach, J. Friction stir welding of DH36 steel. *Sci. Technol. Weld. Join.* **2003**, *8*, 455–461. [CrossRef]
91. Sorensen, C.; Nelson, T. Friction stir welding of ferrous and nickel alloys. *Frict. Stir Weld. Process.* **2007**, 111–121. Available online: https://www.researchgate.net/publication/267856005_Friction_Stir_Welding_of_Ferrous_and_Nickel_Alloys (accessed on 17 February 2020).
92. Perrett, J.; Martin, J.; Peterson, J.; Steel, R.; Packer, S. *Friction Stir Welding of Industrial Steels*; The Minerals, Metals & Materials Society: Pittsburgh, PA, USA, 2011; pp. 65–72.
93. Cater, S.; Martin, J.; Galloway, A.; McPherson, N. Comparison between Friction Stir and Submerged Arc Welding Applied to Joining DH36 and E36 Shipbuilding Steel. In *Friction Stir Welding and Processing VII*; Springer Nature: Berlin, Germany, 2013; pp. 49–58.
94. Thomas, W.M.; Threadgill, P.L.; Nicholas, E.D. Feasibility of friction stir welding steel. *Sci. Technol. Weld. Join.* **1999**, *4*, 365–372. [CrossRef]
95. Lienert, T.J.; Lippold, J.C. Improved weldability diagram for pulsed laser welded austenitic stainless steels. *Sci. Technol. Weld. Join.* **2003**, *8*, 1–9. [CrossRef]
96. Fujii, H.; Cui, L.; Tsuji, N.; Maeda, M.; Nakata, K.; Nogi, K. Friction stir welding of carbon steels. *Mater. Sci. Eng. A* **2006**, *429*, 50–57. [CrossRef]
97. Cui, L.; Fujii, H.; Tsuji, N.; Nogi, K. Friction stir welding of a high carbon steel. *Scr. Mater.* **2007**, *56*, 637–640. [CrossRef]
98. Saeid, T.; Abdollah-zadeh, A.; Assadi, H.; Malek Ghaini, F. Effect of friction stir welding speed on the microstructure and mechanical properties of a duplex stainless steel. *Mater. Sci. Eng. A* **2008**, *496*, 262–268. [CrossRef]

99. Ghosh, M.; Kumar, K.; Mishra, R.S. Analysis of microstructural evolution during friction stir welding of ultrahigh-strength steel. *Scr. Mater.* **2010**, *63*, 851–854. [CrossRef]
100. Miura, T.; Ueji, R.; Fujii, H.; Komine, H.; Yanagimoto, J. Phase transformation behavior of Cr-Mo steel during FSW. *Mater. Des.* **2016**, *90*, 915–921. [CrossRef]
101. Fujii, H.; Ueji, R.; Morisada, Y.; Tanigawa, H. High strength and ductility of friction-stir-welded steel joints due to mechanically stabilized metastable austenite. *Scr. Mater.* **2014**, *70*, 39–42. [CrossRef]
102. Mahoney, M.; Nelson, T.; Sorenson, C.; Packer, S. Friction stir welding of ferrous alloys: Current status. *Mater. Sci. Forum* **2010**, *638–642*, 41–46. [CrossRef]
103. Miura, T.; Ueji, R.; Morisada, Y.; Fujii, H. Enhanced tensile properties of Fe-Ni-C steel resulting from stabilization of austenite by friction stir welding. *J. Mater. Process. Technol.* **2015**, *216*, 216–222. [CrossRef]
104. Fujii, H.; Cui, L.; Nakata, K.; Nogi, K. Mechanical properties of friction stir welded carbon steel joints-Friction stir welding with and without transformation. *Weld. World* **2008**, *52*, 75–81. [CrossRef]
105. Meshram, M.P.; Kodli, B.K.; Dey, S.R. Mechanical Properties and Microstructural Characterization of Friction Stir Welded AISI 316 Austenitic Stainless Steel. *Procedia Mater. Sci.* **2014**, *5*, 2376–2381. [CrossRef]
106. Maltin, C.A.; Nolton, L.J.; Scott, J.L.; Toumpis, A.I.; Galloway, A.M. The potential adaptation of stationary shoulder friction stir welding technology to steel. *Mater. Des.* **2014**, *64*, 614–624. [CrossRef]
107. Bhadeshia, H.K.D.H.; Debroy, T. Critical assessment: Friction stir welding of steels. *Sci. Technol. Weld. Join.* **2009**, *14*, 193–196. [CrossRef]
108. Ragu Nathan, S.; Balasubramanian, V.; Malarvizhi, S.; Rao, A.G. Effect of welding processes on mechanical and microstructural characteristics of high strength low alloy naval grade steel joints. *Def. Technol.* **2015**, *11*, 308–317. [CrossRef]
109. Sabooni, S.; Karimzadeh, F.; Enayati, M.H.; Ngan, A.H.W.; Jabbari, H. *Gas Tungsten arc Welding and Friction Stir Welding of Ultrafine Grained AISI 304L Stainless Steel: Microstructural and Mechanical Behavior Characterization*; Elsevier B.V.: Amsterdam, The Netherlands, 2015; Volume 109, ISBN 8415683111.
110. Meshram, M.P.; Kodli, B.K.; Dey, S.R. Friction Stir Welding of Austenitic Stainless Steel by PCBN Tool and its Joint Analyses. *Procedia Mater. Sci.* **2014**, *6*, 135–139. [CrossRef]
111. Lakshminarayanan, A.K.; Balasubramanian, V.; Salahuddin, M. Microstructure, tensile and impact toughness properties of friction stir welded mild steel. *J. Iron Steel Res. Int.* **2010**, *17*, 68–74. [CrossRef]
112. Li, H.B.; Jiang, Z.H.; Feng, H.; Zhang, S.C.; Li, L.; Han, P.D.; Misra, R.D.K.; Li, J.Z. Microstructure, mechanical and corrosion properties of friction stir welded high nitrogen nickel-free austenitic stainless steel. *Mater. Des.* **2015**, *84*, 291–299. [CrossRef]
113. Wang, D.; Ni, D.R.; Xiao, B.L.; Ma, Z.Y.; Wang, W.; Yang, K. Microstructural evolution and mechanical properties of friction stir welded joint of Fe-Cr-Mn-Mo-N austenite stainless steel. *Mater. Des.* **2014**, *64*, 355–359. [CrossRef]
114. Du, D.; Fu, R.; Li, Y.; Jing, L.; Ren, Y.; Yang, K. Gradient characteristics and strength matching in friction stir welded joints of Fe-18Cr-16Mn-2Mo-0.85N austenitic stainless steel. *Mater. Sci. Eng. A* **2014**, *616*, 246–251. [CrossRef]
115. Kiese, J.; Siemers, C.; Schmidt, C. *A New Class of Oxidation-Resistant, Microstructural-Stabilized and Cold-Workable Titanium Alloys for Exhaust Applications*; The Minerals, Metals & Materials Society: Pittsburgh, PA, USA, 2016; pp. 733–737. ISBN 9781119293668.
116. Saravanan, V.; Rajakumar, S.; Banerjee, N.; Amuthakkannan, R. Effect of shoulder diameter to pin diameter ratio on microstructure and mechanical properties of dissimilar friction stir welded AA2024-T6 and AA7075-T6 aluminum alloy joints. *Int. J. Adv. Manuf. Technol.* **2016**, *87*, 3637–3645. [CrossRef]
117. Vijayavel, P.; Balasubramanian, V.; Sundaram, S. Effect of shoulder diameter to pin diameter (D/d) ratio on tensile strength and ductility of friction stir processed LM25AA-5% SiCp metal matrix composites. *Mater. Des.* **2014**, *57*, 1–9. [CrossRef]
118. Pasta, S.; Reynolds, A.P. Residual stress effects on fatigue crack growth in a Ti-6Al-4V friction stir weld. *Fatigue Fract. Eng. Mater. Struct.* **2008**, *31*, 569–580. [CrossRef]
119. Lee, W.B.; Lee, C.Y.; Chang, W.S.; Yeon, Y.M.; Jung, S.B. Microstructural investigation of friction stir welded pure titanium. *Mater. Lett.* **2005**, *59*, 3315–3318. [CrossRef]
120. Zhang, Y.; Sato, Y.S.; Kokawa, H.; Park, S.H.C.; Hirano, S. Microstructural characteristics and mechanical properties of Ti-6Al-4V friction stir welds. *Mater. Sci. Eng. A* **2008**, *485*, 448–455. [CrossRef]

121. Mironov, S.; Sato, Y.S.; Kokawa, H. Development of grain structure during friction stir welding of pure titanium. *Acta Mater.* **2009**, *57*, 4519–4528. [CrossRef]
122. Knipling, K.E.; Fonda, R.W. Texture development in the stir zone of near-α titanium friction stir welds. *Scr. Mater.* **2009**, *60*, 1097–1100. [CrossRef]
123. Pilchak, A.L.; Juhas, M.C.; Williams, J.C. Microstructural changes due to friction stir processing of investment-cast Ti-6Al-4V. *Metall. Mater. Trans. A Phys. Metall. Mater. Sci.* **2007**, *38*, 401–408. [CrossRef]
124. Rai, R.; De, A.; Bhadeshia, H.K.D.H.; DebRoy, T. Review: Friction stir welding tools. *Sci. Technol. Weld. Join.* **2011**, *16*, 325–342. [CrossRef]
125. Gangwar, K.; Mamidala, R.; Sanders, D.G. Friction stir welding of near α and α + β titanium alloys: Metallurgical and mechanical characterization. *Metals* **2017**, *7*, 565. [CrossRef]
126. Fonda, R.W.; Bingert, J.F. Texture variations in an aluminum friction stir weld. *Scr. Mater.* **2007**, *57*, 1052–1055. [CrossRef]
127. Gangwar, K.; Ramulu, M. Friction stir welding of titanium alloys: A review. *Mater. Des.* **2018**, *141*, 230–255. [CrossRef]
128. Kulkarni, N.; Ramulu, M. Experimental and Numerical Analysis of Mechanical Behavior in Frcition Stir Welded Different Titanium Alloys. In Proceedings of the ASME 2014 International Mechanical Engineering Congress and Exposition, Montreal, QC, Canada, 14–20 November 2014; pp. 1–8.
129. Su, J.; Wang, J.; Mishra, R.S.; Xu, R.; Baumann, J.A. Microstructure and mechanical properties of a friction stir processed Ti-6Al-4V alloy. *Mater. Sci. Eng. A* **2013**, *573*, 67–74. [CrossRef]
130. Mashinini, P.M.; Hattingh, D.G.; Lombard, H. Mechanical Properties and Microstructure of Friction Stir and Laser Beam Welded 3mm Ti6Al4V Alloy. In Proceedings of the World Congress on Engineering, London, UK, 29 June–1 July 2016.
131. Fujii, H.; Sun, Y.; Kato, H.; Nakata, K. Investigation of welding parameter dependent microstructure and mechanical properties in friction stir welded pure Ti joints. *Mater. Sci. Eng. A* **2019**, *527*, 3386–3391. [CrossRef]
132. Edwards, P.; Ramulu, M. Identification of process parameters for friction stir welding Ti-6Al-4V. *J. Eng. Mater. Technol. Trans. ASME* **2010**, *132*, 031006. [CrossRef]
133. Steuwer, A.; Hattingh, D.G.; James, M.N.; Singh, U.; Buslaps, T. Residual stresses, microstructure and tensile properties in Ti-6Al-4V friction stir welds. *Sci. Technol. Weld. Join.* **2012**, *17*, 525–533. [CrossRef]
134. Edwards, P.; Ramulu, M. Fracture toughness and fatigue crack growth in Ti-6Al-4V friction stir welds. *Fatigue Fract. Eng. Mater. Struct.* **2015**, *38*, 970–982. [CrossRef]
135. Liu, H.J.; Zhou, L. Microstructural zones and tensile characteristics of friction stir welded joint of TC4 titanium alloy. *Trans. Nonferrous Met. Soc. China* **2010**, *20*, 1873–1878. [CrossRef]
136. Ebrahimi, M.; Par, M.A. Twenty-year uninterrupted endeavor of friction stir processing by focusing on copper and its alloys. *J. Alloy. Compd.* **2019**, *781*, 1074–1090. [CrossRef]
137. Cederqvist, L.; Sorensen, C.D.; Reynolds, A.P.; Öberg, T. Improved process stability during friction stir welding of 5 cm thick copper canisters through shoulder geometry and parameter studies. *Sci. Technol. Weld. Join.* **2009**, *14*, 178–184. [CrossRef]
138. Cederqvist, L.; Öberg, T. Reliability study of friction stir welded copper canisters containing Sweden's nuclear waste. *Reliab. Eng. Syst. Saf.* **2008**, *93*, 1491–1499. [CrossRef]
139. Cartigueyen, S.; Mahadevan, K. Role of Friction Stir Processing on Copper and Copper based Particle Reinforced Composites–A Review. *J. Mater. Sci. Surf. Eng.* **2015**, *2*, 133–145.
140. Nakata, K. Friction stir welding of copper and copper alloys. *Weld. Int.* **2010**, *55*, 37–41. [CrossRef]
141. Park, H.S.; Kimura, T.; Murakami, T.; Nagano, Y.; Nakata, K.; Ushio, M. Microstructures and mechanical properties of friction stir welds of 60% Cu-40% Zn copper alloy. *Mater. Sci. Eng. A* **2004**, *371*, 160–169. [CrossRef]
142. Çam, G.; Serindağ, H.T.; Çakan, A.; Mistikoglu, S.; Yavuz, H. The effect of weld parameters on friction stir welding of brass plates. *Materwiss. Werksttech.* **2008**, *39*, 394–399. [CrossRef]
143. Machniewicz, T.; Nosal, P.; Korbel, A.; Hebda, M. Effect of FSW Traverse Speed on Mechanical Properties of Copper Plate Joints. *Materials* **2020**, *13*, 1937. [CrossRef]
144. Lee, W.B.; Jung, S.B. The joint properties of copper by friction stir welding. *Mater. Lett.* **2004**, *58*, 1041–1046. [CrossRef]

145. Xue, P.; Xie, G.M.; Xiao, B.L.; Ma, Z.Y.; Geng, L. Effect of heat input conditions on microstructure and mechanical properties of friction-stir-welded pure copper. *Metall. Mater. Trans. A Phys. Metall. Mater. Sci.* **2010**, *41*, 2010–2021. [CrossRef]
146. Khodaverdizadeh, H.; Mahmoudi, A.; Heidarzadeh, A.; Nazari, E. Effect of friction stir welding (FSW) parameters on strain hardening behavior of pure copper joints. *Mater. Des.* **2012**, *35*, 330–334. [CrossRef]
147. Serio, L.M.; Palumbo, D.; De Filippis, L.A.C.; Galietti, U.; Ludovico, A.D. Effect of friction stir process parameters on the mechanical and thermal behavior of 5754-H111 aluminum plates. *Materials* **2016**, *9*, 122. [CrossRef]
148. Costa, J.D.; Ferreira, J.A.M.; Borrego, L.P.; Abreu, L.P. Fatigue behaviour of AA6082 friction stir welds under variable loadings. *Int. J. Fatigue* **2012**, *37*, 8–16. [CrossRef]
149. Salahi, S.; Yapici, G.G. Fatigue Behavior of Friction Stir Welded Joints of Pure Copper with Ultra-fine Grains. *Procedia Mater. Sci.* **2015**, *11*, 74–78. [CrossRef]
150. Surekha, K.; Els-Botes, A. Development of high strength, high conductivity copper by friction stir processing. *Mater. Des.* **2011**, *32*, 911–916. [CrossRef]
151. Liu, H.J.; Shen, J.J.; Huang, Y.X.; Kuang, L.Y.; Liu, C.; Li, C. Effect of tool rotation rate on microstructure and mechanical properties of friction stir welded copper. *Sci. Technol. Weld. Join.* **2009**, *14*, 577–583. [CrossRef]
152. Sahlot, P.; Singh, A.K.; Badheka, V.J.; Arora, A. Friction Stir Welding of Copper: Numerical Modeling and Validation. *Trans. Indian Inst. Met.* **2019**, *72*, 1339–1347. [CrossRef]
153. Xie, G.M.; Ma, Z.Y.; Geng, L. Development of a fine-grained microstructure and the properties of a nugget zone in friction stir welded pure copper. *Scr. Mater.* **2007**, *57*, 73–76. [CrossRef]
154. Cartigueyen, S.; Mahadevan, K. Influence of rotational speed on the formation of friction stir processed zone in pure copper at low-heat input conditions. *J. Manuf. Process.* **2015**, *18*, 124–130. [CrossRef]
155. Cartigueyen, S.; Mahadevan, K. Study of friction stir processed zone under different tool pin profiles in pure copper. *IOSR J. Mech. Civ. Eng.* **2014**, *11*, 6–12. [CrossRef]
156. Ethiraj, N.; Sivabalan, T.; Meikeerthy, S.; Kumar, K.L.V.R.; Chaithanya, G.; Reddy, G.P.K. Comparative study on conventional and underwater friction stir welding of copper plates. In Proceedings of the Int. Conf. Mater. Manuf. Mach. 2019, Sibiu, Romania, 5–7 June 2019; Volume 2128, p. 30003. [CrossRef]
157. Sakthivel, T.; Mukhopadhyay, J. Microstructure and mechanical properties of friction stir welded copper. *J. Mater. Sci.* **2007**, *42*, 8126–8129. [CrossRef]
158. Lin, J.W.; Chang, H.C.; Wu, M.H. Comparison of mechanical properties of pure copper welded using friction stir welding and tungsten inert gas welding. *J. Manuf. Process.* **2014**, *16*, 296–304. [CrossRef]
159. Raju, L.S.; Kumar, A.; Prasad, S.R. Microstructure and mechanical properties of friction stir welded pure copper. *Appl. Mech. Mater.* **2014**, *592–594*, 499–503. [CrossRef]
160. Nagabharam, P.; Srikanth Rao, D.; Manoj Kumar, J.; Gopikrishna, N. Investigation of Mechanical Properties of Friction Stir Welded pure Copper Plates. *Mater. Today Proc.* **2018**, *5*, 1264–1270. [CrossRef]
161. Nia, A.A.; Shirazi, A. Effects of different friction stir welding conditions on the microstructure and mechanical properties of copper plates. *Int. J. Miner. Metall. Mater.* **2016**, *23*, 799–809. [CrossRef]
162. Barmouz, M.; Givi, M.K.B.; Jafari, J. Evaluation of tensile deformation properties of friction stir processed pure copper: Effect of processing parameters and pass number. *J. Mater. Eng. Perform.* **2014**, *23*, 101–107. [CrossRef]
163. Rao, A.N.; Naik, L.S.; Srinivas, C. Evaluation and Impacts of Tool Profile and Rotational Speed on Mechanical Properties of Friction Stir Welded Copper 2200 Alloy. *Mater. Today Proc.* **2017**, *4*, 1225–1229. [CrossRef]
164. Gao, J.; Shen, Y.; Zhang, J.; Xu, H. Submerged friction stir weld of polyethylene sheets. *J. Appl. Polym. Sci.* **2014**, *131*, 1–8. [CrossRef]
165. Arici, A.; Sinmaz, T. Effects of double passes of the tool on friction stir welding of polyethylene. *J. Mater. Sci.* **2005**, *40*, 3313–3316. [CrossRef]
166. Bozkurt, Y. The optimization of friction stir welding process parameters to achieve maximum tensile strength in polyethylene sheets. *Mater. Des.* **2012**, *35*, 440–445. [CrossRef]
167. Aydin, M. Effects of welding parameters and pre-heating on the friction stir welding of UHMW-polyethylene. *Polym.-Plast. Technol. Eng.* **2010**, *49*, 595–601. [CrossRef]
168. Mostafapour, A.; Azarsa, E. A study on the role of processing parameters in joining polyethylene sheets via heat assisted friction stir welding: Investigating microstructure, tensile and flexural properties. *Int. J. Phys. Sci.* **2012**, *7*, 647–654. [CrossRef]

169. Saeedy, S.; Besharati, M.K. Investigation of the effects of critical process parameters of friction stir welding of polyethyleneg. *Proc. Inst. Mech. Eng. Part B J. Eng. Manuf.* **2011**, *225*, 1305–1310. [CrossRef]
170. Mishra, D.; Sahu, S.K.; Mahto, R.P.; Pal, S.K. *Strengthening and Joining by Plastic Deformation*; Springer: Singapore, 2019; ISBN 978-981-13-0377-7.
171. Inaniwa, S.; Kurabe, Y.; Miyashita, Y.; Hori, H. Application of friction stir welding for several plastic materials. In *Proceedings of the 1st International Joint Symposium on Joining and Welding, Osaka, Japan, 6–8 November 2013*; Woodhead Publishing Limited: Cambridge, UK, 2013; Volume 2, pp. 137–142.
172. Rezgui, M.-A.; Trabelsi, A.-C.; Ayadi, M.; Hamrouni, K. Optimization of Friction Stir Welding Process of High Density Polyethylene. *Int. J. Prod. Qual. Eng.* **2011**, *2*, 55–61.
173. Panneerselvam, K.; Lenin, K. Joining of Nylon 6 plate by friction stir welding process using threaded pin profile. *Mater. Des.* **2014**, *53*, 302–307. [CrossRef]
174. Zafar, A.; Awang, M.; Khan, S.R.; Emamian, S. Investigating friction stir welding on thick nylon 6 plates. *Weld. J.* **2016**, *95*, 210s–218s.
175. Khaliel Youssif, M.S.; El-Sayed, M.A.; Khourshid, A.E.F.M. Influence of critical process parameters on the quality of friction stir welded nylon 6. *Int. Rev. Mech. Eng.* **2016**, *10*, 501–507. [CrossRef]
176. Bagheri, A.; Azdast, T.; Doniavi, A. An experimental study on mechanical properties of friction stir welded ABS sheets. *Mater. Des.* **2013**, *43*, 402–409. [CrossRef]
177. Sadeghian, N.; Besharati Givi, M.K. Experimental optimization of the mechanical properties of friction stir welded Acrylonitrile Butadiene Styrene sheets. *Mater. Des.* **2015**, *67*, 145–153. [CrossRef]
178. Pirizadeh, M.; Azdast, T.; Rash Ahmadi, S.; Mamaghani Shishavan, S.; Bagheri, A. Friction stir welding of thermoplastics using a newly designed tool. *Mater. Des.* **2014**, *54*, 342–347. [CrossRef]
179. Mendes, N.; Loureiro, A.; Martins, C.; Neto, P.; Pires, J.N. Morphology and strength of acrylonitrile butadiene styrene welds performed by robotic friction stir welding. *Mater. Des.* **2014**, *64*, 81–90. [CrossRef]
180. Sahu, S.K.; Mishra, D.; Mahto, R.P.; Sharma, V.M.; Pal, S.K.; Pal, K.; Banerjee, S.; Dash, P. Friction stir welding of polypropylene sheet. *Eng. Sci. Technol. Int. J.* **2018**, *21*, 245–254. [CrossRef]
181. Kiss, Z.; Czigány, T. Microscopic analysis of the morphology of seams in friction stir welded polypropylene. *Express Polym. Lett.* **2012**, *6*, 54–62. [CrossRef]
182. Kiss, Z.; Czigány, T. Effect of Welding Parameters on the Heat Affected Zone and the Mechanical Properties of Friction Stir Welded Poly(ethylene-terephthalate-glycol). *J. Appl. Polym. Sci.* **2012**, *125*, 2231–2238. [CrossRef]
183. Panneerselvam, K.; Lenin, K. Investigation on effect of tool forces and joint defects during FSW of polypropylene plate. *Procedia Eng.* **2012**, *38*, 3927–3940. [CrossRef]
184. Panneerselvam, K.; Lenin, K. Effects and Defects of the Polypropylene Plate for Different Parameters in Friction Stir Welding Process. *Int. J. Res. Eng. Technol.* **2013**, *2*, 143–152. [CrossRef]
185. Simoes, F.; Rodrigues, D.M. Material flow and thermo-mechanical conditions during Friction Stir Welding of polymers: Literature review, experimental results and empirical analysis. *Mater. Des.* **2014**, *59*, 344–351. [CrossRef]
186. Squeo, E.A.; Bruno, G.; Guglielmotti, A.; Quadrini, F. Friction Stir Welding of Polyethylene Sheets. The Annals of "Dunarea de Jos" University of Galati Fascicle V. 2009, pp. 241–246. Available online: https://www.researchgate.net/publication/267842113_Friction_stir_welding_of_polyethylene_sheets (accessed on 24 August 2020).
187. Arbegast, W.J. A flow-partitioned deformation zone model for defect formation during friction stir welding. *Scr. Mater.* **2008**, *58*, 372–376. [CrossRef]
188. Rezgui, M.A.; Ayadi, M.; Cherouat, A.; Hamrouni, K.; Zghal, A.; Bejaoui, S. Application of Taguchi approach to optimize friction stir welding parameters of polyethylene. *EPJ Web Conf.* **2010**, *6*, 1–8. [CrossRef]
189. Strand, S.R. Effects of Friction Stir Welding on Polymer Microstructure. Master's Thesis, Ira A. Fulton College of Engineering and Technology, Tempe, AZ, USA, February 2004.
190. Nelson, T.W.; Sorensen, C.D.; John, C.J. Friction Stir Welding of Polymeric Materials. U.S. Patent 681,163,2B2, 2 November 2009.
191. Banjare, P.N.; Sahlot, P.; Arora, A. An assisted heating tool design for FSW of thermoplastics. *J. Mater. Process. Technol.* **2016**, *239*, 83–91. [CrossRef]
192. Laieghi, H.; Alipour, S.; Mostafapour, A. Heat-assisted friction stir welding of polymeric nanocomposite. *Sci. Technol. Weld. Join.* **2020**, *25*, 56–65. [CrossRef]

193. Husain, I.M.; Salim, R.K.; Azdast, T.; Hasanifard, S.; Shishavan, S.M.; Lee, R.E. Mechanical properties of friction-stir-welded polyamide sheets. *Int. J. Mech. Mater. Eng.* **2015**, *10*, 18. [CrossRef]
194. Leng, X.; Yang, W.; Zhang, J.; Ma, X.; Zhao, W.; Yan, J. Designing high-performance composite joints close to parent materials of aluminum matrix composites. *arXiv* **2017**, *34*, 660–663.
195. Hall, I.W.; Kyono, T.; Diwanji, A. On the fibre/matrix interface in boron/aluminum metal matrix composites. *J. Mater. Sci.* **1987**, *22*, 1743–1748. [CrossRef]
196. Vijay, S.J.; Murugan, N. Influence of tool pin profile on the metallurgical and mechanical properties of friction stir welded Al-10wt.% TiB2 metal matrix composite. *Mater. Des.* **2010**, *31*, 3585–3589. [CrossRef]
197. Hassan, A.M.; Qasim, T.; Ghaithan, A. Effect of pin profile on friction stir welded aluminum matrix composites. *Mater. Manuf. Process.* **2012**, *27*, 1397–1401. [CrossRef]
198. Mahmoud, E.R.I.; Takahashi, M.; Shibayanagi, T.; Ikeuchi, K. Effect of friction stir processing tool probe on fabrication of SiC particle reinforced composite on aluminum surface. *Sci. Technol. Weld. Join.* **2009**, *14*, 413–425. [CrossRef]
199. Cavaliere, P.; Cerri, E.; Marzoli, L.; Dos Santos, J. Friction stir welding of ceramic particle reinforced aluminum based metal matrix composites. *Appl. Compos. Mater.* **2004**, *11*, 247–258. [CrossRef]
200. Marzoli, L.M.; Strombeck, A.V.; Dos Santos, J.F.; Gambaro, C.; Volpone, L.M. Friction stir welding of an AA6061/Al_2O_3/20p reinforced alloy. *Compos. Sci. Technol.* **2006**, *66*, 363–371. [CrossRef]
201. Prado, R.A.; Murr, L.E.; Shindo, D.J.; Soto, K.F. Tool wear in the friction-stir welding of aluminum alloy 6061 + 20% Al_2O_3: A preliminary study. *Scr. Mater.* **2001**, *45*, 75–80. [CrossRef]
202. Baxter, S.C.; Reynolds, A.P. Characterization of Reinforcing Particle Size Distribution in a Friction Stir Welded Al-SiC Extrusion. In *Proceedings of the Lightweight Alloys for Aerospace Application*; Jata, K., Lee, E., Frazier, W., Kim, N.J., Eds.; The Minerals, Metals & Materials Society: Pittsburgh, PA, USA, 2001; pp. 283–294.
203. Mahoney, M.W.; Harrigan, W.; Wert, J.A. Friction Stir Welding SiC Discontinuously Reinforced Aluminuium. In Proceedings of the 7th Int. Conf. Joints in Aluminum, Cambridge, UK, 15–17 April 1998; pp. 231–236.
204. Feng, A.H.; Ma, Z.Y. Formation of Cu2FeAl7 phase in friction-stir-welded SiCp/Al-Cu-Mg composite. *Scr. Mater.* **2007**, *57*, 1113–1116. [CrossRef]
205. Feng, A.H.; Xiao, B.L.; Ma, Z.Y. Effect of microstructural evolution on mechanical properties of friction stir welded AA2009/SiCp composite. *Compos. Sci. Technol.* **2008**, *68*, 2141–2148. [CrossRef]
206. Ceschini, L.; Boromei, I.; Minak, G.; Morri, A.; Tarterini, F. Microstructure, tensile and fatigue properties of AA6061/20 vol.%Al2O3p friction stir welded joints. *Compos. Part A Appl. Sci. Manuf.* **2007**, *38*, 1200–1210. [CrossRef]
207. Acharya, U.; Roy, B.S.; Saha, S.C. Torque and force perspectives on particle size and its effect on mechanical property of friction stir welded AA6092/17.5SiC p -T6 composite joints. *J. Manuf. Process.* **2019**, *38*, 113–121. [CrossRef]
208. Kumar, S.; Medhi, T.; Roy, B.S. *Friction Stir Welding of Thermoplastic Composites*; Springer: Singapore, 2019; ISBN 978-981-13-6412-9.
209. Bhushan, R.K.; Sharma, D. Optimization of FSW parameters for maximum UTS of AA6082/SiC/10 P composites. *Adv. Compos. Lett.* **2019**, *28*, 96369351986770. [CrossRef]
210. Jafrey, D.D.; Panneerselvam, K. Study on Tensile Strength, Impact Strength and Analytical Model for Heat Generation in Friction Vibration Joining of Polymeric Nanocomposite Joints Daniel. *Polym. Eng. Sci.* **2016**, *57*, 495–504. [CrossRef]
211. Liu, H.; Hu, Y.; Zhao, Y.; Fujii, H. Microstructure and mechanical properties of friction stir welded AC4A+30vol.%SiCp composite. *Mater. Des.* **2015**, *65*, 395–400. [CrossRef]
212. Wang, D.; Wang, Q.Z.; Xiao, B.L.; Ma, Z.Y. Achieving friction stir welded SiCp/Al–Cu–Mg composite joint of nearly equal strength to base material at high welding speed. *Mater. Sci. Eng. A* **2014**, *589*, 271–274. [CrossRef]
213. Mozammil, S.; Karloopia, J.; Verma, R.; Jha, P.K. Mechanical response of friction stir butt weld Al-4.5%Cu/TiB2/2.5p in situ composite: Statistical modelling and optimization. *J. Alloy. Compd.* **2020**, *826*, 154184. [CrossRef]
214. Narendranath, S.; Chakradhar, D. Effect of FSW on microstructure and hardness of AA6061/SiC/fly ash MMCs. *Mater. Today Proc.* **2018**, *5*, 17866–17872. [CrossRef]
215. Bozkurt, Y.; Uzun, H.; Salman, S. Microstructure and mechanical properties of friction stir welded particulate reinforced AA2124/SiC/25p-T4 composite. *J. Compos. Mater.* **2011**, *45*, 2237–2245. [CrossRef]

216. Acharya, U.; Saha Roy, B.; Chandra Saha, S. A Study of Tool Wear and its Effect on the Mechanical Properties of Friction Stir Welded AA6092/17.5 Sicp Composite Material Joint. *Mater. Today Proc.* **2018**, *5*, 20371–20379. [CrossRef]
217. Ni, D.R.; Chen, D.L.; Xiao, B.L.; Wang, D.; Ma, Z.Y. Residual stresses and high cycle fatigue properties of friction stir welded SiCp/AA2009 composites. *Int. J. Fatigue* **2013**, *55*, 64–73. [CrossRef]
218. Bozkurt, Y.; Boumerzoug, Z. Tool material effect on the friction stir butt welding of AA2124-T4 Alloy Matrix MMC. *J. Mater. Res. Technol.* **2018**, *7*, 29–38. [CrossRef]
219. Fernández, R.; Ibáñez, J.; Cioffi, F.; Verdera, D.; González-Doncel, G. Friction stir welding of 25%SiC/2124Al composite with optimal mechanical properties and minimal tool wear. *Sci. Technol. Weld. Join.* **2017**, *22*, 526–535. [CrossRef]
220. Periyasamy, P.; Mohan, B.; Balasubramanian, V. Effect of heat input on mechanical and metallurgical properties of friction stir welded AA6061-10% SiCp MMCs. *J. Mater. Eng. Perform.* **2012**, *21*, 2417–2428. [CrossRef]
221. Wang, D.; Xiao, B.L.; Wang, Q.Z.; Ma, Z.Y. Evolution of the Microstructure and Strength in the Nugget Zone of Friction Stir Welded SiCp/Al-Cu-Mg Composite. *J. Mater. Sci. Technol.* **2014**, *30*, 54–60. [CrossRef]
222. Salih, O.S.; Ou, H.; Wei, X.; Sun, W. Microstructure and mechanical properties of friction stir welded AA6092/SiC metal matrix composite. *Mater. Sci. Eng. A* **2019**, *742*, 78–88. [CrossRef]
223. Firouzdor, V.; Kou, S. Al-to-Mg friction stir welding: Effect of material position, travel speed, and rotation speed. *Metall. Mater. Trans. A Phys. Metall. Mater. Sci.* **2010**, *41*, 2914–2935. [CrossRef]
224. Mehta, K.P.; Badheka, V.J. A review on dissimilar friction stir welding of copper to aluminum: Process, properties, and variants. *Mater. Manuf. Process.* **2016**, *31*, 233–254. [CrossRef]
225. Miles, M.P.; Melton, D.W.; Nelson, T.W. Formability of friction-stir-welded dissimilar-aluminum-alloy sheets. *Metall. Mater. Trans. A Phys. Metall. Mater. Sci.* **2005**, *36*, 3335–3342. [CrossRef]
226. Simar, A.; Jonckheere, C.; Deplus, K.; Pardoen, T.; De Meester, B. Comparing similar and dissimilar friction stir welds of 2017-6005A aluminum alloys. *Sci. Technol. Weld. Join.* **2010**, *15*, 254–259. [CrossRef]
227. Ipekoglu, G.; Cam, G. Effects of initial temper condition and postweld heat treatment on the properties of dissimilar friction-stir-welded joints between AA7075 and AA6061 aluminum alloys. *Metall. Mater. Trans. A Phys. Metall. Mater. Sci.* **2014**, *45*, 3074–3087. [CrossRef]
228. Ahmadnia, M.; Shahraki, S.; Kamarposhti, M.A. Experimental studies on optimized mechanical properties while dissimilar joining AA6061 and AA5010 in a friction stir welding process. *Int. J. Adv. Manuf. Technol.* **2016**, *87*, 2337–2352. [CrossRef]
229. Sato, Y.S.; Park, S.H.C.; Michiuchi, M.; Kokawa, H. Constitutional liquation during dissimilar friction stir welding of Al and Mg alloys. *Scr. Mater.* **2004**, *50*, 1233–1236. [CrossRef]
230. Yan, J.; Xu, Z.; Li, Z.; Li, L.; Yang, S. Microstructure characteristics and performance of dissimilar welds between magnesium alloy and aluminum formed by friction stirring. *Scr. Mater.* **2005**, *53*, 585–589. [CrossRef]
231. Firouzdor, V.; Kou, S. Formation of liquid and intermetallics in Al-to-Mg friction stir welding. *Metall. Mater. Trans. A Phys. Metall. Mater. Sci.* **2010**, *41*, 3238–3251. [CrossRef]
232. Abdollahzadeh, A.; Shokuhfar, A.; Cabrera, J.M.; Zhilyaev, A.P.; Omidvar, H. The effect of changing chemical composition in dissimilar Mg/Al friction stir welded butt joints using zinc interlayer. *J. Manuf. Process.* **2018**, *34*, 18–30. [CrossRef]
233. Ouyang, J.; Yarrapareddy, E.; Kovacevic, R. Microstructural evolution in the friction stir welded 6061 aluminum alloy (T6-temper condition) to copper. *J. Mater. Process. Technol.* **2006**, *172*, 110–122. [CrossRef]
234. Xue, P.; Xiao, B.L.; Ma, Z.Y. Effect of interfacial microstructure evolution on mechanical properties and fracture behavior of friction-stir-welded Al-Cu joints. *Metall. Mater. Trans. A* **2015**, *46*, 3091–3103. [CrossRef]
235. Zhang, J.; Shen, Y.; Yao, X.; Xu, H.; Li, B. Investigation on dissimilar underwater friction stir lap welding of 6061-T6 aluminum alloy to pure copper. *Mater. Des.* **2014**, *64*, 74–80. [CrossRef]
236. Venkateswaran, P.; Reynolds, A.P. Factors affecting the properties of Friction Stir Welds between aluminum and magnesium alloys. *Mater. Sci. Eng. A* **2012**, *545*, 26–37. [CrossRef]
237. Zhao, Y.Y.; Li, D.; Zhang, Y.S. Effect of welding energy on interface zone of Al-Cu ultrasonic welded joint. *Sci. Technol. Weld. Join.* **2013**, *18*, 354–360. [CrossRef]
238. Liyanage, T.; Kilbourne, J.; Gerlich, A.P.; North, T.H. Joint formation in dissimilar Al alloy/steel and Mg alloy/steel friction stir spot welds. *Sci. Technol. Weld. Join.* **2009**, *14*, 500–508. [CrossRef]

239. Lancaster, J.F. *Metallurgy of Welding*; Woodhead Publishing Limited: Cambridge, UK, 1999; ISBN 978-1-85573-428-9.
240. Fu, B.; Qin, G.; Li, F.; Meng, X.; Zhang, J.; Wu, C. Friction stir welding process of dissimilar metals of 6061-T6 aluminum alloy to AZ31B magnesium alloy. *J. Mater. Process. Technol.* **2015**, *218*, 38–47. [CrossRef]
241. Zettler, R.; Da Silva, A.A.M.; Rodrigues, S.; Blanco, A.; Dos Santos, J.F. Dissimilar Al to Mg alloy friction stir welds. *Adv. Eng. Mater.* **2006**, *8*, 415–421. [CrossRef]
242. Kasai, H.; Morisada, Y.; Fujii, H. Dissimilar FSW of immiscible materials: Steel/magnesium. *Mater. Sci. Eng. A* **2015**, *624*, 250–255. [CrossRef]
243. Peel, M.J.; Steuwer, A.; Withers, P.J.; Dickerson, T.; Shi, Q.; Shercliff, H. Dissimilar friction stir welds in AA5083-AA6082. Part I: Process parameter effects on thermal history and weld properties. *Metall. Mater. Trans. A Phys. Metall. Mater. Sci.* **2006**, *37*, 2183–2193. [CrossRef]
244. Khodir, S.A.; Shibayanagi, T. Friction stir welding of dissimilar AA2024 and AA7075 aluminum alloys. *Mater. Sci. Eng. B Solid-State Mater. Adv. Technol.* **2008**, *148*, 82–87. [CrossRef]
245. Avinash, P.; Manikandan, M.; Arivazhagan, N.; Devendranath, R.K.; Narayanan, S. Friction stir welded butt joints of AA2024 T3 and AA7075 T6 aluminum alloys. *Procedia Eng.* **2014**, *75*, 98–102. [CrossRef]
246. Palanivel, R.; Koshy Mathews, P.; Murugan, N.; Dinaharan, I. Effect of tool rotational speed and pin profile on microstructure and tensile strength of dissimilar friction stir welded AA5083-H111 and AA6351-T6 aluminum alloys. *Mater. Des.* **2012**, *40*, 7–16. [CrossRef]
247. Malarvizhi, S.; Balasubramanian, V. Influences of tool shoulder diameter to plate thickness ratio (D/T) on stir zone formation and tensile properties of friction stir welded dissimilar joints of AA6061 aluminum-AZ31B magnesium alloys. *Mater. Des.* **2012**, *40*, 453–460. [CrossRef]
248. RajKumar, V.; VenkateshKannan, M.; Sadeesh, P.; Arivazhagan, N.; Devendranath Ramkumar, K. Studies on effect of tool design and welding parameters on the friction stir welding of dissimilar aluminum alloys AA 5052-AA 6061. *Procedia Eng.* **2014**, *75*, 93–97. [CrossRef]
249. Jafarzadegan, M.; Abdollah-zadeh, A.; Feng, A.H.; Saeid, T.; Shen, J.; Assadi, H. Microstructure and Mechanical Properties of a Dissimilar Friction Stir Weld between Austenitic Stainless Steel and Low Carbon Steel. *J. Mater. Sci. Technol.* **2013**, *29*, 367–372. [CrossRef]
250. Dehghani, M.; Amadeh, A.; Akbari Mousavi, S.A.A. Investigations on the effects of friction stir welding parameters on intermetallic and defect formation in joining aluminum alloy to mild steel. *Mater. Des.* **2013**, *49*, 433–441. [CrossRef]
251. Morishige, T.; Kawaguchi, A.; Tsujikawa, M.; Hino, M.; Hirata, T.; Higashi, K. Dissimilar welding of Al and Mg alloys by FSW. *Mater. Trans.* **2008**, *49*, 1129–1131. [CrossRef]
252. Masoudian, A.; Tahaei, A.; Shakiba, A.; Sharifianjazi, F.; Mohandesi, J.A. Microstructure and mechanical properties of friction stir weld of dissimilar AZ31-O magnesium alloy to 6061-T6 aluminum alloy. *Trans. Nonferrous Met. Soc. China* **2014**, *24*, 1317–1322. [CrossRef]
253. Azizieh, M.; Sadeghi Alavijeh, A.; Abbasi, M.; Balak, Z.; Kim, H.S. Mechanical properties and microstructural evaluation of AA1100 to AZ31 dissimilar friction stir welds. *Mater. Chem. Phys.* **2016**, *170*, 251–260. [CrossRef]
254. Abe, Y.; Watanabe, T.; Tanabe, H.; Kagiya, K. Dissimilar metal joining of magnesium alloy to steel by FSW. *Adv. Mater. Res.* **2007**, *15–17*, 393–397. [CrossRef]
255. Joshi, G.R.; Badheka, V.J. Microstructures and Properties of Copper to Stainless Steel Joints by Hybrid FSW. *Metallogr. Microstruct. Anal.* **2017**, *6*, 470–480. [CrossRef]
256. Shankar, S.; Chattopadhyaya, S. Friction stir welding of commercially pure copper and 1050 aluminum alloys. *Mater. Today Proc.* **2019**, *25*, 664–667. [CrossRef]
257. Xue, P.; Ni, D.R.; Wang, D.; Xiao, B.L.; Ma, Z.Y. Effect of friction stir welding parameters on the microstructure and mechanical properties of the dissimilar Al-Cu joints. *Mater. Sci. Eng. A* **2011**, *528*, 4683–4689. [CrossRef]

Publisher's Note: MDPI stays neutral with regard to jurisdictional claims in published maps and institutional affiliations.

Article

The Effect of Post-Weld Hot-Rolling on the Properties of Explosively Welded Mg/Al/Ti Multilayer Composite

Marcin Wachowski [1,*], Robert Kosturek [1], Lucjan Śnieżek [1], Sebastian Mróz [2], Andrzej Stefanik [2] and Piotr Szota [2]

[1] Faculty of Mechanical Engineering, Military University of Technology, 00-908 Warsaw, Poland; robert.kosturek@wat.edu.pl (R.K.); lucjan.sniezek@wat.edu.pl (L.Ś.)

[2] Faculty of Production Engineering and Materials Technology, Czestochowa University of Technology, 42-201 Częstochowa, Poland; mroz.sebastian@wip.pcz.pl (S.M.); stefanik.andrzej@wip.pcz.pl (A.S.); szota.piotr@wip.pcz.pl (P.S.)

* Correspondence: marcin.wachowski@wat.edu.pl; Tel.: +48-261-839-245

Received: 21 February 2020; Accepted: 17 April 2020; Published: 19 April 2020

Abstract: The paper describes an investigation of an explosively welded Mg/Al/Ti multilayer composite. Following the welding, the composite was subjected to hot-rolling in three different temperatures: 300 °C, 350 °C and 400 °C, with a total relative strain of 30%. The rolling speed was 0.2 m/s. The investigation of the composite properties involves microhardness analysis and mini-specimen tensile tests of the joints. The composite Mg/Al and Al/Ti bonds in the as-welded state and after rolling in 400 °C were subjected to microstructure analysis using scanning electron (SEM) and transmission electron microscopy (TEM). In the Al/Ti interface, the presence of melted zones with localized intermetallic precipitates has been reported in the as-welded state, and it has been stated that hot-rolling results in precipitation of intermetallic particles from the melted zone. The application of the hot-rolling process causes the formation of a continuous layer in the Mg/Al joint, consisting of two intermetallic phases, Mg_2Al_3 (β) and $Mg_{17}Al_{12}$ (γ).

Keywords: explosive welding; light alloys; hot rolling; composite; microstructure; intermetallic

---

1. Introduction

Light alloy multilayer composites are very promising materials, especially in military applications, due to their high specific strength and ballistic resistance. One of the most interesting is Mg/Al/Ti laminate, characterized by an increasing gradient of hardness. Manufacturing of this material is very problematic, considering the significant differences in properties of components to be joined, such as melting point and ductility. An additional factor is the tendency of the formation of intermetallic compounds between Al/Ti and Mg/Al interfaces [1–8]. Despite the good mechanical properties of intermetallic phases, their brittleness causes further problems during the plastic forming of multilayer composite, including a risk of delamination. An effective method to obtain the Mg/Al/Ti multilayer composite is explosive welding, which is a solid state welding process. Solid state welding of multilayer materials has advantages in accordance with fusion-based welding, i.e., shorter time of process and limited intermetallic formation at the bond zone [9,10]. In the explosive welding process, the energy released during detonation of the high explosive is used to accelerate one metal plate into another, and as a result, the high velocity collision of metal plates occurs, which brings the surfaces of the colliding metals close enough to each other to obtain interaction between their atoms and make the formation of a metallic bond between them possible. The main advantages of this process are the possibility of joining elements of large sizes and of a wide range of base and cladding material thickness without formation of a diffusion zone, which is a potential area of intermetallic compounds presence. Compared to other welding methods, impact welding allows one to obtain high-efficiency joints between a wide variety of similar and/or dissimilar metals without forming intermetallic phases in the bond zones. In the literature, the microstructure and mechanical properties of tri-metal, Ti/Al and Mg/Al layered composites obtained by various methods are widely described [11–16]. Zhang et al. revealed the interface microstructure of explosively welded Ti Gr.2/AA6061/AZ31B composite [17]. A periodic wavy bond without melted zones was observed between Ti and Al plates. A larger wavy interface was revealed in the Mg/Al joint. Fouad presented research results of Ti/AA6081/AZ31 obtained by hot isostatic pressing [18]. An intermetallic layer with a maximum thickness of 10 μm was formed at the AA6081/AZ31 interface and the presence of an Al_2O_3 interface was confirmed. Motevalli and Eghbali investigated the AA1050/cp-Ti/AZ31 laminated composite obtained by accumulative roll bonding [19]. Optimum mechanical properties and microstructural characteristics were achieved after four deformation passes with the strength of the composite at 335.9 MPa. Research results of Wu et al. on Mg/AA5052 laminate fabricated by accumulative roll bonding (ARB) revealed that after the final three cycles of rolling, cracking of the coarse Mg/Al intermetallic compound and rupture of the Al layers occurred [20]. Research on diffusion-welded Mg/Al joints revealed the presence of $Al_{12}Mg_{17}$/Al_3Mg_2 intermetallic compounds [21]. Fronczek et al. revealed that annealing of explosively welded A1050/AZ31/A1050 composite plate results in the formation of the Mg_2Al_3 intermetallic phase localized near to the A1050 plate, $Mg_{17}Al_{12}$, closer to the AZ31 clad and the minor amount of $Mg_{23}Al_{30}$ [22]. Lazurenko et al. investigated the multilayer cp-Ti/Al-1Mn aluminum alloy composite produced by explosive welding and revealed the formation of Al_3Ti and AlTi phases, metastable phase Al_5Ti_3, disordered $AlTi_3$, amorphous structures and ordered solid solutions [23]. Although explosive welding allows to significantly limit the formation of intermetallic compounds on a Mg/Al/Ti multilayer material interface, the composite itself is subjected to the further technological operations after welding; usually plastic forming (including hot forming) in order to obtain a specific shape [24]. The application of hot forming after the explosive welding process can result in the formation of intermetallic phases in the considered bond zones that influence the mechanical properties of the whole composite. While the microstructure and mechanical properties of laminated composites obtained by explosive welding are widely described in the literature, information about research on the Mg/Al/Ti laminate, obtained by the method of explosive welding with subsequent hot-forming, is not presented. Therefore, it is important to understand the effect of the post-weld hot-forming process on the microstructure, especially the formation of diffusion zones. The aim of this paper is to investigate the influence of post-weld hot-rolling on the properties of the composite, especially the Mg/Al and Al/Ti interfaces, which are the potential areas of intermetallic compounds formation.

2. Materials and Methods

For the research, a Mg/Al/Ti multilayer composite was produced by explosive welding. The process of explosive welding was carried out in the company ZTW EXPLOMET sp.j. Gałka, Szulc (Opole, Poland). The plates used to obtain the investigated composite had dimensions equal to 1200 × 800 mm. The type of explosive used during the process was ANFO (ammonium nitrate fuel oil). The thickness of the explosive layer was 50 mm, the stand-off distances for both Al/Ti and Mg/Al were 5 mm and the initial position angle between plates was 0°. Detonation velocity was equal to 3200 m/s. Before joining, surfaces of the joined materials were polished and cleaned by acetone. The charge initiation point was located in the center of the shorter side of the plate. The schematic of the explosive connection system is shown in Figure 1.

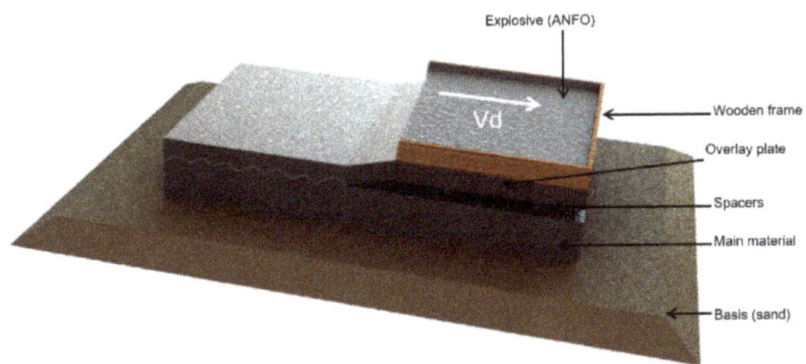

Figure 1. Scheme of explosive welding process, Vd—detonation velocity.

Obtaining an AZ31/AA2519/Ti6Al4V laminate requires the use of an additional AA1050 aluminum layer between AZ31/AA2519 and AA2519/Ti6Al4V plates. The following plates were used to make the laminate: magnesium alloy AZ31 (5 mm thick), aluminum alloy AA2519 cladded on both sides with AA1050 (4 mm thick in total) and titanium alloy Ti6Al4V (5 mm thick). The chemical composition of the components is shown in Table 1. The chemical composition of the materials was established by the materials supplier. The multilayer material was investigated in the state after the explosive welding and after applying the forming process—hot rolling, performed in three different temperatures: 300 °C, 350 °C and 400 °C, with a total relative strain of 30%. For laboratory testing, a laboratory rolling mill with a working roll diameter of 300 mm was used. The rolling speed was 0.2 m/s. The investigation of the composite properties involves a microhardness distribution analysis, performed with load of 0.98 N was applied for 10 s, and mini-specimen tensile testing conducted on Instron 8802 MTL universal testing machine (Instron, Warsaw, Poland). Space between microhardness test indentations was equal to 150 μm. The scheme of the used mini-specimen is presented in Figure 2. The tensile strength was measured separately for Ti6Al4V/AA1050/AA2519 and AZ31/AA1050/AA2519 joints. For obtaining the tensile strength values, 3 mini-samples of Ti6Al4V/AA1050/AA2519 material for each state (as-welded, hot-rolled 300 °C, 350 °C, 400 °C) were tested (equal to 12 samples). The same case was for AZ31/AA1050/AA2519 mini-samples (equal to 12 samples as well). The samples were cut from the central part of the plate in order to avoid the delaminated or weakened areas of the joint that are characteristic of the edges and corners of the plates being joined. In addition, the test plates were cut at a certain angle to each other, as the impact wave generated by the explosion propagated radially through the material. Therefore, for a representative microstructure, the test plates should be cut in a direction parallel to the direction of the impact wave propagation.

Table 1. Chemical composition of the laminate components.

AZ31	**Al**	**Zn**	**Mn**	**Si**	**Cu**	**Ca**	**Fe**	**Mg**
	2.50<	0.60<	0.20	0.10	0.050	0.040	0.0050	Rest
AA1050	**Fe**	**Si**	**Cu**	**Mg**	**Mn**	**Zn**	**Ti**	**Al**
	0.4	0.25<	0.05<	0.18	0.05<	0.07<	0.05<	Rest
AA2519	**Fe**	**Si**	**Cu**	**Mg**	**Zr**	**Sc**	**Ti**	**Al**
	0.08	0.06	5.77	0.18	0.2	0.36	0.04	Rest
Ti6Al4V	**O**	**V**	**Al**	**Fe**	**H**	**C**	**N**	**Ti**
	<0.20	3.5	5.5	<0.30	<0.0015	<0.08	0.05<	Rest

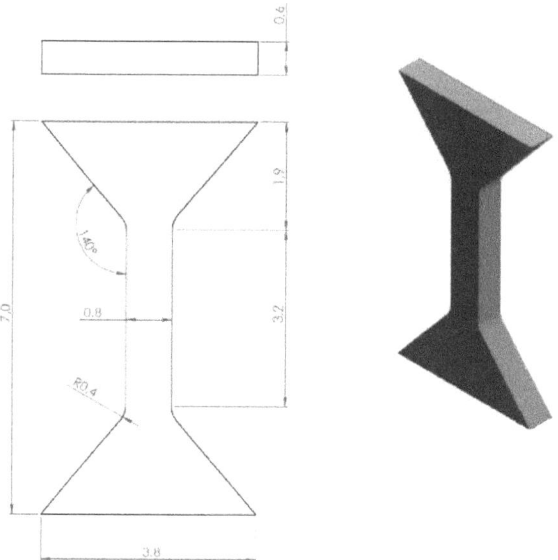

Figure 2. The scheme of mini-specimen for tensile test.

The composite Mg/Al and Al/Ti bonds in the as-welded state and after rolling at 400 °C were subjected to microstructure analysis using scanning electron (SEM) (JEOL, Warsaw, Poland) and transmission electron microscopy (TEM) (JEOL, Warsaw, Poland). Microscopic examinations were carried out on samples whose surfaces were oriented parallel to the direction of propagation of the shock wave during the joining process and parallel to the rolling direction. A wire EDM was used to cut the samples. High cutting accuracy was achieved, and the heat of cut was avoided on the surface layer of the materials. Because the hot rolling process can contribute to significant changes in the microstructure, not only of the magnesium alloy but also of aluminum and titanium, and can also cause significant structural changes within the bond zone, to understand the phenomena occurring during explosive bonding and hot rolling, advanced microscopic methods were used. In order to visualize the microstructure of the composite components and the morphology of the transition layers, the obtained material was investigated using scanning electron microscope JEOL JSM-6610 equipped with an X-ray energy dispersion (EDX) spectrometer and backscattered electron detector (BSE) (JEOL, Warsaw, Poland). As a part of metallographic sample preparation, samples were cut along the axial direction using a precision diamond saw and then mounted in resin, ground with the abrasive paper of 80, 320, 600, 1200, 2400 and 4000 gradations and polished using diamond pastes (3 μm and 1 μm gradation). In selected joints, a precise microstructural analysis was carried out. For this purpose, samples were cut for observation on the transmission electron microscope JEOL JEM-1200. The identification of phases present in individual layers of multilayer material was carried out using the electron diffraction technique (SAED) (JEOL, Warsaw, Poland). A light microscopy investigation was performed using Olympus LEXT OLS4100 microscope (OLYMPUS, Warsaw, Poland).

3. Results and Discussion

After the joining process, the AZ31/AA1050/AA2519/AA1050/Ti6Al4V multilayer composite was obtained (Figure 3).

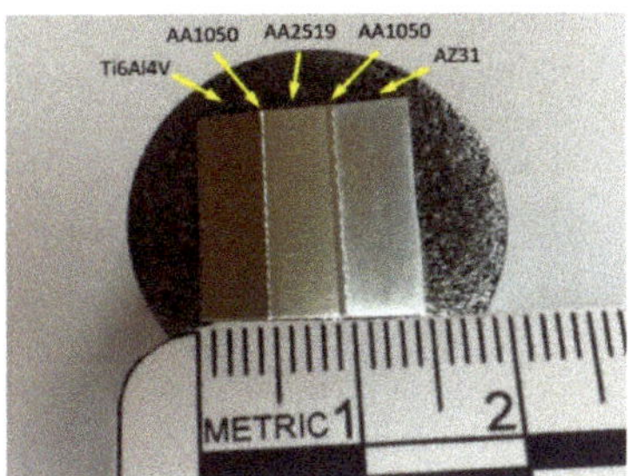

Figure 3. Cross-section of the obtained multilayer composite.

In this investigation, the four different samples have been taken under investigation: one in the as-welded state and three after hot-rolling at different temperatures. The designation of samples, together with their descriptions, are presented in Table 2.

Table 2. The designation of the samples.

Designation	State of the Sample
EXW	Mg/Al/Ti composite in the as-welded state
300C	Mg/Al/Ti composite subjected to the post-weld hot-rolling at 300 °C
350C	Mg/Al/Ti composite subjected to the post-weld hot-rolling at 350 °C
400C	Mg/Al/Ti composite subjected to the post-weld hot-rolling at 400 °C

In the first part of the investigation, the microhardness measurements were performed in order to establish the microhardness distribution for each sample (Figure 4). It is note-worthy that in the bonded areas in some cases it is difficult to obtain an accordance with each layer in all samples; this is an effect of small differences in the width of materials after the explosive welding process. The material affected by this phenomenon in the most visible way is the AA1050 interlayer, due to its softness. As can observed in the microhardness distribution, the hot-rolling process increases the microhardness for AZ31 and AA2519 components. The total relative strain was constant (30%), and only the rolling temperature was a factor which influenced the final properties of the composite. In the case of the 300C sample, it can be stated that the highest value of strain hardening was obtained as the result of the post-weld hot-rolling process. Rolling at a higher temperature of 350C and 400C results in slightly lower values of microhardness for both AZ31 and AA2519. This is probably the result of involving recrystallization processes in the higher temperature range [25,26]. Ti6Al4V is characterized by the relatively higher spread of microhardness values and this tendency maintains after applying of hot-rolling, due to the high recrystallization temperature of this alloy [27]. The highest strain hardening of AZ31 and AA2519 has been achieved after rolling at 300 °C, with reported values of microhardness of about 80 HV0.1 and 105 HV0.1, respectively. For the titanium component, the spread of results covers values from 300 HV0.1 to 370 HV0.1.

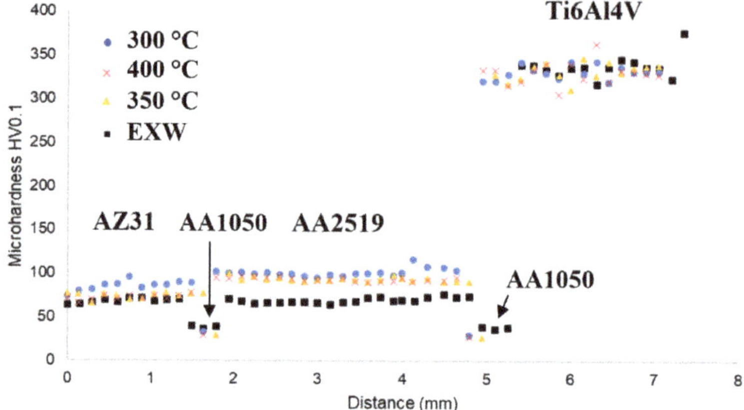

Figure 4. The distribution of microhardness in the investigated composite in the as-welded state (EXW) and after hot-rolling (300C, 350C, 400C).

Now, we establish the basic mechanical properties that the tensile tests of mini-specimens were carried on. Mini-specimens were cut from Mg/Al and Al/Ti joints separately for all analyzed samples. The obtained results of the tensile test for Mg/Al and Al/Ti joints are presented in Figures 5 and 6, respectively. A summary of the tensile tests and the standard deviation of mechanical properties is presented in Table 3.

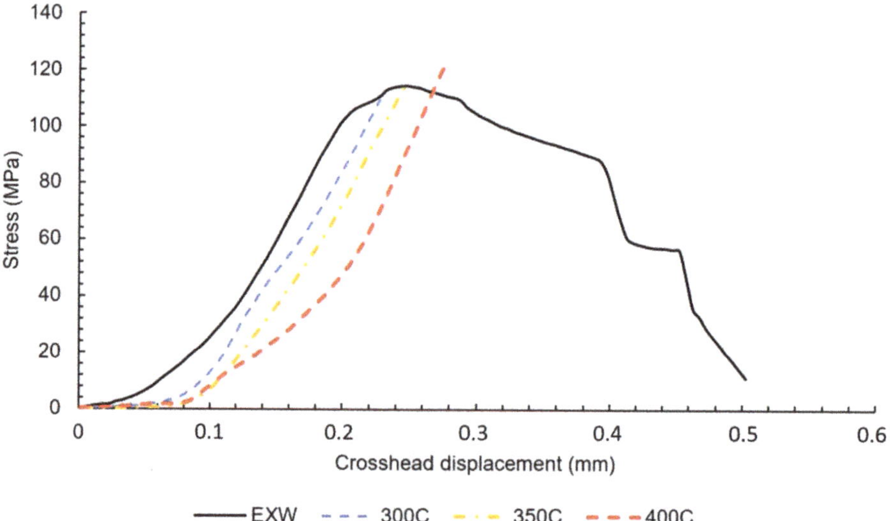

Figure 5. The tensile curves for Mg/Al interface in the as-welded state (EXW) and after hot-rolling (300C, 350C, 400C).

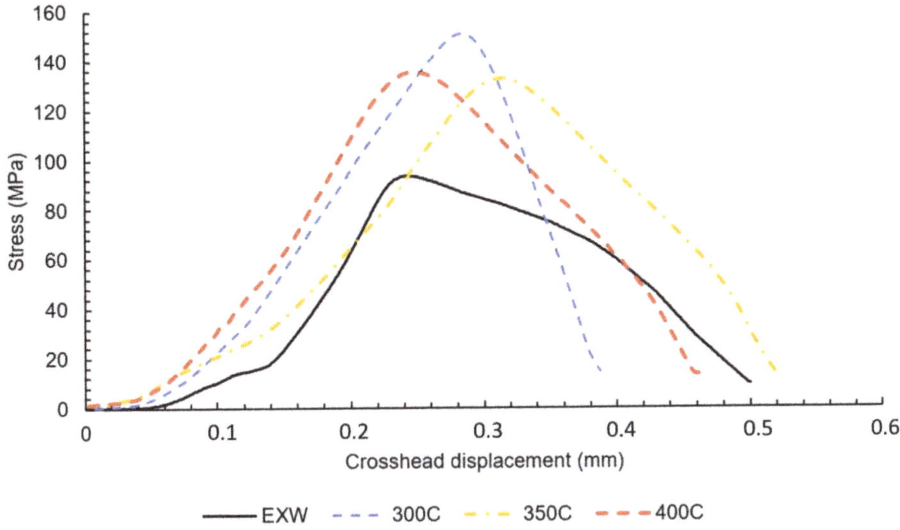

Figure 6. The tensile curves for Al/Ti interface in the as-welded state (EXW) and after hot-rolling (300C, 350C, 400C).

Table 3. Tensile tests results with standard deviation.

	Ti6Al4V/AA1050/AA2519		AZ31/AA1050/AA2519	
	Rm (MPa)	**Standard Deviation**	**Rm (MPa)**	**Standard Deviation**
As-welded	101	7.15	114	6.22
Hot-rolled 300 °C	154	2.87	109	3.12
Hot-rolled 350 °C	132	3.50	115	3.30
Hot-rolled 400 °C	135	4.25	125	5.02

The analysis of the obtained curves allows one to draw a conclusion that post-weld hot rolling affects both joints in a different way. The Mg/Al interface shows a strong correlation of tensile strength and rolling temperature. The sample in the as-welded state is characterized by tensile strength about 114 MPa. The curves obtained for samples after hot-rolling are characterized by the lack of plastic deformation. It has to be taken into consideration that the higher value of elongation for the as-welded sample could not be an effect of plastic deformation but also crack opening or delamination. Although hot-rolling caused a noticeable decrease in plasticity, the post-weld processing at 400 °C given a positive effect in the highest tensile strength of 125 MPa. At the same time, the Al/Ti interface is characterized by significant differences in tensile strength, with 101 MPa value reported for sample in the as-welded state, and 154 MPa for the sample after hot-rolling in 300 °C. The samples 350 °C and 400 °C have intermediate values of tensile strength—132 MPa and 135 MPa, respectively. In Ti6Al4V/AA1050/AA2519 mini-samples independently on the state of the material, the failure occurs on Ti6Al4V/AA1050 interface, which indicates the crucial role of this bond in the composite cohesion. The Mg/Al fracture surfaces of samples EXW (Figure 7A) and 400C (Figure 7B) are presented below.

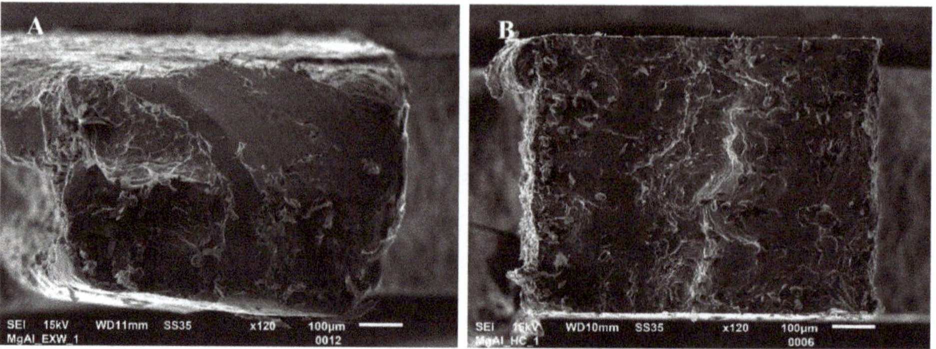

Figure 7. The fracture surface of Mg/Al mini-specimen in the as-welded state (**A**) and after hot-rolling at 400 °C (**B**).

The fractography investigation revealed the brittle mechanism of failure for samples before and after hot-rolling. However, the failure of material after treatment was proceeded through the diffusion layer, which was confirmed by EDX investigation. The results of the element distribution on the fracture surface of Mg/Al mini-specimen after hot-rolling at 400 °C for aluminum (Figure 8A) and magnesium alloy (Figure 8B) are presented below. Obtained maps show that the surface of the fractured sample is characterized by the presence of both aluminum and magnesium. This result suggests that failure occurs in the diffusion layer. In all AZ31/AA1050/AA2519 mini-samples independently on the state of the material, the failure occurs on the AZ31/AA1050 interface, which indicates on lower strength properties of AZ31/AA1050 than AA1050/AA2519 independently on the presence of thick or thin intermetallic phase layer [28].

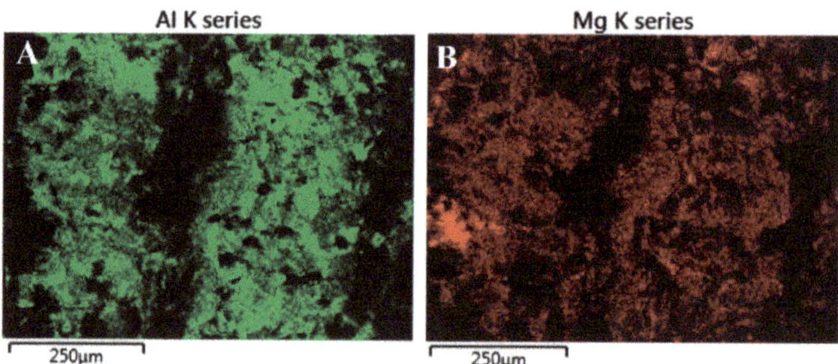

Figure 8. The results of the element distribution on the fracture surface of Mg/Al mini-specimen after hot-rolling at 400 °C: aluminum (**A**) and magnesium (**B**).

The Al/Ti fracture surfaces of samples EXW (Figure 9A) 400C (Figure 9B) are presented below. The analysis of the obtained fracture surfaces shows that despite the similar character of fracture, the sample subjected to the hot-rolling exhibits a slightly higher participation of ductile fracture.

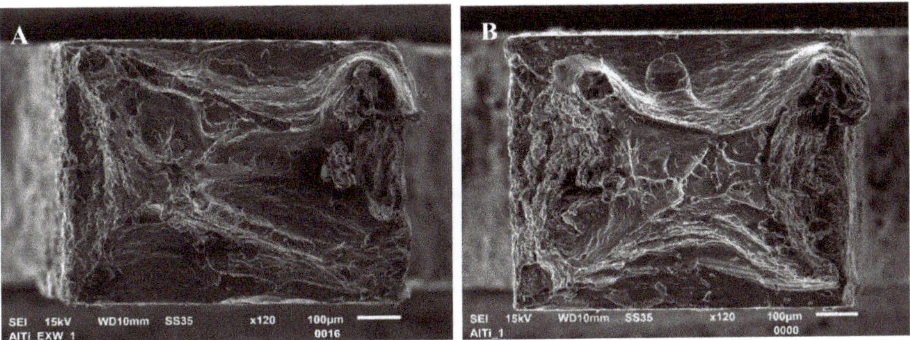

Figure 9. The fracture surface of Al/Ti mini-specimen in the as-welded state (**A**) and after hot-rolling at 400 °C (**B**).

For microstructure analysis, two samples have been chosen: in the as-welded state and after hot-rolling at 400 °C. Light microscopy images of investigated interfaces are presented in Figure 10.

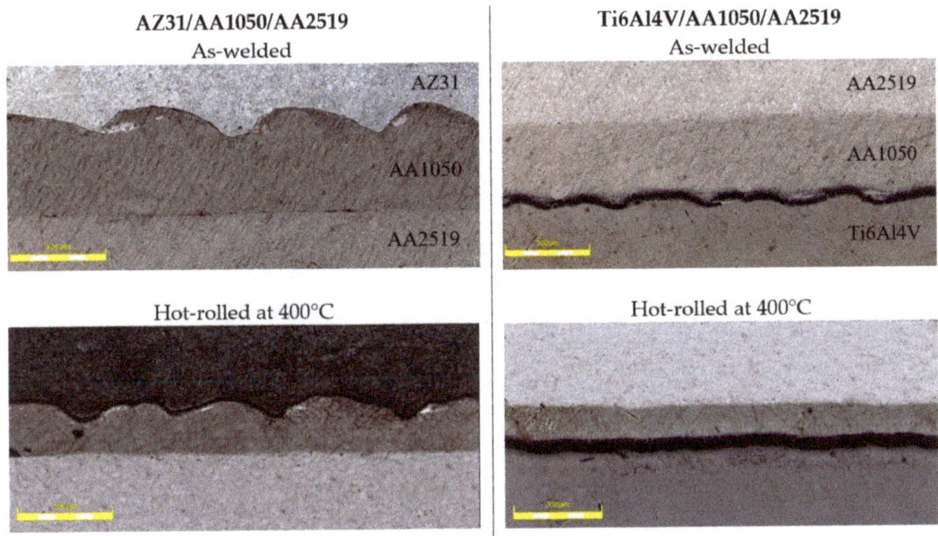

Figure 10. Light microscopy images of investigated interfaces (500 µm scale bar).

The AZ31/AA1050 joint is a potential area of intermetallic phase growth from the Al-Mg system. Analyzing the phase equilibrium system of the main alloying elements of the considered joint, three intermetallic compounds can be distinguished: Mg_2Al_3 (β), $Mg_{17}Al_{12}$ (γ) and $Mg_{23}Al_{30}$ (R or ε). An explosive welding process can result in the formation of vortex in the joint line, characterized by the local mixing of joint materials [29]. Similarly to the bimetallic joint, the vortex is also the area with potential for intermetallic growth. In the AZ31/AA150 joint not subjected to thermo-plastic treatment, no formation of intermetallic phases was observed, which was confirmed by the results of SEM tests (Figure 11A) and analysis of the Mg and Al elements distribution on the sample surface—EDX (Figure 11B). An observed flat joint was observed free of any defects in the form of voids, delamination and melted areas, which proves that the joining process was carried out correctly.

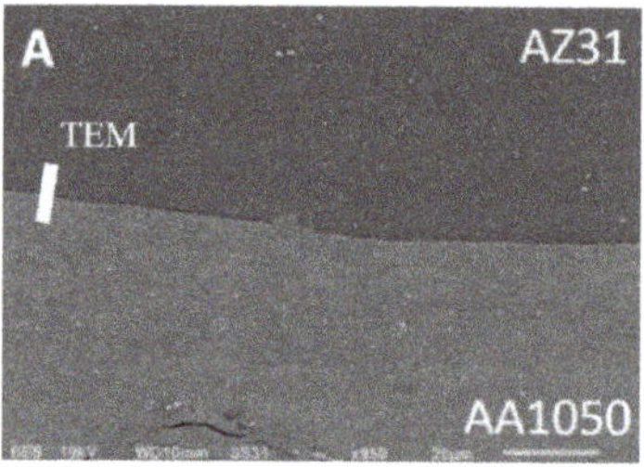

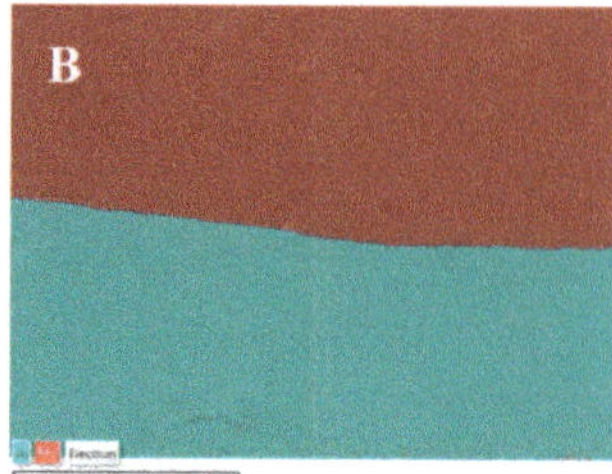

Figure 11. AZ31/AA1050 joint in as-welded state: (**A**) SEM results, (**B**) Mg and Al elements distribution on the sample surface.

In order to confirm the absence of intermetallic phases and melted zones at the Mg-Al boundary, a phase analysis was carried out using the SAED electron diffraction technique (Figure 12). The TEM results also revealed the occurrence of high plastic deformation in the AZ31 material near the joint and the presence of equiaxial ultrafine grains in the microstructure of AA1050 material with their size about 1 µm.

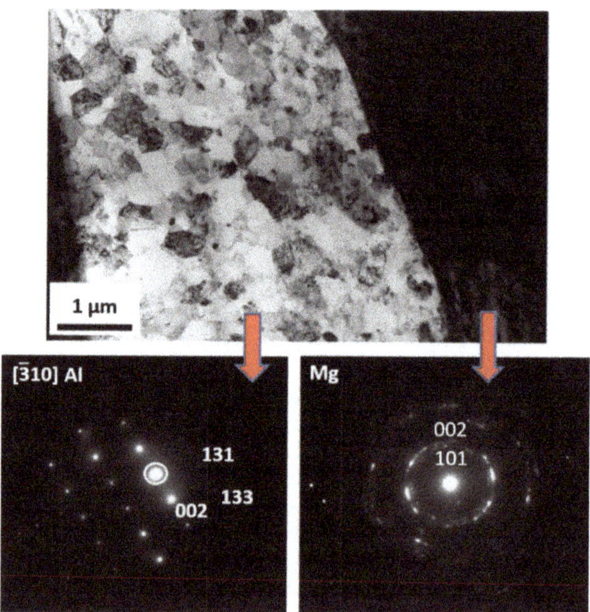

Figure 12. TEM image of AZ31/AA1050 joint microstructure in as-welded state with selected area diffraction patterns (SAED). TEM lamella position in Figure 11A.

Scanning and transmission electron microscopy investigation of the AZ31/AA1050 joint after the rolling process allows one to conclude that, due to the diffusion of aluminum and magnesium in the joint area during hot rolling, a continuous layer consisting of two intermetallic phases is formed. The microstructure of the characteristic joint zone is presented in Figure 13. Due to the short time of the explosive welding process and small zone impact, the cooling rates of joined materials allowed one to limit the formation of the intermetallic compounds. At the same time, the hot rolling process promotes diffusion changes within the joints, which may result in the formation of intermetallic phases. The severe plastic deformations of the joint zone decreases the amount of energy necessary to initiate heat-activated phenomena, including the formation of new compounds. In the case of the Mg/Al joint, the formed diffusion zone had a layered character with two sublayers of different compounds. From the aluminum alloy side, the Mg_2Al_3 (β) phase is formed and the $Mg_{17}Al_{12}$ (γ) phase from the magnesium alloy was confirmed by TEM investigation results (Figure 14). The results indicated the significant influence of the hot rolling process on the intermetallic compounds formation on the joint interface.

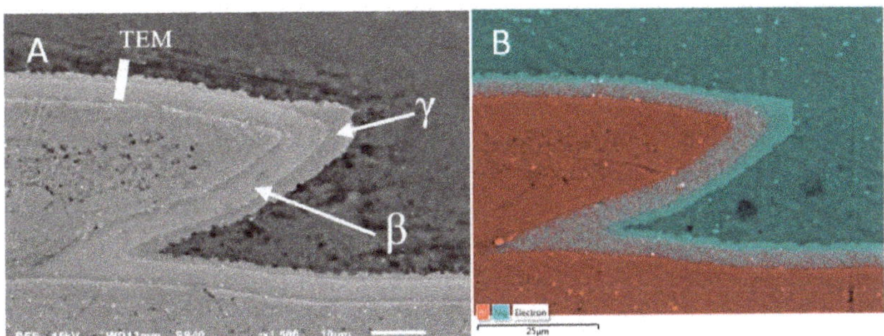

Figure 13. Microstructure of AZ31/AA1050 joint after hot-rolling in 400 °C: (**A**) SEM results, (**B**) Mg and Al elements distribution on the sample surface.

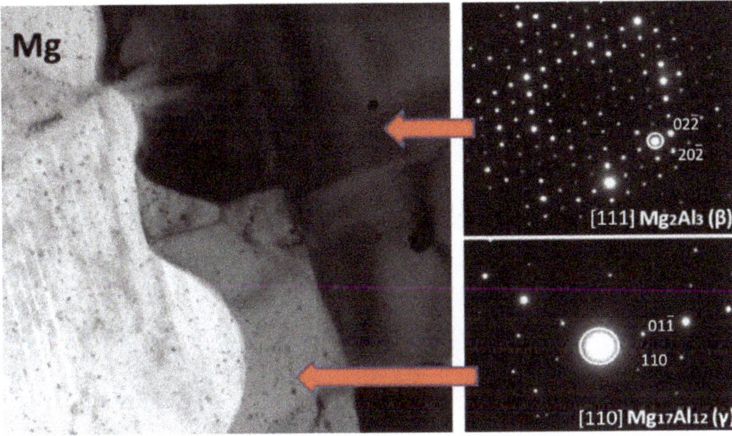

Figure 14. TEM image of AZ31/AA1050 joint microstructure after hot-rolling at 400 °C with selected area diffraction patterns (SAED). TEM lamella position at Figure 13A.

It should also be noted that there are disproportions in the thickness of the individual intermetallic phases (maximum 4.5 μm for β and 9 μm for γ), resulting in differences in their growth rates (Figure 13). In order to confirm the presence of intermetallic phases β and γ at the Mg-Al limit, a phase analysis was performed using the SAED electron diffraction technique (Figure 14). The higher-resolution observation revealed a lack of cracks or delamination between intermetallic layers and base materials. The TEM investigation revealed coarsened equiaxed grains in the microstructure of AZ31.

Similarly, as in the case of Al/Mg bonds, the joint AA1050/Ti6Al4V is also a potential area of intermetallic phase growth. Analyzing the titanium-aluminum phase balance system, four intermetallic compounds can be identified: Ti_3Al (α2), TiAl (γ), $TiAl_2$ and $TiAl_3$. In contrast to the stoichiometric $TiAl_3$ and $TiAl_2$ phases, the Ti_3Al and TiAl phases are present in a wide range of aluminum contents. The research results of the as-welded AA1050/Ti6Al4V joint revealed the presence of the melted zone with localized intermetallic particles (Figure 15A). The analysis of Ti and Al elements distribution on the sample surface—EDX allows to observe the mixture character of the melted zone (Figure 15B). Element compositions corresponding to the spectrums in Figure 15B are presented in Figure 16.

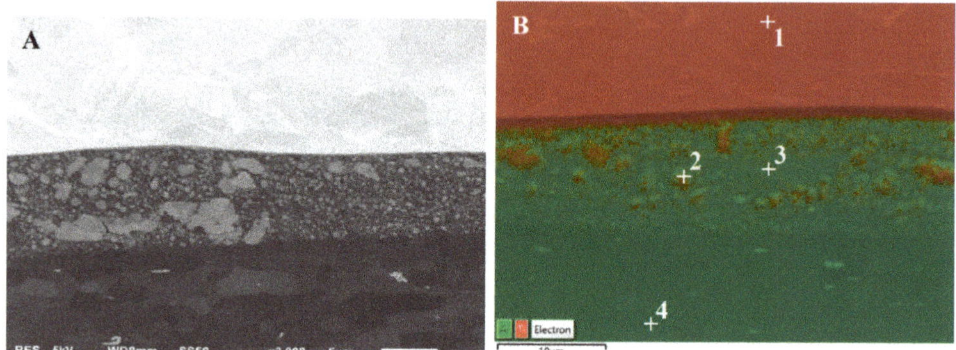

Figure 15. Microstructure of AA1050/Ti6Al4V joint in as-welded state: (**A**) SEM results, (**B**) Ti and Al elements distribution on the sample surface.

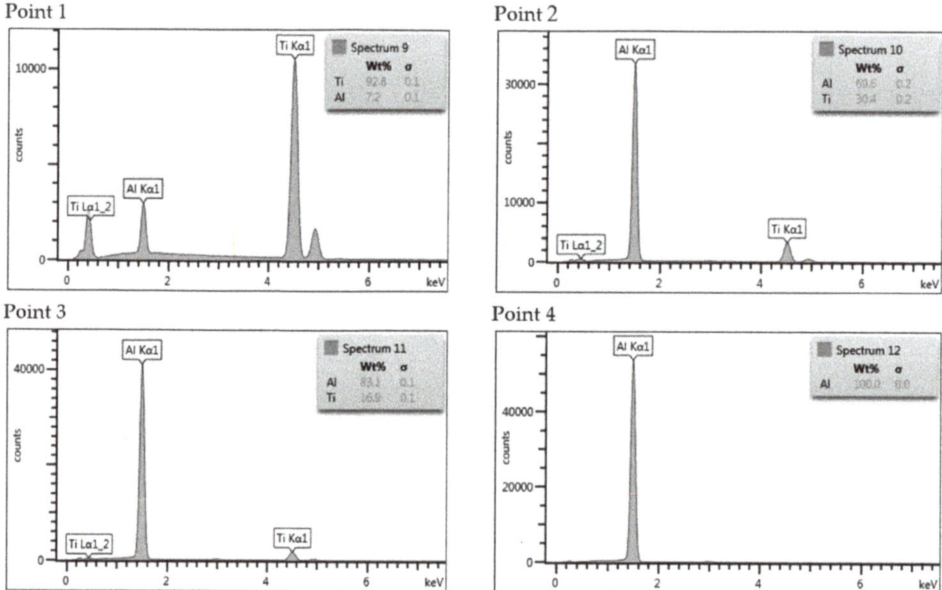

Figure 16. Element compositions corresponding to the spectrums in Figure 15B.

The results of the research on the material after hot rolling at a temperature of 400 °C revealed the formation of an ultrafine precipitation in the melted zone (Figure 17). The diffusion zone in hot-rolled Ti/Al joint has a significantly different character than in Mg/Al, consisting of a mixture of both materials, together with the ultrafine dispersion of $TiAl_3$ intermetallic precipitates. This specific microstructure suggests that intermetallic phase forms a Ti/Al solid solution during hot rolling.

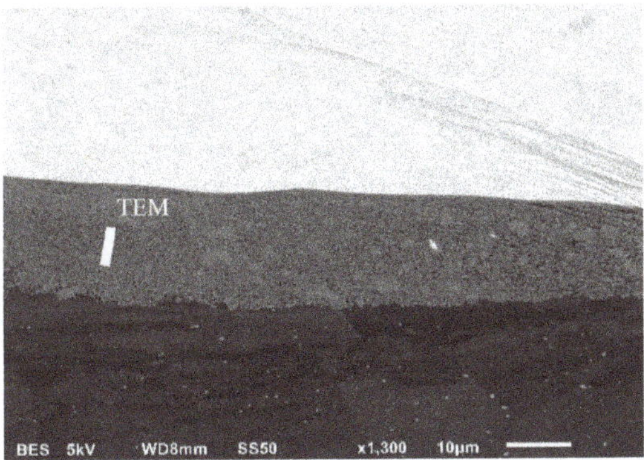

Figure 17. SEM image of the microstructure of AA1050/Ti6Al4V joint hot-rolled at 400 °C.

The $TiAl_3$ intermetallic phase is most likely from a thermodynamic point of view. In the case of Al/Ti, intermetallic growth is in the nature of precipitation rather than layered growth, like in the case of the Mg/Al interface. The presence of $TiAl_3$ was confirmed by TEM observation and the results of the SAED electron diffraction technique (Figure 18). Investigations revealed ultrafine precipitates of the $TiAl_3$. The size of the precipitates was approximately 100 nm.

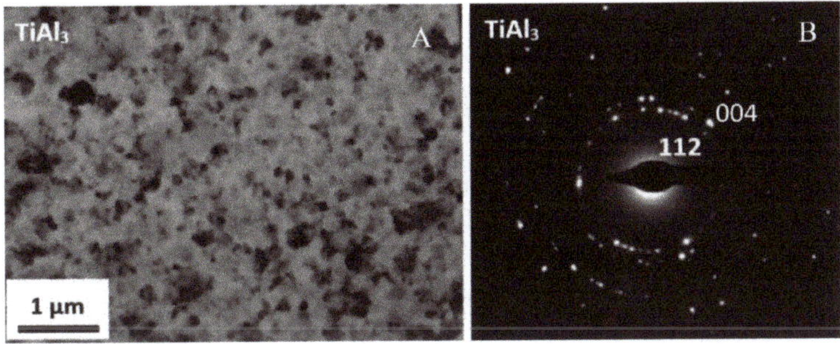

Figure 18. TEM image of AA1050/Ti6Al4V joint microstructure in hot-rolled state (**A**), with selected area diffraction patterns (SAED) (**B**). TEM lamella position in Figure 17.

4. Conclusions

The analysis of the results shows that explosive welding allows to obtain Mg/Al/Ti multilayer composite, characterized by lack of any imperfections in joints, such as delamination, voids or cracks. The application of the post-weld hot-rolling process results in the diffusion of elements under the influence of temperature and plastic deformation. Hot-rolling at 300 °C results in the highest reported level of strain hardening of AZ31 and AA2519 components. The curves obtained in the tensile test of mini-specimens allow one to draw a conclusion that post-weld hot rolling affects both joints in a different way. The Mg/Al interface shows an accordance of tensile strength and rolling temperature, with the highest value of 125 MPa reported for rolling at 400 °C. At the same time, the Al/Ti interface is characterized by significant differences in tensile strength, with 93 MPa value reported for sample in the as-welded state and 135 MPa for the sample after hot-rolling at 400 °C. The microstructure analysis revealed that hot-rolling affects the microstructure of both joints. In the case of the Mg/Al interface, the growth of continuous intermetallic layers and, for the Al/Ti interface, the precipitation of $TiAl_3$ intermetallic from solid solution (melted zone), have been reported.

Author Contributions: Conceptualization, M.W., R.K. and S.M.; methodology, M.W., R.K. and S.M.; formal analysis, M.W., R.K. and L.Ś.; Data Curation, L.Ś., A.S.; P.S.; investigation, M.W., R.K., S.M., A.S. and P.S.; writing—original draft preparation, M.W. and R.K.; writing—review and editing, S.M. and L.Ś., visualization, M.W.; supervision, L.Ś. All authors have read and agreed to the published version of the manuscript.

Funding: This research was funded by Polish Ministry of National Defence, grant number: PBG/13-998.

Conflicts of Interest: The authors declare no conflict of interest.

References

1. Chang, H.; Zheng, M.Y.; Gan, W.M.; Wu, K.; Maawad, E.; Brokmeier, H.G. Texture evolution of the Mg/Al laminated composite fabricated by the accumulative roll bonding. *Scr. Mater.* **2009**, *61*, 717–720. [CrossRef]
2. Yang, S.; Bao, J. Microstructure and Properties of 5083 Al/1060 Al/AZ31 Composite Plate Fabricated by Explosive Welding. *J. Mater. Eng. Perform.* **2018**, *27*, 1177–1184. [CrossRef]
3. Zhang, N.; Wang, W.; Cao, X.; Wu, J. The effect of annealing on the interface microstructure and mechanical characteristics of AZ31B/AA6061 composite plates fabricated by explosive welding. *Mater. Des.* **2015**, *65*, 1100–1109. [CrossRef]
4. Yan, Y.B.; Zhang, Z.W.; Shen, W.; Wang, J.H.; Zhang, L.K.; Chin, B.A. Microstructure and properties of magnesium AZ31B–aluminum 7075 explosively welded composite plate. *Mater. Sci. Eng. A-Struct.* **2010**, *527*, 2241–2245. [CrossRef]
5. Fronczek, D.M.; Chulist, R.; Litynska-Dobrzynska, L.; Lopez, G.A.; Wierzbicka-Miernik, A.; Schell, N.; Szulc, Z.; Wojewoda-Budka, J. Microstructural and Phase Composition Differences Across the Interfaces in Al/Ti/Al Explosively Welded Clads. *Metall. Mater. Trans. A* **2017**, *48*, 4154–4165. [CrossRef]
6. Wierzba, A.; Mróz, S.; Szota, P.; Stefanik, A.; Mola, R. The influence of the asymmetric arb process on the properties of al-mg-al multi-layer sheets. *Arch. Metall. Mater.* **2015**, *60*, 2821–2825. [CrossRef]
7. Kosturek, R.; Wachowski, M.; Śnieżek, L.; Gloc, M. The Influence of the Post-Weld Heat Treatment on the Microstructure of Inconel 625/Carbon Steel Bimetal Joint Obtained by Explosive Welding. *Metals* **2019**, *9*, 246. [CrossRef]
8. Čížek, L.; Ostroushko, D.; Mazancova, E.; Szulc, Z.; Molak, R.; Wachowski, M. Structure and Properties of Sandwich material steel Cr13Ni10+ Ti after explosive cladding. *Metall. J.* **2009**, *62*, 6.
9. Oliveira, J.P.; Ponder, K.; Brizes, E.; Abke, T.; Edwards, P.; Ramirez, A.J. Combining resistance spot welding and friction element welding for dissimilar joining of aluminum to high strength steels. *J. Mater. Process. Tech.* **2019**, *273*, 116192. [CrossRef]
10. Lazurenko, D.V.; Bataev, I.A.; Mali, V.I.; Esikov, M.A.; Bataev, A.A. Effect of Hardening Heat Treatment on the Structure and Properties of a Three-Layer Composite of Type 'VT23—08ps—45KhNM' Obtained by Explosion Welding. *Met. Sci. Heat Treat.* **2019**, *60*, 651–658. [CrossRef]
11. Szachogłuchowicz, I.; Hutsaylyuk, V.; Śnieżek, L. Low cycle fatigue properties of AA2519–Ti6Al4V laminate bonded by explosion welding. *Eng. Fail. Anal.* **2016**, *69*, 77–87. [CrossRef]
12. Luo, C.; Liang, W.; Chen, Z.; Zhang, J.; Chi, C.; Yang, F. Effect of high temperature annealing and subsequent hot rolling on microstructural evolution at the bond-interface of Al/Mg/Al alloy laminated composites. *Mater. Charact.* **2013**, *84*, 34–40. [CrossRef]
13. Chen, M.C.; Hsieh, H.C.; Wu, W. The evolution of microstructures and mechanical properties during accumulative roll bonding of Al/Mg composite. *J. Alloy. Compd.* **2006**, *416*, 169–172. [CrossRef]
14. Hutsaylyuk, V.; Śnieżek, L.; Chausov, M.; Torzewski, J.; Pylypenko, A.; Wachowski, M. Cyclic deformation of aluminium alloys after the preliminary combined loading. *Eng. Fail. Anal.* **2016**, *69*, 66–76. [CrossRef]
15. Xu, L.; Cui, Y.Y.; Hao, Y.L.; Yang, R. Growth of intermetallic layer in multi-laminated Ti/Al diffusion couples. *Mater. Sci. Eng. A Struct.* **2006**, *435*, 638–647. [CrossRef]
16. Thiyaneshwaran, N.; Sivaprasad, K.; Ravisanka, B. Work hardening behavior of Ti/Al-based metal intermetallic laminates. *Int. J. Adv. Manuf. Technol.* **2017**, *93*, 361–374. [CrossRef]
17. Zhang, T.T.; Wang, W.X.; Zhou, J.; Cao, X.Q.; Yan, Z.F.; Wei, Y.; Zhang, W. Investigation of Interface Bonding Mechanism of an Explosively Welded Tri-Metal Titanium/Aluminum/Magnesium Plate by Nanoindentation. *Jom-J. Min. Met. Mater. Soc.* **2017**, *70*, 504–509. [CrossRef]
18. Fouad, Y. Characterization of a high strength Al-alloy interlayer for mechanical bonding of Ti to AZ31 and associated tri-layered clad. *Alex. Eng. J.* **2014**, *53*, 289–293. [CrossRef]

19. Motevalli, P.D.; Eghbali, B. Microstructure and mechanical properties of Tri-metal Al/Ti/Mg laminated composite processed by accumulative roll bonding. *Mater. Sci. Eng. A Struct.* **2015**, *628*, 135–142. [CrossRef]
20. Wu, K.; Chang, H.; Maawad, E.; Gan, W.M.; Brokmeier, H.G.; Zheng, M.Y. Microstructure and mechanical properties of the Mg/Al laminated composite fabricated by accumulative roll bonding (ARB). *Mater. Sci. Eng. A Struct.* **2010**, *527*, 3073–3078. [CrossRef]
21. Dietrich, D.; Nickel, D.; Krause, M.; Lampke, T.; Coleman, M.P.; Randle, V. Formation of intermetallic phases in diffusion-welded joints of aluminium and magnesium alloys. *J. Mater. Sci.* **2011**, *46*, 357–364. [CrossRef]
22. Fronczek, D.M.; Chulist, R.; Litynska-Dobrzynska, L.; Kac, S.; Schell, N.; Kania, Z.; Szulc, Z.; Wojewoda-Budka, J. Microstructure and kinetics of intermetallic phase growth of three-layered A1050/AZ31/A1050 clads prepared by explosive welding combined with subsequent annealing. *Mater. Des.* **2017**, *130*, 120–130. [CrossRef]
23. Lazurenko, D.V.; Bataev, I.A.; Mali, V.I.; Bataev, A.A.; Maliutina, I.N.; Lozhkin, V.S.; Esikov, M.A.; Jorge, A.M.J. Explosively welded multilayer Ti-Al composites: Structure and transformation during heat treatment. *Mater. Des.* **2016**, *102*, 122–130. [CrossRef]
24. Mróz, S.; Stefanik, A.; Szota, P.; Kwapisz, M.; Wachowski, M.; Śnieżek, L.; Gałka, A.; Szulc, Z. Numerical and experimental modeling of plastic deformation the multi-layer Ti/Al/Mg materials. *Arch. Metall. Mater.* **2019**, *64*, 1361–1368. [CrossRef]
25. Zhang, X.; Ma, F.; Zhang, W.; Li, X. Kinetics of Dynamic Recrystallization in AA2024 Aluminum Alloy. *Mod. Appl. Sci.* **2014**, *8*, 47–52. [CrossRef]
26. Guo, L.; Fujita, F. Influence of rolling parameters on dynamically recrystallized microstructures in AZ31 magnesium alloy sheets. *J. Magnes. Alloy.* **2015**, *3*, 95–105. [CrossRef]
27. Yumeng, L.; Jinxu, L.; Shukui, L.; Xingwang, C. Effect of Hot-rolling Temperature on Microstructure and Dynamic Mechanical Properties of Ti-6Al-4V Alloy. *Rare Met. Mater. Eng.* **2018**, *47*, 1333–1340. [CrossRef]
28. Zhang, W.; Ao, S.; Oliveira, J.P.; Li, C.; Zeng, Z.; Wang, A.; Luo, Z. On the metallurgical joining mechanism during ultrasonic spot welding of NiTi using a Cu interlayer. *Scr. Mater.* **2020**, *178*, 414–417. [CrossRef]
29. Lee, T.; Nassiri, A.; Dittrich, T.; Vivek, A.; Daehn, G. Microstructure development in impact welding of a model system. *Scr. Mater.* **2020**, *178*, 203–206. [CrossRef]

Article

Influence of Impact Velocity on the Residual Stress, Tensile Strength, and Structural Properties of an Explosively Welded Composite Plate

Aleksander Karolczuk [1], Krzysztof Kluger [1,*], Szymon Derda [1], Mariusz Prażmowski [1] and Henryk Paul [2]

1 Department of Mechanics and Machine Design, Opole University of Technology, Mikołajczyka 5, 45-271 Opole, Poland; a.karolczuk@po.edu.pl (A.K.); szymon.derda@doktorant.po.edu.pl (S.D.); m.prazmowski@po.edu.pl (M.P.)

2 Institute of Metallurgy and Materials Science, Polish Academy of Sciences, Reymonta 25, 30-0 59 Kraków, Poland; h.paul@imim.pl

* Correspondence: k.kluger@po.edu.pl

Received: 20 May 2020; Accepted: 9 June 2020; Published: 12 June 2020

Abstract: This study aimed to analyze the effect of the impact velocity of a Zr 700 flyer plate explosively welded to a Ti Gr. 1/P265GH bimetallic composite on the residual stress formation, structural properties, and tensile strength. The residual stresses were determined by the orbital hole-drilling strain-gauge method in a surface layer of Zr 700 in as-received and as-welded conditions. The analysis of the tensile test results based on a force parallel to interfaces was used to propose a model for predicting the yield force of composite plates. Compressive residual stresses found in the initial state of the Zr 700 plate were transformed to tensile stresses on the surface layer of the welded Zr 700 plate. A higher impact velocity resulted in higher tensile stresses in the Zr 700 surface layer. To increase the resistance of the composite plate to stress-based corrosion cracking, a lower value of impact velocity is recommended in the welding process.

Keywords: explosive welding; residual stress; orbital hole-drilling strain-gauge method; prediction of tensile yield force; explosive cladding; Zr 700

---

1. Introduction

Metallic composites belong to a group of materials in which the multilayer structure of different metallic alloys provides special functional properties [1]. Explosive cladding is one of the manufacturing processes used to produce multilayer metallic composites. It involves the energy of detonation to accelerate a flyer plate that, as a result, collides with a base plate [2,3]. High-velocity impact leads to the formation of a strong bond between colliding plates. Low-density materials, such as aluminum and magnesium alloys, can be explosively welded to steels and applied in the ship building and automotive industries [4,5]. A copper layer [6,7] within a composite multilayer material provides excellent electrical conductivity, and so-called reactive materials such as titanium [8,9], zirconium, and niobium exhibit corrosive resistance in an aggressive environment. This, in turn, predisposes them to wide usage in design applications in process equipment in particular [10]. A composition of tungsten foil and cooper layers is applicable for thermonuclear reactors [11] since it offers high resistance to heat loads and irritation.

Zirconium exhibits outstanding performance as a material for use in highly corrosive environments with a broad range of chemical media and temperatures [10,12,13]. This allows the material to be used in the chemical process industry for heat exchangers cooled with seawater and other pieces of process equipment as well as in nuclear-fuel-reprocessing plants. To reduce the financial cost of producing the process equipment, a zirconium alloy could be used as a relatively thin layer in cladded plates. According to Banker [10], replacing the solid structure with a zirconium wall of a thickness above 20 mm by a multilayer plate could have cost-effective benefits. Explosive welding is categorized within the solid-state process and is one of the ways to achieve a high-quality connection between dissimilar materials.

The parameters of the explosive welding process determine the quality of the bond and the mechanical properties of the cladded plates. For higher explosive welding parameters, locally melted areas may be formed, which can involve brittle intermetallic compounds and shrinkage cracks [14,15]. Therefore, all parameters must be adequately selected to achieve optimal efficiency of the process. Such parameters include the physical properties of the welded materials, parameters of the explosive material, and the geometry of the welding setup. The explosively induced joint is inevitably associated with large material deformation, a significant temperature gradient, and rapid phase change in the surroundings of the impact zone. Therefore, residual stresses are locked-in stresses in the explosively welded plates [16,17]. The residual stress state may have a considerable effect on material performance. A compressive residual stress state is known to increase fatigue strength [18], and can be a cause of dimension instability during cutting or other manufacturing processes. In contrast, a tensile residual stress state tends to be undesirable as it accelerates crack growth and can induce stress-based corrosion cracking [19,20]. Nagano et al. [21] noted that a passive oxide film in pure zirconium and its alloys ruptured because of stress in corrosive environments depending on temperature and HNO_3 concentration. Several other studies have reported on the stress-based corrosion cracking in zirconium and its alloys [22–25]. A high residual stress gradient can result in rapid delamination of welded plates within a few seconds after the collision of plates in the case of improper welding parameters. Even if the bond survives, the geometrical stability of the welded plate during cutting can be lost. In the case of the application of multilayer plates in processing equipment, the possible failure of the corrosive resistance layer is unacceptable as it would lead to undetected corrosion because of the release of corrosive compounds throughout the backing material. To reduce the failure probability of the corrosive resistance layer, the residual stresses in the near-surface layer should be as low as possible, with a preferably compressive character. However, the problem of the influence of explosive welding parameters on residual stress states in the zirconium layer of composite plates was not profoundly analyzed.

The present study is aimed at analyzing the influence of impact velocity on the tensile strength, the structural properties of composite plates, and the generation of residual stress in the flyer plate (a plate with explosive charge) made of Zr 700 alloy. Two impact velocities were achieved by altering the stand-off distance by using a fixed quantity of the explosive charge.

Presently, limited research has been devoted to the determination of residual stress in composite structures obtained through the explosive welding process. Due to the wide range of analyzed composite structures and the use of different methods to determine residual stresses, unambiguous conclusions on the state of residual stress in explosively welded multilayer structures cannot be deduced. In most cases, tensile residual stresses were detected in the surface layer of the flyer plate [16,26–31]. Limited cases with compressive residual stresses were found [17,32,33]. The hole-drilling strain-gauge method seems to be the most relevant, as it is included in the ASTM standard [34]. In addition, both the limitation and accuracy of this method are well established and are available in the literature.

In this study, the residual stresses were determined by employing the hole-drilling strain-gauge method recommended by the ASTM standard [34]. The stresses were determined in the Zr 700 layer for two plates welded under different impact velocities. Additionally, the residual stresses in the Zr 700 plate before welding were determined and used as the reference value.

The characteristic features of explosively welded materials include a wavy character of the interface between joint metals and locally melted areas. These features of the interface were described by measuring the sizes of the melted areas as well as the height and length of the interface wave. The data were obtained through the microhardness measurement across welded plates. In addition, tensile tests were conducted with force applied in the parallel direction to the interface. Moreover, a model for predicting the yield force of the composite plate is proposed.

2. Experiment

2.1. Materials in As-Delivered Condition

A composite structure consisting of three layers made of P265GH pressure vessel steel, Ti Gr. 1, and zirconium Zr 700 was manufactured in the explosive welding process. Tables 1 and 2 summarize the chemical composition and mechanical properties of the materials examined in this study, respectively. The tensile tests were conducted to estimate the presented mechanical properties.

Table 1. Chemical composition of materials in as-delivered conditions [35].

Materials	Chemical Composition (wt %)											
P265GH	Mn 0.959	Si 0.260	C 0.147	Al 0.051	Ni 0.030	Cr 0.022	P 0.011	Nb 0.008	S 0.006	Mo 0.005	N 0.004	Fe Balance
Zr 700	O 0.067		Fe 0.060		C 0.004		N <0.002		H <0.0003		Zr + Hf Balance	
Ti Gr. 1	O 0.070		F 0.020		C 0.020		N <0.010		H 0.010		Ti Balance	

Table 2. Basic mechanical properties of applied materials.

Material	E, (GPa)	ν, (-)	R_{p02}, (MPa)	R_m, (MPa)	A, (%)
Zr 700	101	0.38	216	269	35
Ti Gr. 1	109	0.37	251	325	46
P265GH	193	0.29	268	391	41

E is the elastic modulus, ν is Poisson's ratio, A is the elongation, R_{p02} is the yield strength, and R_m is the tensile strength.

The microstructure of the materials under as-delivered conditions is illustrated in Figure 1.

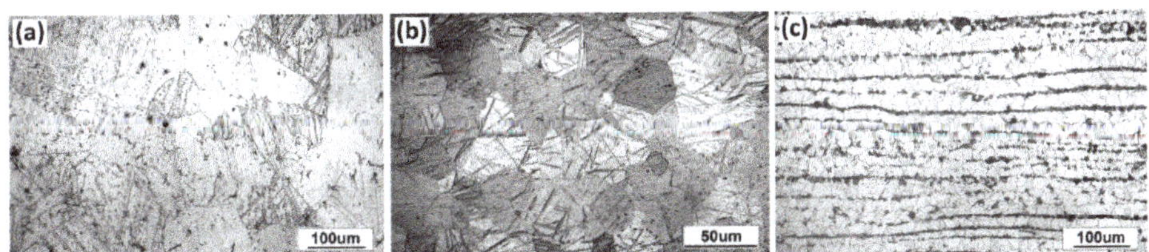

Figure 1. Microstructure of materials in as-delivered conditions: (**a**) Zr 700 alloy, (**b**) Ti Gr. 1 alloy, and (c) P265GH steel.

The microstructure of Zr 700 plate in the section perpendicular to the rolling direction is presented in Figure 1a. The material was characterized by the structure of the α-phased grains sized between 70 and 170 µm. The microstructure of the material used in the interlayer, i.e., Ti Gr. 1 alloy, consisted of α-phased equiaxed grains sized between 20 and 40 µm (Figure 1b). Figure 1c presents the P265GH carbon steel microstructure characterized by an equiaxed structure of mid-sized grains of 4–11 and 10–20 µm for pearlite and ferrite, respectively. As shown, a band structure composed of fine grains of pearlite was observed; this is a typical phenomenon for materials after the hot-forming process.

2.2. Explosive Welding Process

The explosive welding process was conducted by High Energy Technology Works "Explomet" (Opole, Poland). Consequently, two multilayer plates were produced with different welding parameters. In both cases, the flyer plate made of Zr 700 with a thickness of 10 mm was cladded to the preliminary welded bimetallic plate composed of a 2-mm Ti Gr. 1 layer and a 14-mm P265GH steel layer. The plates with dimensions of 300 mm × 500 mm were welded in parallel by applying an explosive charge of ammonites, with NH_4NO_3 (High Energy Technology Works "Explomet", Opole, Poland) as the main component. The detailed composition of the explosive charge was not provided by the supplier of the composite plates. The applied explosive charge resulted in a detonation velocity of $v_D = 2500$ m/s, which was measured using a fiber optic system [36]. The welding processes for the studied plates have different values of stand-off distance δ (which is the initial distance between the flyer plate, Zr 700, and the basic Ti Gr. 1–P265GH bimetal). For the first plate, labeled as B3, the stand-off distance was δ = 10 mm, whereas it was δ = 15 mm for the second plate (B4). The application of different δ values resulted in different impact velocities v_P estimated using the Deribas formula [37,38]. The summarized welding parameters are presented in Table 3.

Table 3. Explosive welding parameters.

Plate	Flyer	Thickness, (mm)	Detonation Velocity, v_D, (mm)	Stand-off Distance, δ, (mm)	Impact Velocity v_p, (m/s)
B3	Zr 700	10	2500	10	425
B4	Zr 700	10	2500	15	468

Both the welded plates underwent the flattening process and ultrasonic examination [39], revealing no discontinuities except near the ignition and narrow area at the plate edges (approximately 20 mm from the edge). Figure 2 shows the locations of the ignition point and samples used for residual stress measurement and microstructural analysis.

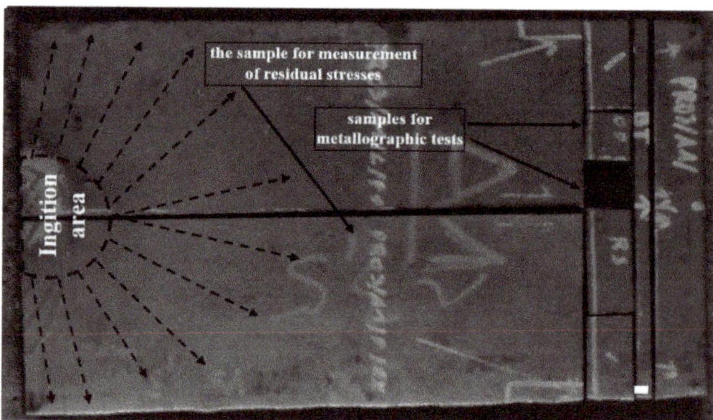

Figure 2. Explosively welded plate with the marked ignition area and samples for residual stress estimation and microstructural analysis.

2.3. Residual Stress Estimation

Residual stress was identified using the incremental hole-drilling strain-gage method recommended in the ASTM E837-13a standard [34], and the experiments were conducted on the surface of the flyer Zr 700 plate. Additionally, for referential analysis, residual stresses in the Zr 700 plate in the as-delivered condition were estimated.

For measuring the relieved strains during incremental hole drilling, a Vishay RS-200 device (Vishay Precision Group, Malvern, PA, USA) with a pneumatic turbine was used in orbital drilling mode. This method appears to be beneficial in several ways, one of which is the considerable improvement of drilling conditions [40]. In addition, a Vishay milling cutter (Vishay Precision Group, Malvern, PA, USA) was used for the drilling and producing holes with the final diameter of approximately 1.9 mm. The holes were drilled in incremental steps of 0.05 mm to a total depth of 1 mm. This study employed a three-element type-A [34] strain rosette gauge of FRS-2 (TML Lab. Company, Tokyo, Japan), paired with a multi-channel signal-acquisition device (P3 Strain Indicator and Recorder, Vishay Precision Group, Malvern, PA, USA) (with resolution of strain measurement equal to 10^{-6}). Four measurement points were located in the middle part of plate B3 and two on plate B4. The distance between the points varied from 30 to 70 mm. Additional residual stress estimation for the Zr 700 plate under the as-delivered condition was based on two measurement points.

2.4. Structural Properties

The process of explosive welding is characterized by the formation of an interface between the periodic deformation and the interfacial wave [3] (Figure 3a).

Figure 3. (**a**) Interfacial wave of plate B4, and (**b**) structural properties of the wavy interface: length of the welded line (L, wave height (H), wavelength (n), and melted area (P).

The geometrical parameters of the wave depend on the process parameters and physical properties of the welded materials [7,41–43]. These parameters could be correlated with mechanical properties of the composite structure [7,44,45]. For some welding systems, a melted area can be formed in the vortex of the collision zone. In this study, the following geometrical quantities were measured: wave height H, wavelength n, and melted area P along the length of welded line L. Figure 3b illustrates the measurement of the geometrical parameters of the samples cut out from both welded plates by using the digital optical method according. The total length of welded line L varies between 15 and 18 mm, including 8–12 points for wavelength and wave height determination for the Zr 700–Ti Gr. 1 interface and 20–37 points for the Ti Gr. 1–P265GH interface. Analyzed data were collected to calculate the mean and standard deviation for each parameter. Following a previous study [7], the authors calculated a parameter that describes the averaged amount of melted area: equivalent melted thickness, $EMT = P/L$.

In addition to the geometrical properties, microhardness distribution was included as a structural property. The microhardness distribution provides information on induced material hardening due to severe plastic deformation that occurred during the impact of two plates. The distribution of Vickers microhardness (HV) under the 50-G load was measured along three lines perpendicular to interfaces with the distance of 0.06 mm between points in the vicinity of the welded zone. The final results were obtained as the mean values of the three measurements (three lines) supplemented with error bars representing standard deviations. The HV hardness of the as-delivered plates was also measured for comparison.

2.5. Mechanical Test

The tensile test was designed to verify the strength of the cross-section perpendicular to the interfaces. Figure 4 presents the geometry of the specimen used for the tensile test.

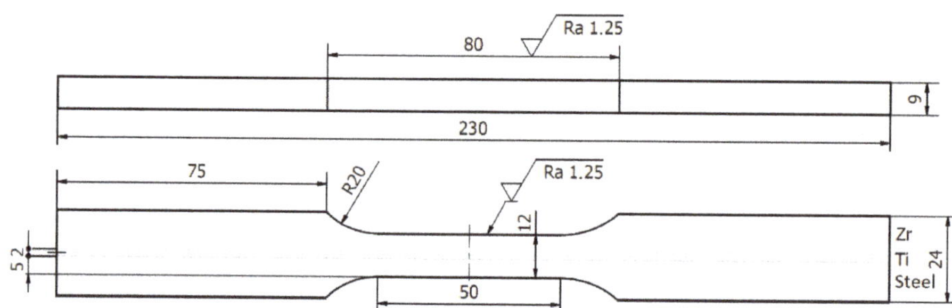

Figure 4. Geometry of specimen for the tensile test (units in mm).

The test facilitated the determination of force F_{p02} for the 0.2% offset of plastic strain, ultimate force F_m, and elongation to rupture A. The obtained values for plate B4 and its composite plates can be compared. These results can be verified using theoretical values calculated based on the mechanical properties of materials in the as-delivered condition.

3. Calculation, Results, and Discussion

3.1. Residual Stresses

The relieved strain components recorded during the incremental drilling were evaluated and recalculated with respect to residual principal stresses by using the Eval 7 software (Sint Technology). This software is fully compliant with the international ASTM E837-13a standard [34] for residual stress measurement using the hole-drilling strain-gage method. The software also allows evaluation of the uncertainty of stress calculation [46]. The following accuracies of material and drilling parameters were found: Young's modulus = ±3 GPa, Poisson ratio = ±0.01, strain = ±0.6 μm/m, strain gauge factor = ±1%, hole diameter = ±0.04 mm, and hole depth = ±0.025 mm. A uniform stress field was assumed throughout the calculations. The obtained maximum (σ_1) and minimum (σ_2) principal stresses may be considered as average values along the 1-mm hole depth, i.e., near the surface layer of Zr 700 plate. The results of the residual stresses determined in the Zr 700 layers of composite plates B3 and B4 as well as the reference stresses determined for the Zr 700 plate before welding are presented in Figure 5.

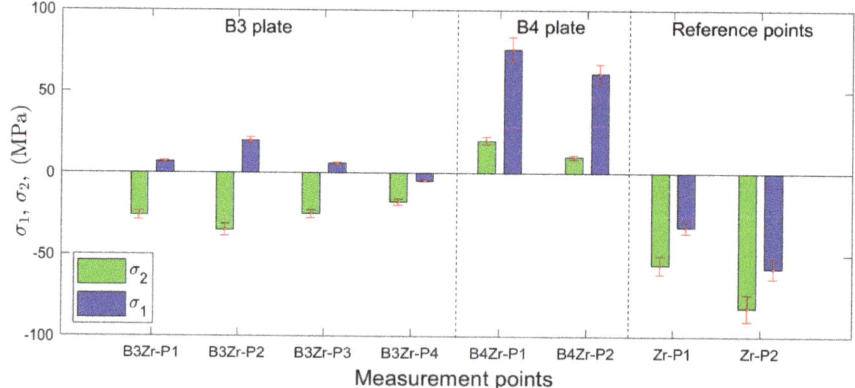

Figure 5. Residual stresses identified in the Zr 700 layer of B3 (points P1–P4), B4 (points P1 and P2), and Zr 700 plate before welding (reference points P1 and P2).

The principal residual stresses determined for the as-delivered condition in the Zr 700 plate represent the state of the residual stress before welding (reference points in Figure 5). This biaxial compressive stress state with component values between −82 ± 8 MPa and −33 ± 4 MPa was transformed into the biaxial tensile stress state (10 ± 1 MPa and 76 ± 8 MPa) for composite plate B4. In contrast, composite plate B3, produced with a lower impact velocity of $V_p = 425$ m/s as compared to plate B4 ($V_p = 468$ m/s), indicated lower values of residual stresses, ranging between −35 ± 4 MPa and 20 ± 2 MPa. The performed manufacturing processes had a considerable effect on the residual stress state. The initial compressive stress state, which was beneficial in terms of stress-based corrosion cracking, became more tensile with the increasing magnitude of impact velocity.

3.2. Structural Properties

Figure 6 presents the morphologies of the interfacial waves for composite plates B3 and B4. As shown, the interfacial waves of the Zr 700–Ti Gr. 1 joint in both the plates did not exhibit melted areas (*EMT* = 0 µm) but showed higher values for the wavelength and wave height than those observed for the Ti Gr. 1–P265GH joint. The measured parameters are presented in Figure 7. Compared to plate B3, the higher impact velocity (by 10%) for plate B4 resulted in higher wavelength (by 31%) and wave height (by 14%) at the Zr 700–Ti Gr. 1 interface. At the Ti Gr. 1–P265GH interface, the wave parameters were nearly equal (within the standard deviation band). The *EMT* values were low [45], at approximately 3 and 5 µm for plates B3 and B4, respectively.

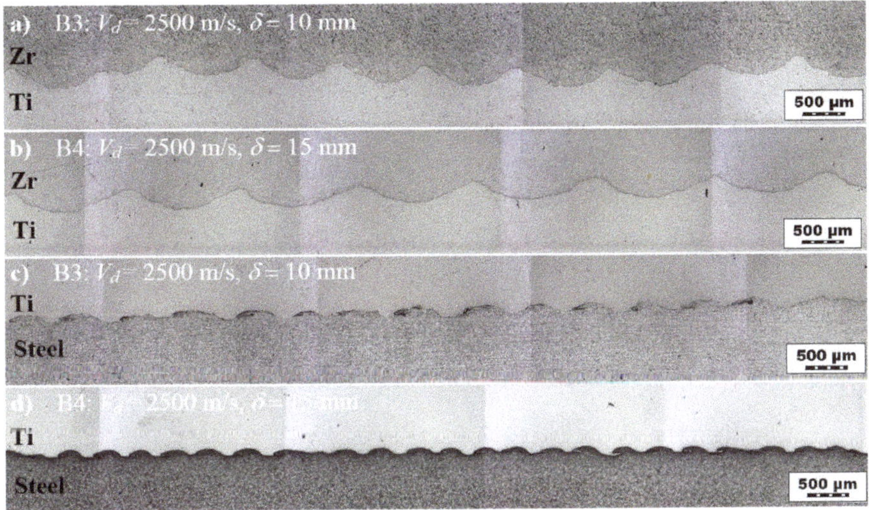

Figure 6. Morphology of interfacial waves for composite plates B3 and B4. Zr–Ti interface for (**a**) plate B3 and (**b**) plate B4. Ti–steel interface for (**c**) plate B3 and (**d**) plate B4.

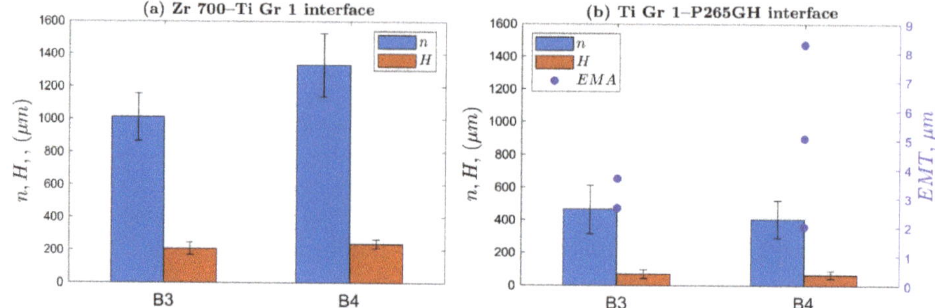

Figure 7. Wavelengths n and wave heights H for (**a**) the Zr 700–Ti Gr. 1 interface and (**b**) the Ti Gr. 1–P265GH interface with an equivalent thickness of melted area (*EMT*).

The melted areas at the Ti Gr. 1–P265GH interface were localized mainly in the vortex formed as a result of fluid–structure interaction (Figure 8.). The melted areas comprise a new phase with mixed chemical composition of Ti and Fe (Figure 9). In both plates B3 and B4, multiple microcracks (Figures 8 and 9a) were detected in the melted areas; these were probably formed owing to severe residual stresses generated during the rapid cooling rate (shrinkage cracks [14]). Furthermore, grain deformation was observed to intensify closer to the weld line.

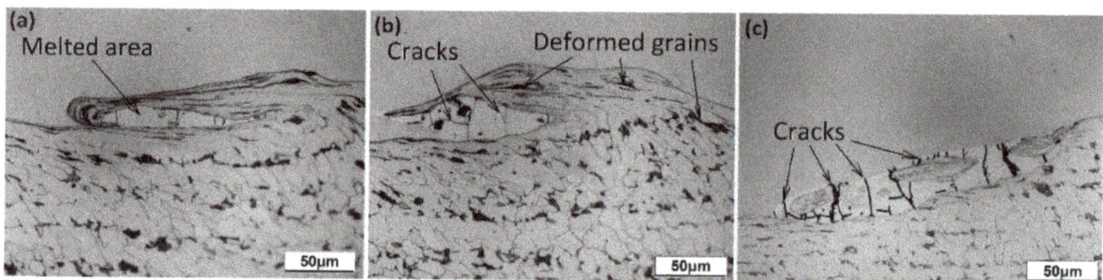

Figure 8. Microstructure with grain deformation and microcracks observed in melted areas of (**a**) plate B3 and (**b**,**c**) plate B4 in different locations.

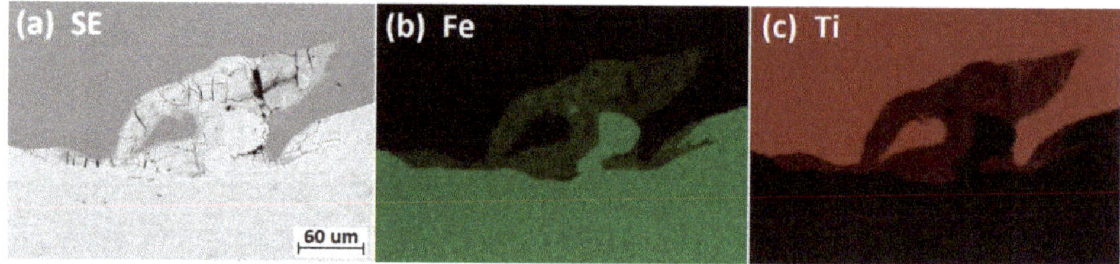

Figure 9. (**a**) SEM (Scanning Electron Microscope) and (**b**,**c**) EDX (Energy Dispersive X-Ray Analysis) maps showing the distribution of Fe and Ti at the Ti Gr. 1–P265GH interface for plate B4.

The microhardness distributions presented in Figure 10 exhibited an increase in the plate hardness in close vicinity of interfaces. In both plates B3 and B4, the distributions of microhardness were similar within the standard-deviation error bands. The highest value of microhardness of 250 $HV_{0.05}$ was detected in the Ti Gr 1–P265GH interface; it exceeded the hardness of steel by approximately 40%.

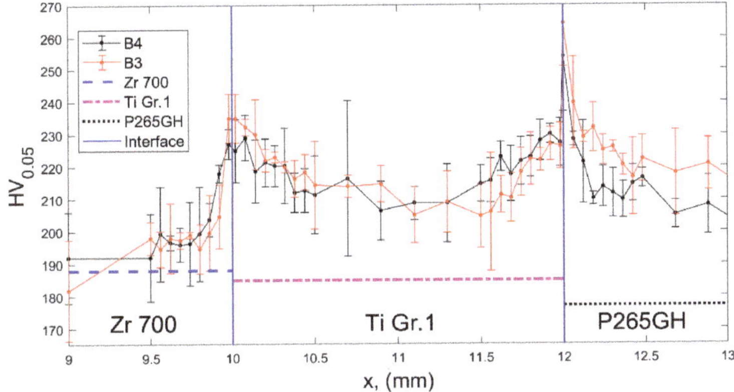

Figure 10. Vickers microhardness distribution across the composite plates, with horizontal lines representing hardness of materials in the as-delivered conditions.

3.3. Mechanical Test

Figure 11 presents the results of tensile tests represented as strain–force curves. The curves for the specimens made of composite plates B3 and B4 differ insignificantly.

The estimated curve parameters of yield force F_{p02}, ultimate force F_m, elongation A, and equivalent Young's modulus E_{eq} were calculated, as shown in Table 4. The equivalent Young's modulus was determined for the linear range of the curve as force F divided by total cross-section area A and strain ε, $E_{eq} = F/(A\varepsilon)$. Additionally, the fractures of both plates are presented in Figure 11.

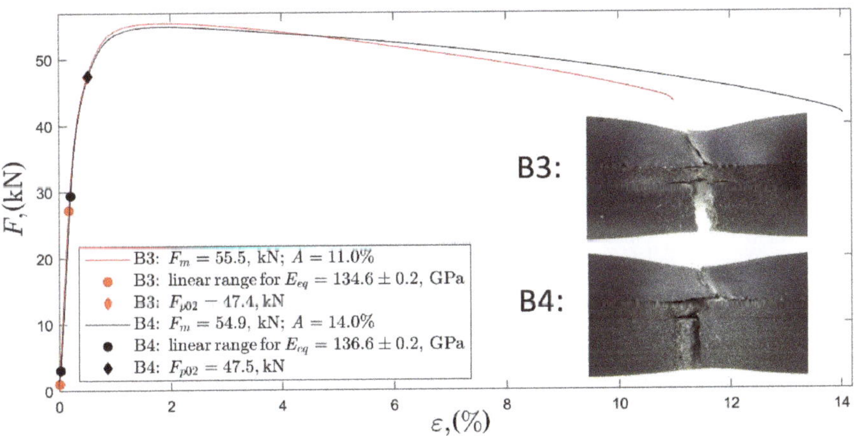

Figure 11. Strain–force curves of tensile test conducted on composite plates B3 and B4.

Table 4. Results of tensile tests on composite plates B3 and B4.

Plate	F_{p02}, (kN)	F_m, (kN)	A, (%)	E_{eq}, (GPa)
B3	47.4	55.5	11	134.6
B4	47.5	54.9	14	136.6

The strain–force curve was estimated theoretically to quantify the obtained results with respect to the possible increase of strength properties [47] in relation to material properties in the as-delivered condition. The following assumptions were made in the calculation model: (i) the cross-section of the specimen (Figure 4) comprised uniform strain distribution and (ii) all the layers comprised the uniaxial stress state. This model applies constitutive empirical relations of $\sigma^{Zr} - \varepsilon^{Zr}$, $\sigma^{Ti} - \varepsilon^{Ti}$, and $\sigma^{St} - \varepsilon^{St}$ of materials in the as-delivered condition (tensile tests). The recorded strain signal, ε, for the composite plate was used to estimate the stress in each layer through the following empirical constitutive relations: $\sigma^{Zr}\left(\varepsilon = \varepsilon^{Zr}\right)$, $\sigma^{Ti}\left(\varepsilon = \varepsilon^{Ti}\right)$, and $\sigma^{St}\left(\varepsilon = \varepsilon^{St}\right)$. The stresses σ^{Zr}, σ^{Ti}, and σ^{St} estimated through the force balance equation in each layer were used to determine force value F:

$$F = \sum_{i=1}^{3} \sigma^i A^i = \sigma^{Zr} A^{Zr} + \sigma^{Ti} A^{Ti} + \sigma^{St} A^{St}, \quad (1)$$

where A_{Zr}, A_{Ti}, and A_{St} are the initial cross-sectional areas of the Zr 700, Ti Gr 1, and P265GH layers, respectively. The empirical strain–stress curves for the material in the as-welded condition are presented in Figure 12a.

The model implemented the initial cross-sectional areas, and thus its applicability was limited to a small strain regime (assumed to be less than 1%). The slope of the calculated curve in the linear range (up to ~20 kN; Figure 12b) is in accordance with the slope of the experimental curves for both plates B3 and B4. Computed yield force $F_{p02} = 25.7$ kN was considerably lower than its empirical value of $F_{p02} = 47.5$ kN. The increase in the yield force by approximately 85% was due to material hardening in the vicinity of the interfaces. The research presented in [48–50] showed that the yield strength proportionally increased with the hardness of different types of steels and zirconium alloys [51]. The yield force F_{p02} of explosively welded plates is proposed to be predicted according to the proportional increase in yield stress R_{p02} as the function of hardness rate:

$$F_{p02} = \sum_{i=1}^{3} c^i r^i_{HV} R^i_{p02} A^i = c^{Zr} r^{Zr}_{HV} R^{Zr}_{p02} A^{Zr} + c^{Ti} r^{Ti}_{HV} R^{Ti}_{p02} A^{Ti} + c^{St} r^{St}_{HV} R^{St}_{p02} A^{St}, \quad (2)$$

where c^i represents the proportionality factors for each layer ($i = [Zr, Ti, St]$), and $r^i_{HV} = HV^i_h / HV^i_0$ represents the hardness rates (hardness of the hardened material HV^i_h divided by the hardness of the material in the initial state HV^i_0). The hardness of HV^i_h was calculated as the average hardness of each layer (5 mm + 2 mm + 5 mm). The calculated hardness rates were as follows: $r^{Zr}_{HV} = 0.97$, $r^{Ti}_{HV} = 1.16$, and $r^{St}_{HV} = 1.25$. The proportionality factors were assumed to be equal, i.e., $c^{Zr} = c^{Ti} = c^{St} = c$. Proportionality factor $c = 1.58$ was identified by fitting the experimental and calculated (Equation (2)) yield forces.

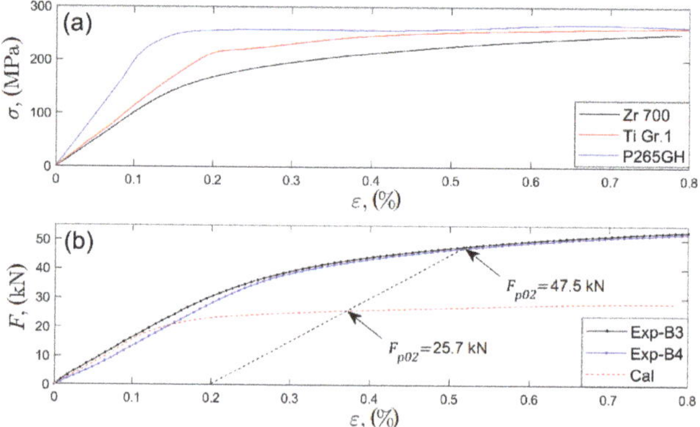

Figure 12. (**a**) Recorded strain stress curves for materials in as-delivered condition. (**b**) Experimental force–strain curves for plates B3 and B4 (Exp-B3 and Exp-B4) and the calculated curve.

The process of explosive welding includes problems of phase transformation, shock impact, plastic deformation, and fluid–structure interaction. A thorough analysis of the process requires sophisticated multiphysics numerical modeling [52–55], which is nowadays narrowed to the prediction of jet formation, morphology of the interfacial wave, and weldability. Residual stress formation has not yet been numerically simulated. However, based on the knowledge of the explosive welding process, a general scheme of residual stress generation can be proposed. During the shock of impact of the Zr 700 plate with the Ti Gr. 1 layer, a large hydrostatic compressive stress due to inertial forces could be formed in the contact region. The increase of local temperature reduces yield stress of both materials in contact, allowing the occurrence of plastic deformation, which is sometimes manifested by the adiabatic shear bands [8,56,57]. More intensive, larger, and plastically deformed regions appear in a material with lower resistance to plastic yield. In most cases, the material with a lower yield strength and higher elongation is selected as the flyer. In this case, larger and more intensive plastic deformations occur in the flyer region in the vicinity of impact, and more elastic potential energy is stored in the base plate. When the pressure of explosive gas is released, the elastically deformed base plate below the interface expands and compresses the plastically deformed layer of the flyer plate. The induced compressive residual stresses in the vicinity of the interface must be balanced by the tensile residual stresses formed in the outer layer of the flyer plate (as a result of the spring-back effect [3]). The locally melted areas solidify and form a new phase; these areas are the sources of thermal residual stresses, which deviate from the general field of stresses described. The higher the differences in Young's modulus and thermal expansion coefficients, the higher the residual stresses. Stress relief with respect to heat treatment is possible only for materials with close values of thermal expansion coefficients.

In this study, the difference in the impact velocities by approximately 10% resulted in insignificant differences in microhardness distribution and tensile yield force of specimens with reduced thickness (5 mm + 2 mm + 5 mm). In addition, a notable increase of the wavelength by approximately 31% was observed. In contrast to most structural properties, the residual stresses in a 1-mm-thick Zr 700 surface layer exhibited profound sensitivity to the applied impact velocities. The initial compressive residual stress state was significantly transformed towards the tensile type, reaching 76 ± 8 MPa for higher impact velocities.

4. Conclusions

The main conclusions of the study are summarized as follows:

- The compressive residual stress, which was initially present in the Zr 700 flyer plate, decreased in the explosive welding process, resulting in a tensile type with an increase in impact velocity.
- To protect the composite plate from stress-based corrosion cracking, a lower value of the impact velocity is recommended.
- The experimental yield force of composite specimens is around 85% higher than the yield force of combined properties of materials in the as-delivered condition.
- The experimentally estimated residual stresses could be used to verify the numerical method applied in modeling of the explosive welding process.
- In addition, a simple model based on microhardness measurement for yield force prediction of the composite plate was proposed. However, the model needs further verification.

Author Contributions: Conceptualization, Writing, Original Draft, Methodology, A.K.; Investigation, K.K.; Investigation, Data Curation, S.D.; Investigation, M.P.; Writing—Review and Editing, Funding acquisition, H.P. All authors have read and agreed to the published version of the manuscript.

Funding: This study was financed by the National Centre for Research and Development, Poland, contract no Techmastrateg2/412341/8/NCBR/2019.

Acknowledgments: The authors acknowledge the support of Z. Szulc and High Energy Technology Works 'Explomet' (Opole, Poland) for delivery of the composite plates.

Conflicts of Interest: The Authors declare no conflict of interest.

References

1. Wang, H.; Wang, Y. High-Velocity Impact Welding Process: A Review. *Metals* **2019**, *9*, 144. [CrossRef]
2. Findik, F. Recent developments in explosive welding. *Mater. Des.* **2011**, *32*, 1081–1093. [CrossRef]
3. Blazynski, T.Z. *Explosive Welding, Forming and Compaction*; Blazynski, T.Z., Ed.; Applied Science Publishers Ltd.: London, UK; New York, NY, USA, 1983; ISBN 9789401197533.
4. Kaya, Y. Microstructural, Mechanical and Corrosion Investigations of Ship Steel-Aluminum Bimetal Composites Produced by Explosive Welding. *Metals* **2018**, *8*, 544. [CrossRef]
5. Wachowski, M.; Kosturek, R.; Śnieżek, L.; Mróz, S.; Stefanik, A.; Szota, P. The Effect of Post-Weld Hot-Rolling on the Properties of Explosively Welded Mg/Al/Ti Multilayer Composite. *Materials* **2020**, *13*, 1930. [CrossRef] [PubMed]
6. Kaya, Y. Investigation of Copper-Aluminium Composite Materials Produced by Explosive Welding. *Metals* **2018**, *8*, 780. [CrossRef]
7. Paul, H.; Skuza, W.; Chulist, R.; Miszczyk, M.; Gałka, A.; Prażmowski, M.; Pstruś, J. The Effect of Interface Morphology on the Electro-Mechanical Properties of Ti/Cu Clad Composites Produced by Explosive Welding. *Met. Mater. Trans. A* **2019**, *51*, 750–766. [CrossRef]
8. Wachowski, M.; Gloc, M.; Ślęzak, T.; Plocinski, T.; Kurzydłowski, K.J. The Effect of Heat Treatment on the Microstructure and Properties of Explosively Welded Titanium-Steel Plates. *J. Mater. Eng. Perform.* **2017**, *26*, 945–954. [CrossRef]
9. Kim, J.-G.; Park, J.; An, B.-S.; Lee, Y.H.; Yang, C.-W.; Kim, J.-G. Investigation of Zirconium Effect on the Corrosion Resistance of Aluminum Alloy Using Electrochemical Methods and Numerical Simulation in an Acidified Synthetic Sea Salt Solution. *Materials* **2018**, *11*, 1982. [CrossRef]
10. Banker, J.G.; Barberis, P.; Dean, S.W. Explosion Cladding: An Enabling Technology for Zirconium in the Chemical Process Industry. *J. ASTM Int.* **2010**, *7*, 103050. [CrossRef]
11. Zhou, Q.; Feng, J.; Chen, P. Numerical and Experimental Studies on the Explosive Welding of Tungsten Foil to Copper. *Materials* **2017**, *10*, 984. [CrossRef]
12. Samardžić, I.; Kožuh, Z.; Mateša, B. Structural analysis of three-metal explosion joint: Zirconium-titanium-steel. *Metalurgija* **2010**, *49*, 119–122.
13. Kalavathi, V.; Bhuyan, R.K. A detailed study on zirconium and its applications in manufacturing process with combinations of other metals, oxides and alloys—A review. *Mater. Today Proc.* **2019**, *19*, 781–786. [CrossRef]
14. Prasanthi, T.; Sudha, R.C.; Saroja, S. Explosive cladding and post-weld heat treatment of mild steel and titanium. *Mater. Des.* **2016**, *93*, 180–193. [CrossRef]
15. Manikandan, P.; Hokamoto, K.; Raghukandan, K.; Chiba, A.; Deribas, A.A. The effect of experimental parameters on the explosive welding of Ti and stainless steel. *Sci. Technol. Energ. Mater.* **2005**, *66*, 370–374.
16. Karolczuk, A.; Kowalski, M.; Kluger, K.; Zok, F. Identification of Residual Stress Phenomena Based on the Hole Drilling Method in Explosively Welded Steel-Titanium Composite. *Arch. Met. Mater.* **2014**, *59*, 1119–1123. [CrossRef]
17. Karolczuk, A.; Paul, H.; Szulc, Z.; Kluger, K.; Najwer, M.; Kwiatkowski, G. Residual Stresses in Explosively Welded Plates Made of Titanium Grade 12 and Steel with Interlayer. *J. Mater. Eng. Perform.* **2018**, *27*, 4571–4581. [CrossRef]
18. Lavigne, O.; Kotousov, A.; Luzin, V. Microstructural, Mechanical, Texture and Residual Stress Characterizations of X52 Pipeline Steel. *Metals* **2017**, *7*, 306. [CrossRef]
19. Ghosh, S.; Rana, V.P.S.; Kain, V.; Mittal, V.; Baveja, S. Role of residual stresses induced by industrial fabrication on stress corrosion cracking susceptibility of austenitic stainless steel. *Mater. Des.* **2011**, *32*, 3823–3831. [CrossRef]
20. Wu, H.; Li, C.; Fang, K.; Zhang, W.; Xue, F.; Zhang, G.; Wang, X. Effect of residual stress on the stress corrosion cracking in boiling magnesium chloride solution of austenite stainless steel. *Mater. Corros.* **2018**, *69*, 1572–1583. [CrossRef]
21. Nagano, H.; Kajimura, H.; Yamanaka, K. Corrosion resistance of zirconium and zirconium-titanium alloy in hot nitric acid. *Mater. Sci. Eng. A* **1995**, *198*, 127–134. [CrossRef]

22. Yamamoto, M.; Kato, C.; Ishijima, Y.; Kano, Y.; Ebina, T. Stress Corrosion Cracking of Zirconium in Boiling Nitric Acid Solution at Potentiostatic Condition. *ECS Trans.* **2009**, *16*, 101–108. [CrossRef]
23. Ishijima, Y.; Kato, C.; Motooka, T.; Yamamoto, M.; Kano, Y.; Ebina, T. Stress Corrosion Cracking Behavior of Zirconium in Boiling Nitric Acid Solutions at Oxide Formation Potentials. *Mater. Trans.* **2013**, *54*, 1001–1005. [CrossRef]
24. Beavers, J.A.; Griess, J.C.; Boyd, W. Stress corrosion cracking of zirconium in hot nitric acid. *Corrosion* **1981**, *37*, 292–297. [CrossRef]
25. Cox, B. Environmentally-induced cracking of zirconium alloys—A review. *J. Nucl. Mater.* **1990**, *170*, 1–23. [CrossRef]
26. Sedighi, M.; Honarpisheh, M. Experimental study of through-depth residual stress in explosive welded Al–Cu–Al multilayer. *Mater. Des.* **2012**, *37*, 577–581. [CrossRef]
27. Sedighi, M.; Honarpisheh, M. Investigation of cold rolling influence on near surface residual stress distrubution in explosive welded multilayer. *Strength Mater.* **2012**, *44*, 693–698. [CrossRef]
28. Yasheng, W.; Hongneng, C.; Ninxu, M. Measurement of residual stresses in a multi-layer explosive welded joint with successive milling technique. *Strain* **1999**, *35*, 7–10. [CrossRef]
29. Varavallo, R.; Moreira, V.D.M.; Paes, V.; Brito, P.; Olivas, J.; Pinto, H.C. Residual Stresses of Explosion Cladded Composite Plates of ZERON 100 Superduplex Stainless Steel and ASTM SA516-70 Carbon Steel. *Adv. Mater. Res.* **2014**, *996*, 500–505. [CrossRef]
30. Mateša, B.; Kozak, D.; Stoić, I.; Samardžić, I. The influence of heat treatment by annealing on clad plates residual stresses. *Metalurgija* **2012**, *51*, 229–232.
31. Karolczuk, A.; Kluger, K.; Kowalski, M.; Żok, F.; Robak, G. Residual Stresses in Steel-Titanium Composite Manufactured by Explosive Welding. *Mater. Sci. Forum* **2012**, *726*, 125–132. [CrossRef]
32. Taran, Y.; Balagurov, A.M.; Sabirov, B.M.; Evans, A.; Davydov, V.; Venter, A.M. Residual stresses in a stainless steel—titanium alloy joint made with the explosive technique. *J. Phys. Conf. Ser.* **2012**, *340*, 012105. [CrossRef]
33. Mateša, B.; Kožuh, Z.; Dunđer, M.; Samardzic, I. Determination of clad plates residual stresses by x-ray diffraction method. *Teh. Vjesn. Tech. Gaz.* **2015**, *22*, 1533–1538. [CrossRef]
34. ASTM E837—13a Standard Test. *Method for Determining Residual Stresses by the Hole-Drilling Strain-Gage Method*; ASTM International: West Conshohocken, PA, USA, 2013.
35. Prazmowski, M.; Paul, H. Experimental investigations of the bonding zone in the explosive welding of a differently structured steel-zirconium platers. *J. Mach. Eng.* **2019**, *19*, 99–110. [CrossRef]
36. Tete, A.D.; Deshmukh, A.; Yerpude, R. Velocity of detonation (VOD) measurement techniques—Practical approach. *Int. J. Eng. Technol.* **2013**, *2*, 259–265. [CrossRef]
37. Deribas, A.A. Acceleration of metal plates by a tangential detonation wave. *J. Appl. Mech. Tech. Phys.* **2000**, *41*, 824–830. [CrossRef]
38. Manikandan, P.; Hokamoto, K.; Deribas, A.A.; Raghukandan, K.; Tomoshige, R. Explosive Welding of Titanium/Stainless Steel by Controlling Energetic Conditions. *Mater. Trans.* **2006**, *47*, 2049–2055. [CrossRef]
39. ASTM International. *A578M-17 AA Standard Specification for Straight-Beam Ultrasonic Examination of Rolled Steel Plates*; ASTM International: West Conshohocken, PA, USA, 2010; Volume 1, pp. 1–5.
40. Nau, A.; Scholtes, B. Evaluation of the High-Speed Drilling Technique for the Incremental Hole-Drilling Method. *Exp. Mech.* **2012**, *53*, 531–542. [CrossRef]
41. Sun, Z.; Shi, C.; Xu, F.; Feng, K.; Zhou, C.; Wu, X. Detonation process analysis and interface morphology distribution of double vertical explosive welding by SPH 2D/3D numerical simulation and experiment. *Mater. Des.* **2020**, *191*, 108630. [CrossRef]
42. Sun, Z.; Shi, C.; Fang, Z.; Shi, H. A dynamic study of effect of multiple parameters on interface characteristic in double-vertical explosive welding. *Mater. Res. Express* **2020**, *7*, 016541. [CrossRef]
43. Greenberg, B.A.; Ivanov, M.A.; Inozemtsev, A.V.; Pushkin, M.S.; Patselov, A.M.; Besshaposhnikov, Y.R. Comparative characterisation of interfaces for two- and multi-layered Cu-Ta explosively welded composites. *Compos. Interfaces* **2019**, *27*, 705–715. [CrossRef]
44. Karolczuk, A.; Kowalski, M.; Bański, R.; Żok, F. Fatigue phenomena in explosively welded steel–titanium clad components subjected to push–pull loading. *Int. J. Fatigue* **2013**, *48*, 101–108. [CrossRef]
45. Prażmowski, M.; Paul, H.; Zok, F. The Effect of Heat Treatment on the Properties of Zirconium—Carbon Steel Bimetal Produced By Explosion Welding. *Arch. Met. Mater.* **2014**, *59*, 1143–1149. [CrossRef]

46. Scafidi, M.; Valentini, E.; Zuccarello, B. Error and Uncertainty Analysis of the Residual Stresses Computed by Using the Hole Drilling Method. *Strain* **2010**, *47*, 301–312. [CrossRef]
47. Boroński, D.; Kotyk, M.; Maćkowiak, P.; Śnieżek, L. Mechanical properties of explosively welded AA2519-AA1050-Ti6Al4V layered material at ambient and cryogenic conditions. *Mater. Des.* **2017**, *133*, 390–403. [CrossRef]
48. Pavlina, E.J.; Van Tyne, C.; Van Tyne, C. Correlation of Yield Strength and Tensile Strength with Hardness for Steels. *J. Mater. Eng. Perform.* **2008**, *17*, 888–893. [CrossRef]
49. Busby, J.T.; Hash, M.C.; Was, G.S. The relationship between hardness and yield stress in irradiated austenitic and ferritic steels. *J. Nucl. Mater.* **2005**, *336*, 267–278. [CrossRef]
50. Cahoon, J.R.; Broughton, W.H.; Kutzak, A.R. The determination of yield strength from hardness measurements. *Met. Trans.* **1971**, *2*, 1979–1983.
51. Lodh, A.; Pant, P.; Kumar, G.; Krishna, K.V.M.; Tewari, R.; Samajdar, I. Orientation-dependent solid solution strengthening in zirconium: A nanoindentation study. *J. Mater. Sci.* **2019**, *55*, 4493–4503. [CrossRef]
52. Bataev, A.; Tanaka, S.; Zhou, Q.; Lazurenko, D.V.; Jorge, A.M.; Bataev, A.; Hokamoto, K.; Mori, A.; Chen, P. Towards better understanding of explosive welding by combination of numerical simulation and experimental study. *Mater. Des.* **2019**, *169*, 107649. [CrossRef]
53. Zhang, Z.L.; Feng, D.L.; Liu, M. Investigation of explosive welding through whole process modeling using a density adaptive SPH method. *J. Manuf. Process.* **2018**, *35*, 169–189. [CrossRef]
54. Feng, J.; Chen, P.; Zhou, Q.; Dai, K.; An, E.; Yuan, Y. Numerical simulation of explosive welding using Smoothed Particle Hydrodynamics method. *Int. J. Multiphys.* **2017**, *11*, 315–325. [CrossRef]
55. Mahmood, Y.; Dai, K.; Chen, P.; Zhou, Q.; Bhatti, A.; Arab, A. Experimental and Numerical Study on Microstructure and Mechanical Properties of Ti-6Al-4V/Al-1060 Explosive Welding. *Metals* **2019**, *9*, 1189. [CrossRef]
56. Yang, Y.; Wang, B.; Hu, B.; Hu, K.; Li, Z. The collective behavior and spacing of adiabatic shear bands in the explosive cladding plate interface. *Mater. Sci. Eng. A* **2005**, *398*, 291–296. [CrossRef]
57. Gloc, M.; Wachowski, M.; Plocinski, T.; Kurzydlowski, K.J. Microstructural and microanalysis investigations of bond titanium grade1/low alloy steel st52-3N obtained by explosive welding. *J. Alloy. Compd.* **2016**, *671*, 446–451. [CrossRef]

Article

Numerical and Metallurgical Analysis of Laser Welded, Sealed Lap Joints of S355J2 and 316L Steels under Different Configurations

Hubert Danielewski [1,*], Andrzej Skrzypczyk [1], Marek Hebda [2], Szymon Tofil [1], Grzegorz Witkowski [1], Piotr Długosz [3] and Rastislav Nigrovič [4]

1 Faculty of Mechatronics and Mechanical Engineering, Kielce University of Technology, Tysiąclecia Państwa Polskiego 7, 25-314 Kielce, Poland; tmaask@tu.kielce.pl (A.S.); tofil@tu.kielce.pl (S.T.); gwitkowski@tu.kielce.pl (G.W.)
2 Faculty of Materials Engineering and Physics, Cracow University of Technology, Warszawska 24, 31-155 Cracow, Poland; mhebda@pk.edu.pl
3 Lukasiewicz Network-Cracow Institute of Technology, Zakopiańska 73, 30-418 Cracow, Poland; piotr.dlugosz@kit.lukasiewicz.gov.pl
4 Faculty of Mechanical Engineering, ŽilinskáUniverzita v ŽilineUniverzitná 1, 010 26 Žilina, Slovakia; rastislav.nigrovic@snop.eu
* Correspondence: hdanielewski@tu.kielce.pl

Received: 18 November 2020; Accepted: 16 December 2020; Published: 20 December 2020

Abstract: This paper presents the results of laser welding of dissimilar joints, where low-carbon and stainless steels were welded inthe lap joint configuration. Performed welding of austenitic and ferritic-pearlitic steels included a sealed joint, where only partial penetration of lower material was obtained. The authors presented acomparative study of the joints under different configurations. The welding parameters for the assumed penetration were estimated via anumericalsimulation. Moreover, a stress–strain analysis was performed based on theestablished model. Numerical analysis showed significant differences in joint properties, therefore, further study was conducted. Investigation of the fusion mechanism in the obtained joints wascarried out using electron dispersive spectroscopy (EDS) and metallurgical analysis. The study of the lap joint under different configurations showed considerable dissimilarities in stress–strain distribution and relevant differences in the fusion zone structure. The results showed advantages of using stainless steel as the upper material of a microstructure, and uniform chemical element distribution and stress analysis is considered.

Keywords: laser beam welding; sealed lap joints of dissimilar materials; austenitic and ferritic-pearlitic steels; numerical simulation; microstructure analysis

1. Introduction

A number of analytical, numerical, and empirical studies have shown the potential of laser beam welding (LBW). First commercially applied in 1970 [1–3], LBW has become one of the most widely studied techniques of joining metallic parts.A focused laser beam causes rapid vaporization and ionizationof the metal. The keyhole effectallows lap welding [4,5]. The weld bead geometry and the joint properties are related to the process parameters and the properties of the welded materials [6,7]. Some process parameters depend on the laser type used (wavelength, transverse mode, pulse/continuous-wave operating mode), the laser machine system (spot size, single or multi-spot optics, focal length), and the programmed parameters (emission frequency, output power, welding speed, focal point position). The material properties, such as thermal conductivity, specific heat capacity, latent heat, and thermal diffusivity, affect the welding dynamics andoutcome [8]. The laser beam penetration is related to the surface reflectivity and the ionization effect of the metalvapor, therefore, the intensity of plasma generation varies for different materials [9]. The beam penetration through two materials is a complex phenomenon, moreover, in dissimilar lap joints, the results of the welding process depend on the welded material configuration.

Numerous studies of laser welding technology, including lap welding, are being carried out. Many of those studies focus on joining special application materials such as titanium, aluminum and nickel-based alloys, zinc-coated steels, or a combination of these materials [10,11]. However, as theresearchersfocus on one configuration type, there is a lack of publications showinghow joint properties change under different configurations. Therefore, the authors presented acomparative study of sealed lap joints with partial penetration, where low-carbon and stainless steels are welded alternately using a laser beam [12]. The joints are intended for use in pipeline components and large crude oil storage tanks, where high joint quality and strength are critical characteristics [13,14].

Nowadays, the development of computing power makes it possible to perform advanced calculations of welding processes, based on the finite element method (FEM) and computer-aided design (CAD) geometry, divided into finite elements (FEs). By solving the heat transfer problem, using differential equations, many physical phenomena, including phase transformation and different heat transfer mechanisms, can be taken into account [15–17]. Nevertheless, in numerical simulations, some parameters cannot be used in the model directly as input data, and some simplificationsare necessary [18]. Simple numerical models of laser welding provide a thermal solution where the fusion zone (FZ) and the heat-affected zone (HAZ) dimensions can becalculated. Nevertheless, a more advanced thermo-mechanical solution, a stress–strain analysis, can be carried out. The simulation of a lap joint is complex, and the complexity of the process increases when materials with different thermophysical material properties are to be joined. Realistic results can be obtained by developing an accurate heat source (HS) model andadequate boundary conditions [19,20].

The welding of dissimilar materials is problematic, especially when a sealed lap joint is considered. Many publications report the study of joint properties based on process parameters, while neglecting other aspects [21]. In this paper, the authors proposed a new approach to the problem, by analyzing the joint depending on the configuration of welded materials. On the basis of a numerical simulation, welding parameters were estimated, and stress–strain analysis was performed. The structure of the welds wasstudiedusing metallographic and electron dispersive spectroscopy (EDS) analyses. The researchshowedadvantages and disadvantages of austenitic and ferritic-pearlitic steels welded under different configurations ina sealed lap joint.

2. Material and Experimental Design

2.1. Materials

The materials used in the experiment were 4 mm thick steel sheets in grades S355J2 (according to EN 10025-2 [22]) and 316L (according to ASTM A240 [23]), with dimensions of 50 mm × 20 mm. The S355J2 steel is a commonly used, unalloyed, low-carbon construction steel with a ferritic-pearlitic structure. The other material is austenitic stainless steel 316L, with a high content of chromium and nickel. Both steels are characterized as materials of good weld ability, however, the differences in the chemical composition (specified in Table 1) affect the welding process and the properties of the joints.

Table 1. Chemical composition of materials used.

Material	Elements (wt %)										
	C	Mn	Si	P	S	Cr	Ni	Al	Fe	Cu	CEV
S355J2	0.17	1.6	0.02	0.017	0.011	0.02	-	0.05	98.1	0.06	0.45%
316L	0.018	1.57	0.48	0.04	0.002	16.7	11.2	-	balance	-	-

The materials used have different thermophysical properties that are not constant and change with temperature (Figures 1 and 2). This phenomenon affects the laser beam absorption, which increases with material temperature during welding and plays a significant role in the process dynamics. Figures 1 and 2 show the material database with the modified specific heat used to model the latent heat effect, calculated using JMatPro (Sentesoftware, Guildford, Surrey, UK) included in the Simufact Welding 8 software library (MSC Software Company, Hamburg Germany).

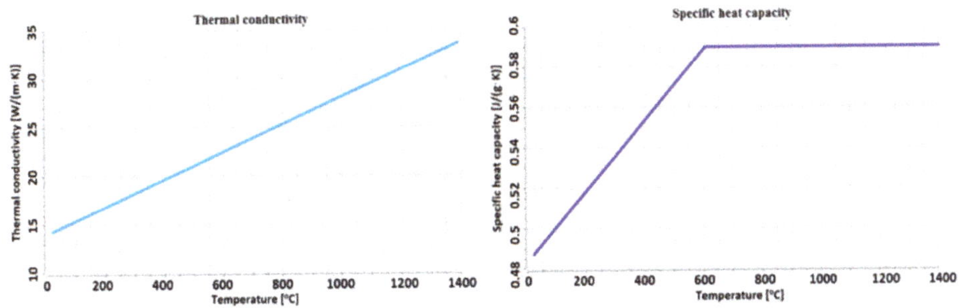

Figure 1. The temperature-dependent thermo-physical properties of 316L steel.

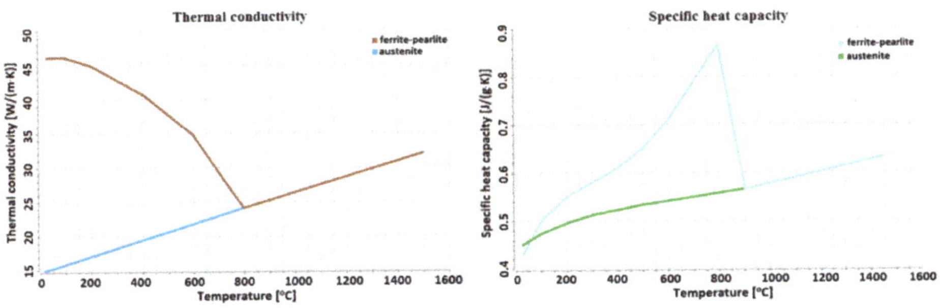

Figure 2. The temperature-dependent thermophysical properties of S355J2 steel.

The thermophysical properties for the abovementioned materials have a dissimilar range. For austenitic steel, the thermal conductivity uniformly increases in the range of 14 to 34 W/(m·K). Moreover, the specific heat capacity has linear characteristics in the range of 0.49 to 0.59 J/(g·K), and from 600 °C it is linearhorizontal. However, for low carbon steel, these properties are related to the material phase, and for the austenitic phase, the linear dependence of thermal conductivity in the range from 15 to 33 W/(m·K) can be observed. On the other hand, for the ferrite and pearlite phase, the thermal conductivity decreases from 46 W/(m·K), until it reaches 800 °C, where it overlaps within austenite conductivity. In S355J2 steel, for the austenitic phase, specific heat capacity increases from 0.3 to 0.62 J/(g·K), with a characteristic similar to linear for the ferrite and pearlite phases, and specific heat capacity rapidly increases to 800 °C, reaching the value of 0.86 J/(g·K), and then rapidly decreases; from 900 °C, it overlaps within the austenite value.

2.2. Numerical Simulation Procedure

For a programming simulation of the welding process, particularly LBW, calibration of the HS model is required and, therefore, at the preliminary stage, by comparing the results of trial welding with the simulation, an accurate model was obtained [24]. Softwarecommonly used for welding simulationsare based on solving heat-mass flow phenomena (ANSYS with the Fluent module (Ansys Inc., Southpointe, Canonsburg, PA, USA), software with an additional welding module (Abaqus), and software dedicated towelding applications, such as SYSWELD and Simufact Welding [25,26]. From the aforementioned software, Simufact Welding was used for estimating welding parameters, and performing stress and strain analysis. The selected program, based on Marc solver (MSC Software

Company, Hamburg Germany), is software dedicated to welding applications, and provides sufficient accuracy of results with a relatively quick calculation time. Simulations of conventional arc welding use the double-ellipsoid Goldak model, however, simulations of laser welding are based on volumetric heat sources (conical and cylindrical), with uneven power distribution (Gaussian parameter). Moreover, some methods use a combination of HSs, with a conical (1) source for simulating the keyhole effect, and a disc-shaped source for simulating laser beam absorption by the material surface (Figure 3) [27,28].

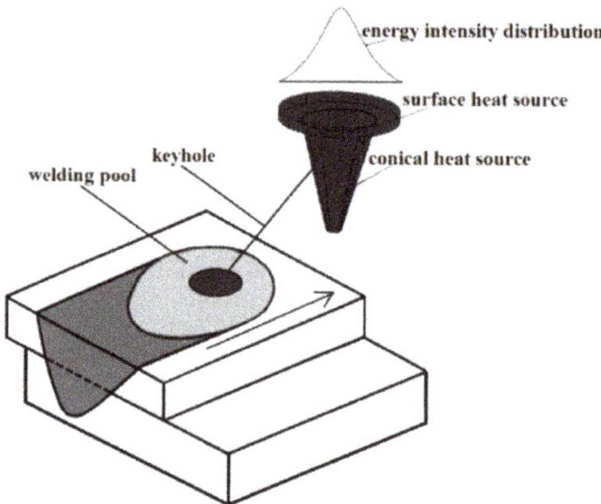

Figure 3. Heatsource model of laser lap joint welding.

Conical volumetric heat source with the Gaussian distribution can be described by the following equation:

$$Q(x,y,z) = Q_0 \exp\left(-\frac{x^2+y^2}{\left(r_i + \frac{r_t - r_i}{z_t - z_i}(z - z_i)\right)^2}\right) \tag{1}$$

where Q_0—the maximum heat flux density in a volumetric heat source, $r_t - r_i$—the dimensions of the upper and lower conical radius, $z_t - z_i$—the depth of the conical heat source, x, y, z—the heat source coordinates.

Solving the governing heat equation based on Fourier's law for three-dimensional heat conduction (2), witha partial differential equation in a nonlinear form, is done with the following equation:

$$\rho c(T)\frac{\partial T}{\partial t} = \frac{\partial}{\partial x}\left(k(T)\frac{\partial T}{\partial x}\right) + \frac{\partial}{\partial y}\left(k(T)\frac{\partial T}{\partial y}\right) + \frac{\partial}{\partial z}\left(k(T)\frac{\partial T}{\partial z}\right) + q_v \tag{2}$$

where *c(T)*—temperature-dependent specific heat capacity; *k(T)*—temperature-dependent thermal conductivity; q_v—volumetric internal energy; x, y, z—space coordinates; T—temperature; ρ —density; and t—time.

The conical heat sourcecan be described as follows:

$$q_l(x,\ y,\ z) = \frac{9\eta_l P_l e^3}{\pi(e^3-1)(z_t - z_i)\left(r_t^2 + r_t r_i + r_i^2\right)} \exp\left(-\frac{3\left[(x - vt)^2 + y^2\right]}{\left(r_t - (r_t - r_i)\frac{z_t - z}{z_t - z_i}\right)}\right) \tag{3}$$

where q_l—heat flux, $z_t - z_i$ —z coordinates (heat source depth), $r_t - r_i$—upper and lower conical radius, e—natural logarithm, $r_t - (r_t - r_i)\frac{z_t - z}{z_t - z_i}$—linear decrease in distribution along theconical heat source, P_l—laser power, η_l—laser heat source efficiency.

Thermal conductivity, specific heat, and emissivity in the heat transfer analysis depend on the temperature, however, the used model is based on the assumption that mass density is constant.

By an extrapolating or interpolating procedure, the temperature-dependent properties are averaged. Latent heat is related to phase transformation from solid to liquid metal, or vice versa. Phase change modeling is a complex process, however, using simplification, where latent heat is uniformly released in the solid–liquid range, it can be calculated [29,30].

Due to the convection–diffusion effect involved, the Petro–Galerkin model with nodal velocity vectors was used, which can be described as follows:

$$\frac{\partial T}{\partial t} + v \cdot \nabla T = \nabla \cdot (\kappa \nabla T) + Q \tag{4}$$

where v—nodal velocity vector, T—temperature, κ—diffusion tensor, Q—source term.

The surface energy is determined by calculating the thermal energy that affects the boundary conditions, including thermochemical ablation. These phenomena affect the convective heat flux, and the mass and the enthalpy flow are related to molecular diffusion. The surface energy is correlated with the heat transfer to the material by conduction through a heat-mass flow towards the surface as a result of the evaporation of the material.

Specimens with a size of 50 mm × 20 mm were used for the FE model. Three-dimensional solid hexahedral finite elements were used while meshing. The sizes of the elements were determined by adjusting the resolution and accuracy of the temperature distribution in the regions of severe thermal gradients. A mesh convergence study was performed, within FE sizes of 1.0000, 0.7500, 0.5000, 0.2500, 0.1250, and 0.0625 mm. For 0.0625 and 0.1250 mm, no relevant differences in weld geometry and heat distribution were observed, however, there were some differences between 0.1250 and 0.2500 mm. Therefore, the nominal FE size was set as 0.2500 mm, and in an area where a temperature exceeding 400 °C may occur (Figure 4, region 4), an FE refinement was performed, and the FE sizewas set as 0.1250 mm. Two 4 mm thick sheets (Figure 4, elements 1 and 2) were oriented in a lap configuration and fixed in three-dimensional space by additional elements (Figure 4, element 3). The welding trajectory was set in the center of the refinement area (9.5 mm from the sheet edge).

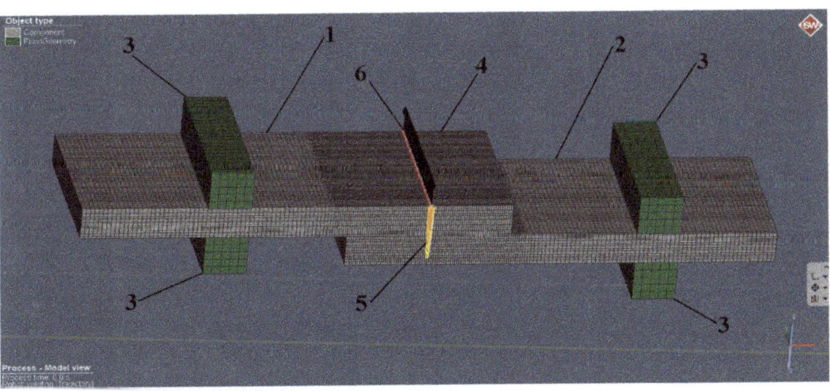

Figure 4. Laser welding finite element (FE) model with defined elements: 1—topsheet 2—bottom sheet, 3—fixed bearings,4—refined FE area, 5—heat sources, 6—welding trajectory.

The geometries of the heat sources (Figure 3) (3), are related to the used welding optics. In this case, the focal length was equal to 270 mm with a focal point diameter of 0.3 mm. Nevertheless, the HS geometry calibration for obtaining more accurate results was required. Therefore, a trial calibration weld at the speed of 1 m/min and output power equal to 4 kW was performed (to achieve a keyhole effect). A comparison with the simulation results showed a small discrepancy (width of the face of the weld, with anerror less than 15%), and so the heat source geometry was adjusted until the error was less than 1% [31–33]. According to the preliminary calibration performed, a conical heat source with a depth equal to 7 mm, upper radius r_t equal to 0.4 mm, and lower radius r_i equal to 0.2 mm, as well as a disc-shaped heat source, with a radius equal to 0.5 mm and depth equal to 0.05 mm, were programmed. The volume heat fraction defined the laser power division between the conical and disc HS and for laser keyhole welding, the value was set as 0.95, which means that 95% of the total power was assigned to the conical HS. Moreover, the Gaussian distribution parameter for the disc HS was defined as 1.0 and for the conical HS as 2.8, and they are related to TEM01* (CO_2 laser, transverse mode). The materials used in the simulation were structural steel ofgrade S355J2 and 316L austenitic stainless steel. Laser welding with a keyhole effect makes it possible to obtain deep material penetration, however, for a CO_2 laser, the metal surface has high reflectivity and, therefore, HS efficiency for S355J2 was assumed as 0.6, and for 316L as 0.7. The programmed efficiencies are related to laser beam interaction with the materials used, where the reflectivity and the ionization effect have a significant influence on those aspects [34–36].

The simulation of dissimilar, sealed, lap joint laser welding, for two configurations of the top material, was carried out using the Simufact Welding software. Simufact solves a heat transfer problem in solid material, considering the convection–radiation effect, however, no mixing effect of welded material can be taken into account. Figure 5a shows a keyhole in the cross-section during the lap welding, and Figure 5b shows the top hat of the keyhole, during the formation of the face of the weld. The results are presented for the temperature scale in the range of 20 to 3070 °C (from normal condition to the boiling point).

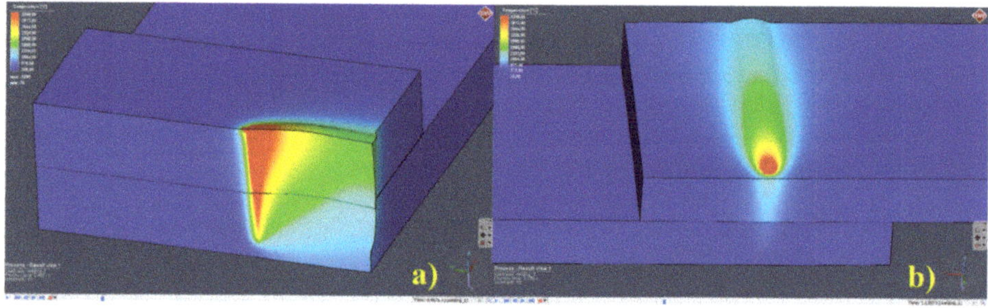

Figure 5. Simulation of laser lap joint welding using the deep penetration mode: (**a**) view of the keyhole effect in the cross-section, (**b**) top view of a moving keyhole, and weld formation.

The first joint configuration assumes low-carbon S355J2 steel placed on the top, and the second configuration assumes the opposite: S355J2 steel on the bottom and 316L stainless steel on the top. The research assumed obtaining a sealed lap joint, with partial weld penetration in the bottom sheet [37,38]. Numerical simulations of laser welding were performed at a constant speed of 1 m/min and variable output power in the range of 3 to 6 kW, which increased with each subsequent step by the value of 0.5 kW until the assumed penetration was obtained. The heat source operating time, based on the HS speed and the trajectory length, was equal to 1.2 s, however, the full simulation time, including cooling, was programmed as 30 s. According to the performed simulations, the obtained weld bead geometry and stress–strain distribution (according to the determined measurement points) were studied.

2.3. Experimental Welding Procedure

The first step of the experimental research was the HS validation. For this purpose, a preliminary test according to the procedure described in Section 2.2 was carried out. Based on the aforementioned procedure, calibration of the HS geometry and the efficiency coefficient was carried out. The parameters estimated in the numerical simulation, welding speed equal to 1 m/min and output power 6 kW, were used to perform trial welds with a TrumpfTruFlow6000 CO_2 CW laser (wavelength 10.6 μm)

(Trumpf Group, Ditzingen, German), using welding optics with the focal length of 270 mm and the spot diameter equal to 0.3 mm. This type of laser has worse parameters than fiber or disc lasers, however, further planned research requires the use of single and twin spot optics for an extended weld area, which are available for this type of laser. To shield the welding zone, helium (5.0) was used, with the flow rate equal to 20 l/min conveyed coaxially. The laser beam was focused on the surface of the top plate in a PA (flat) position and the welding trajectory was established based on the boundary conditions of the simulation [39]. Using single pass welding, lap joints in two configurations were obtained: the 1st with the low-carbon steel placed on the top (and the 316L sheet at the bottom), and the 2nd with the 316L steel placed on the top (and the S355J2 sheet at the bottom). Welding procedure qualification was performed according to PN-EN ISO 15609-4: 2009 [40] and the joint quality level was specified according to PN-EN ISO 13919-1 [41]. The results of the simulation and the metallographic analysis of the trial joints were studied.

2.4. Microstructure Analysis

Validation of the simulation results is related to a weld bead build analysis, therefore, measurements of characteristic geometries were carried out. Specimens for further analysis were prepared by cutting them in half (according to the cross-section of the weld), polishing, and etching with Adler reagent. A metallographic analysis of weld structures according to PN-EN ISO 17639 [42], using a Hirox KH-8700 (Hirox Co Ltd., Tokyo, Japan) confocal digital microscope and a scanning electron microscope JSM-7100F (JEOL Ltd., Tokyo, Japan), was carried out. By visual tests and weld bead build evaluation, welding defect detection and inclusion analysis wasperformed. Investigation of the weld uniformity by element distribution analysis, using a JEOL scanning electron microscope with an electron dispersive X-ray spectroscopy analyzer, was carried out [43].

3. Results

3.1. Global Observation

Due to the significant mismatch of thermal conductivities, specific heat capacity, and surface absorptivities in the welded materials, the FZ and HAZ are different. Moreover, a numerical simulation analysis showed further differences. The research aimed to obtain sealed lap joints, with a bottom sheet welded approximately to 3/4ths of its thickness. A macroscopic analysis showed only partial material mixing in the bottomsheet (Figure 6). The decision of which configuration to choose for better properties is complex and requires an extended thermalstress–strain numerical analysis, as well as micro- and macro-structure examinations. The preliminary visual tests (VTs) showed a good weld build, lack of defects, a convex face of the weld, and assumed partial penetration of the bottom plate was obtained, therefore, according to therequirement of the PN-EN ISO 13919-1 standard, a B quality level was obtained [44,45].

In the obtained welds beads, some separate lap joint characteristic regions were indicated, where: S1—root of the weld, S2—intersheet area, S3—uniform mixing of top plate weld area, S4—HAZ region (Figure 6). Potentially non-uniform mixing of fused materials was detected, which can affect the joint strength and corrosion resistance and can lead to micro-cracking. Therefore, a microstructure analysis, extended to include a numerical stress–strain study for lap welds evaluations, was carried out [46,47].

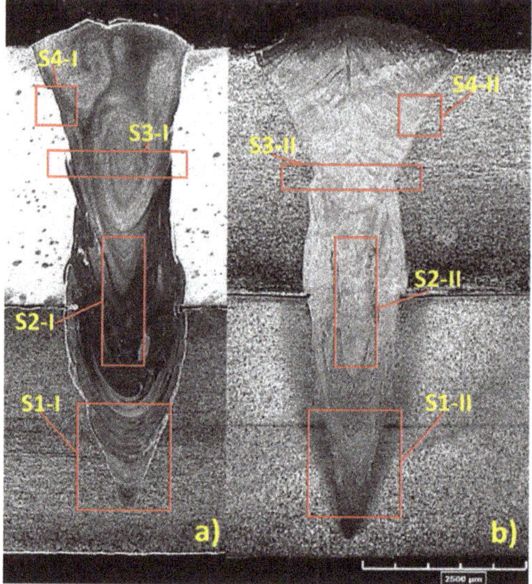

Figure 6. The build of welds with the region defined for further investigation, (**a**) 1st joint configuration (S355J2 steel on the top), (**b**) 2nd joint configuration (316L steel on the top).

3.2. Numerical Simulation

The weld bead geometry based on the solid–liquid range was analyzed. On the basis of the programmed HS parameters and established boundary conditions, a numerical simulation of laser lap welding was performed at a constant speed and varying output power. The total thickness of the materials positioned in the lap configuration was equal to 8 mm. As sealed joints were to be obtained, the partial penetration of the weld was up to 7 mm at an output power equal to 6 kW. The differences in the welded materials, especially the absorption coefficient, the thermal conductivity, and the vaporization–ionization effect, made it necessary to change the HS efficiency for each simulation configuration. Therefore, the aforementioned HS efficiency for the low-carbon steel was adopted as 0.6 and for the stainless steel as 0.7 [48,49].

For the 1st configuration, aweld depth equal to 7.1 mm, a weld face width equal to 2.87 mm, and a width of the overlap region equal to 1.92 mm were calculated, with experimental values equal to 7.16 mm, 2.51 mm, and 1.83 mm, respectively (Figure 7). The calculation results for configuration 2 showed similar values, where the depth was equal to 7.15 mm, the weld face width to 3.1 mm, and the width of the overlap region to 1.7 mm. The experimental values were equal to 7.3 mm, 3.15 mm, and 1.59 mm, respectively (Figure 8). Some differences in the weld build were observed. The weld obtained in the 2nd configuration had a typical U-groove weld build, with a wider face of the weld, however, the weld waist was narrower. The 1st configuration showed a narrower face of the weldand a wider weld waist.

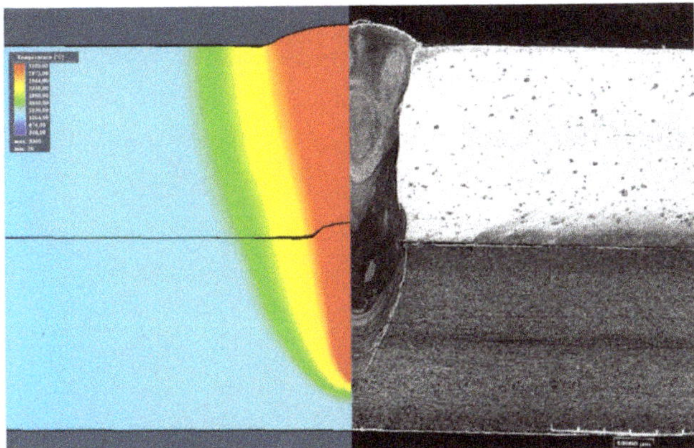

Figure 7. Comparison of the simulation and the trial joint weld build—1st configuration (S355J2 steel on the top).

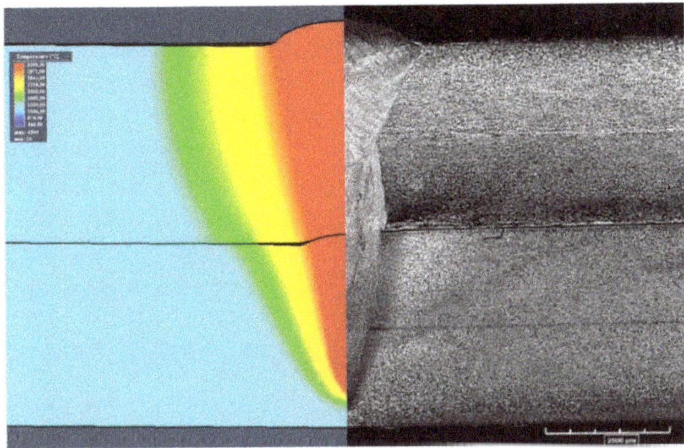

Figure 8. Comparison of the simulation and the trial joint weld build—2nd configuration (316L steel on the top).

Figures 7 and 8 compare the simulation and experimental results. The predicted weld bead geometries were compared to the measurements obtained from the trial joints [50]. The results showed good agreement between the predicted and measured characteristic dimensions of weld geometries.

The studied weld bead build showed some differences but no welding defects or incorrect weld structures were observed. Therefore, in order to decide which configuration gives better results, a further study, starting from a stress–strain distribution analysis, was carried out.

Considering the stress and strain analysis, it is relevant to study the variability of those phenomena in time. Therefore, measurement points, located across determined lines (characteristic joints areas), as shown in Figure 9, were used to perform the analysis, referred to the welding and cooling stages [51–53].

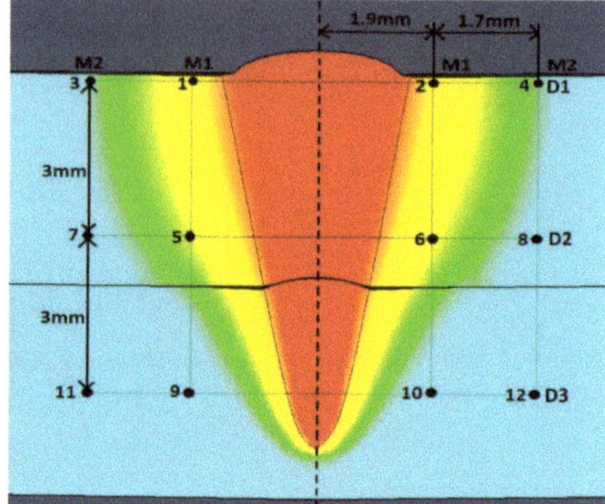

Figure 9. Defined points and lines for stress–strain numerical analysis, selected in critical joint areas.

The measurement points (1–12) were defined along the M1 and M2 measurement lines—parallel to the weld axis, where M1 is the inside line (near fusion line) and M2 is the outside line (near the HAZ line), and D1, D2, and D3 lines are perpendicular to the weld axis, where D1 is the top line, D2 is the central line, and D3 is the bottom line of weld profile. Based on the defined measurement points and lines, a stress–strain analysis for the 1st and 2nd configurations was carried out (Figures 10–14).

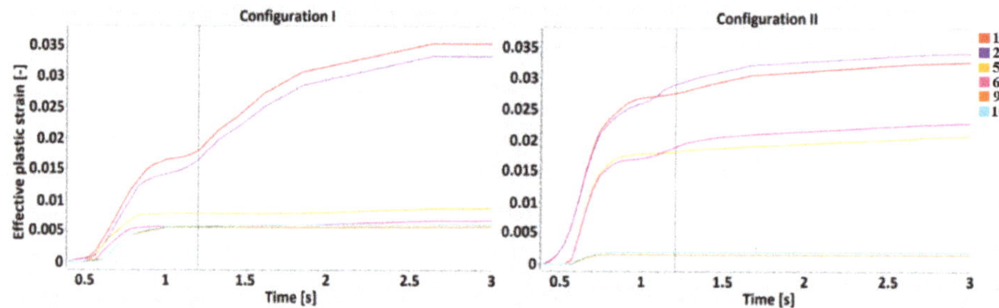

Figure 10. The effective plastic strain curve for the 1st (S355J2 steel on the top) and 2nd (316L steel on the top) configurations along the M1 line.

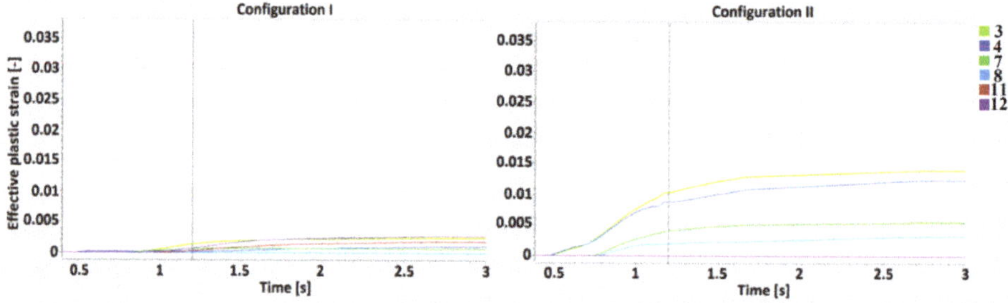

Figure 11. The effective plastic strain curve for the 1st (S355J2 steel on the top) and 2nd (316L steel on the top) configurations along the M2 line.

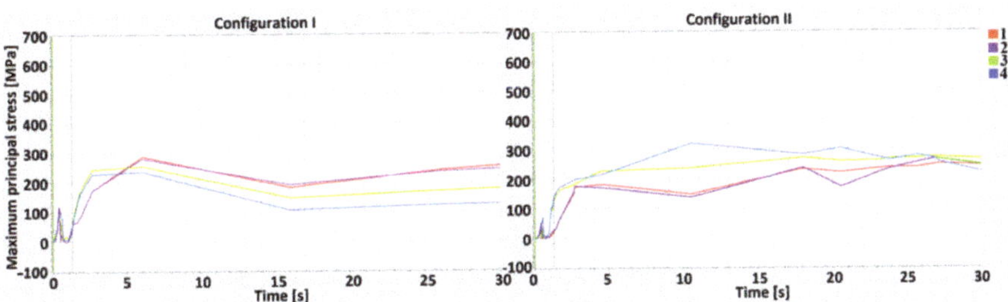

Figure 12. The maximum principal stress curve for the 1st (S355J2 steel on the top) and 2nd (316L steel on the top) configuration along the D1 line—top.

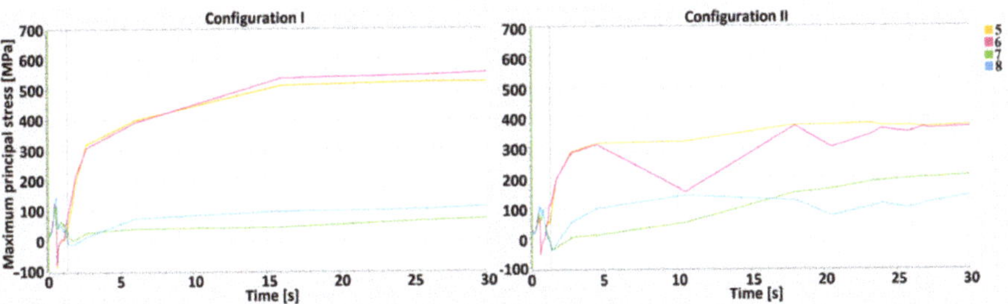

Figure 13. The maximum principal stress curve for the 1st (S355J2 steel on the top) and 2nd (316L steel on the top) configuration along the D2 line—central.

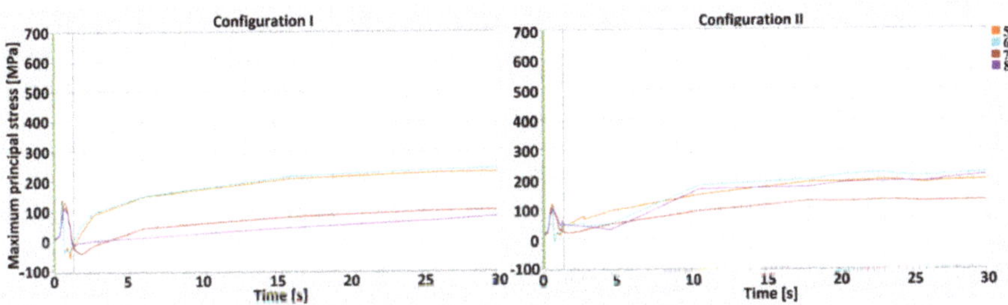

Figure 14. The maximum principal stress curve for the 1st (S355J2 steel on the top) and 2nd (316L steel on the top) configuration along the D3 line—bottom.

Effective plastic strain is a monotonically increasing scalar value which is calculated incrementally, as a function of the plastic component of the rate of the deformation tensor. The effective plastic strain increases whenever the material is actively yielding [54]. An analysis of the effective plastic strain results is shown below, according to the defined M-lines. Welding time alone (1.2 s) during the performed simulation is shown by the perpendicular line.

The tensorial strain values are not monotonically increasing as they reflect the current, the total (elastic+plastic) state of deformation, primarily in the HAZ and across the fusion line where, due to the constituent thermal expansion mismatch, the phase transformation is the most intense. The value of effective plastic strain is the integral of stepwise increments of plastic deformation continuing for a period of time; therefore, the analyzed values are shown according to the welding time. The results show the highest effective plastic strain value across the M1 line and are similar for both joint configurations, however, in the 2nd case, the curve is sharper. The values obtained across the M2 line show greater differences and, in the 2nd configuration, they were almost 4.7 times greater than in the 1st configuration. While the values for the M1 line obtained at the surface are similar (points 1 and 2), in the central zone of the 1st configuration, they are equal to the bottom zone values (points 5 and 6). For the 2nd configuration, three separate regions of effective plastic strain can be identified. The computation times of the welding process (without cooling) were equal to 1.2 s, therefore, for these specific points, some changes in the plots can be observed [55–57].

A maximum principal stress (MPS) analysis, concerning values changing in time, was performed as well (Figures 12–14). The points selected for the MPS analysis were related to the weld depth (D-lines). By performing a distribution analysis of the maximum principal stress, the areas of stress concentrations can be identified. These particular areas show where crack propagation could start when critical material parameters are reached. An analysis of the results can shown whether in particular joints configurations, the MPS did not exceed the critical values, disqualifying one of the lap joints [58,59].

Analysis of MPS showed higher stress concentrations across the D2 line, particularly in points 5 and 6 (in the middle of the HAZ). The greatest calculated values, equal to 540 MPa, occurred in the 1st configuration, where low-carbon steel is located on the top. In the 2nd configuration, the highest stress occurred across the D2 line as well, however, it did not exceed 400 MPa. At the surface (D1 line), the maximum principal stress exceeded 300 MPa only in point 4. Across the D3 line, the maximum principal stress is similar in both configurations, with slight differences in austenitic steel (in 1st configuration). An increase in the maximum principal stress in D1 and D2 occurred after the welding, and MPS values were related to the movement of the HS, and increased rapidly during the cooling stage, after the heat source was turned off [60–63]. One must keep in mind that the measurement points were located on the sectional plane in the middle of the welding trajectory.

3.3. Metallographic Analysis

Important differences between stress–strain distribution showed significant dissimilarity in joints properties, however, considering only simulation, it is hard to define which configuration provides better results, and further analysis is required. Therefore, the microstructure of characteristic weld areas, chosen alloyed elements distribution (Figure 21), and the precipitates in the fusion zone were studied and the results are shown below.

Figure 15 shows the microstructure of the base material (BM). The structures shown are typical for the used materials, with a ferritic-pearlitic microstructure of low-carbon S355J2 steel (Figure 15a) and an austenitic structure of stainless 316L steel (Figure 15b). The HAZ areas (Figure 16) in both materials are different. In 316L steel, the HAZ is narrow, with grain refining along the fusion line, and an increase in the volume fraction of ferrite δ. The HAZ identified in S355J2 steel is much wider and is made up of three separate regions: overheated zone (OZ), normalization zone (NZ), and partial recrystallization zone (RZ) [64,65]. The HAZ was investigated according to the S3 region (Figure 6).

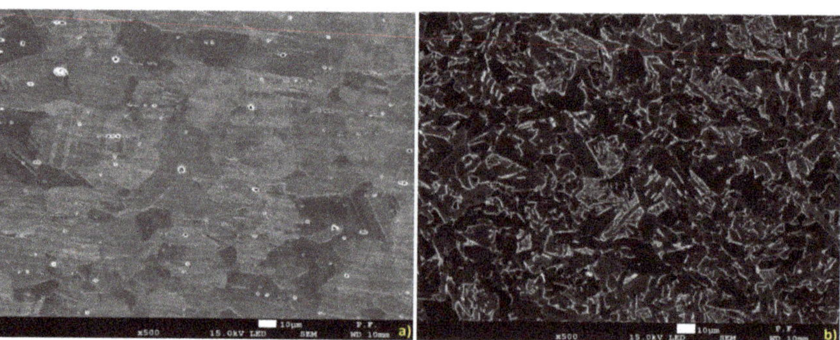

Figure 15. The microstructure of welded base materials: (**a**) 316L steel, (**b**) S355J2 steel.

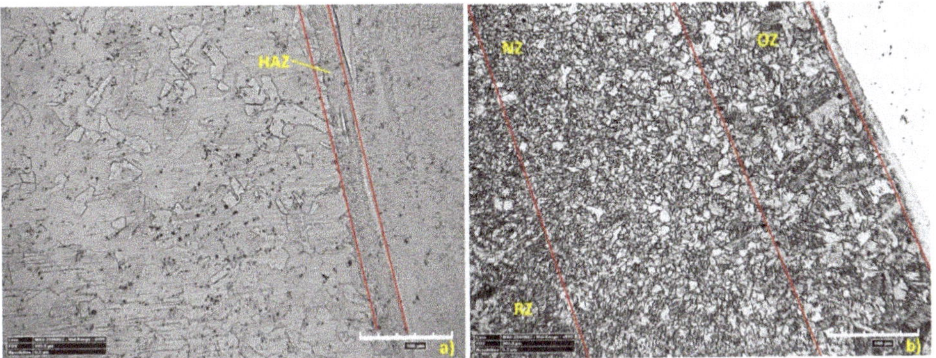

Figure 16. The microstructure of the heat-affected zone (HAZ): (**a**) 316L steel, (**b**) S355J2 steel.

The weld composition is related to the mixture of the low-carbon and stainless steel alloying components, and the weld structure is a result of the solidification process of this newly formed material (Figure 17).

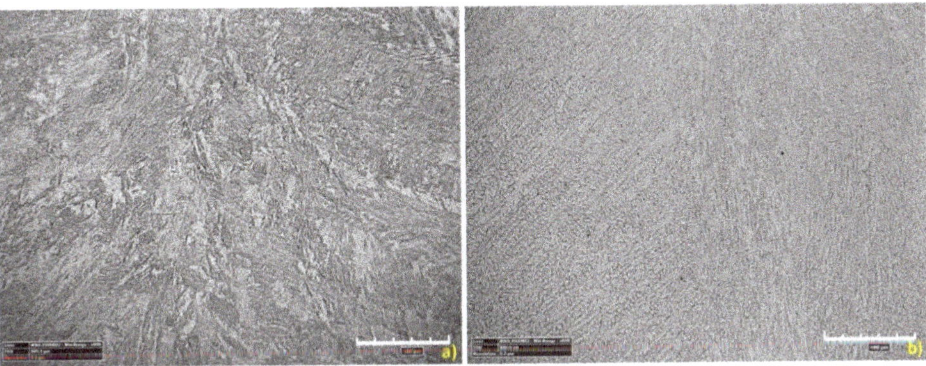

Figure 17. The weld structure approximately in the middle of the upper plate: (**a**) 1st configuration (S355J2 steel on top), (**b**) 2nd configuration (316L steel on top).

The obtained structure has a dendritic build, however, in the case where low-carbon steel is the topsheet, the structure is composed of coarse-grained dendrites and acicular ferrite content. Meanwhile, for the configuration where stainless steel is the topsheet, the observed structure has a typical pillar dendritic build. In the 1st case, an austenitic-ferritic structure can be observed, however, in the 2nd case, there is only an austenitic structure. Moreover, fusion zones in the bottom sheet showed regions with non-uniform structures (Figure 18), according to region S1 (Figure 6) [66,67].

The SEM analysis showed differences in the weld structures, where separate areas inside the FZ were identified (Figure 19A,B). In area A, a pillar-dendritic structure was observed, while in area B, a dendritic cellular structure dominated.

In the root of the weld (region S1), where the weld penetration was achieved, some differences were observed. The bottom weld area in austenitic steel had a structure similar to that of the BM (Figure 20a), while in ferritic-pearlitic steel, the structure was similar to that of the HAZ (Figure 20b).

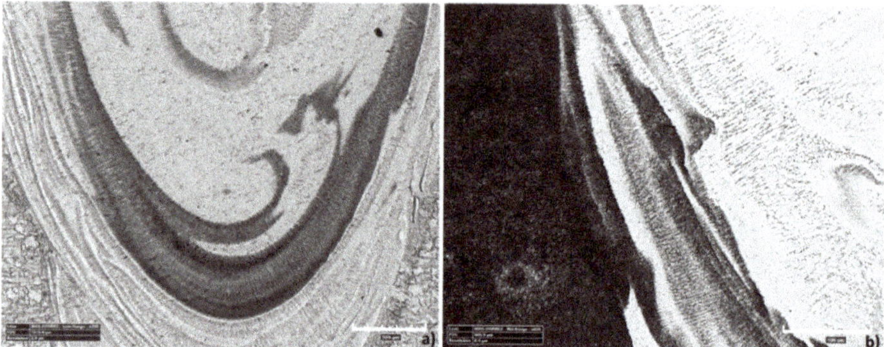

Figure 18. Dissimilarity in the weld structure, down plate: (**a**) 1st configuration (S355J2 steel on top), (**b**) 2nd configuration (316L steel on top).

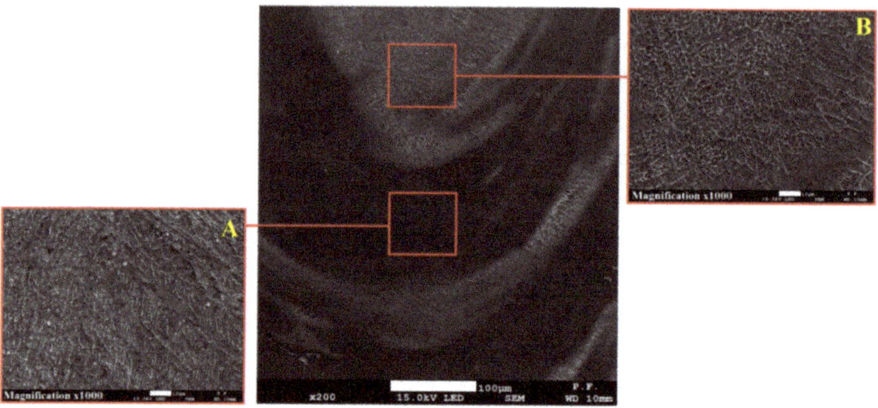

Figure 19. The weld structure approximately in the middle of the down plate, 2nd configuration (316L on top), with identified differences in structure (central picture) and the enlarged areas: (**A**)—pillar dendritic structure, (**B**)—cellular dendritic structure.

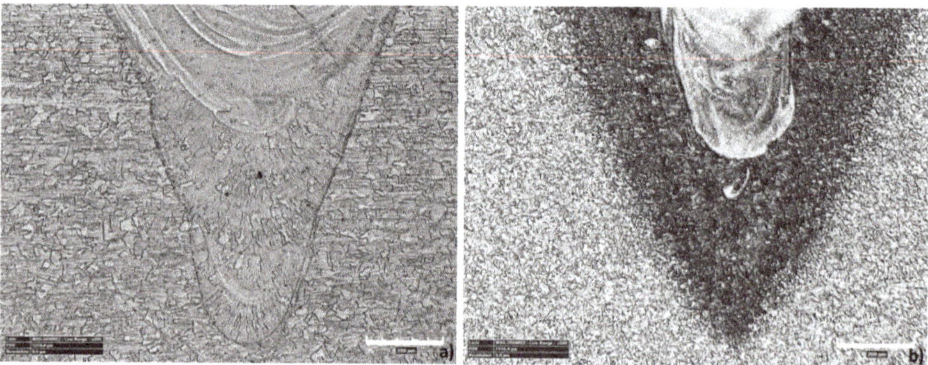

Figure 20. The fusion line of the weld in the down plate: (**a**) 1st configuration (S355J2 steel on top), (**b**) 2nd configuration (316L steel on top).

The non-uniform fusion zone structurewas confirmed through the EDS analysis. Therefore, according to the defined regions (presented in Figure 6), important differences in chemical compositionwere studied (Table 1) based on iron, nickel, and chromium distributions in the weld cross-section (Figure 21).

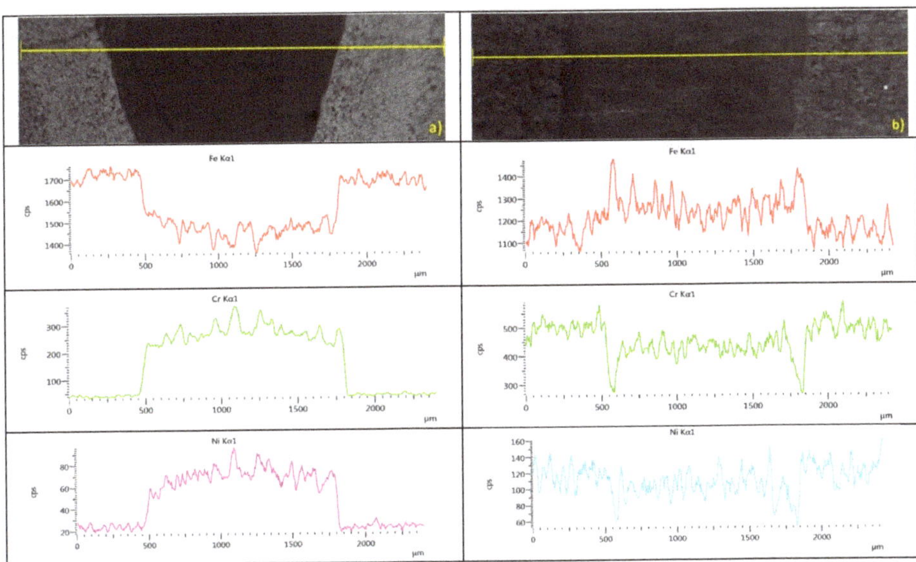

Figure 21. Iron, chromium, and nickel distribution according to S3 region: (**a**) 1st configuration, (**b**) 2nd configuration.

The uniform distribution in the topsheet FZ proves the high mixture factor in this region, but the distribution in the bottomsheet FZ is more problematic (Figure 22).

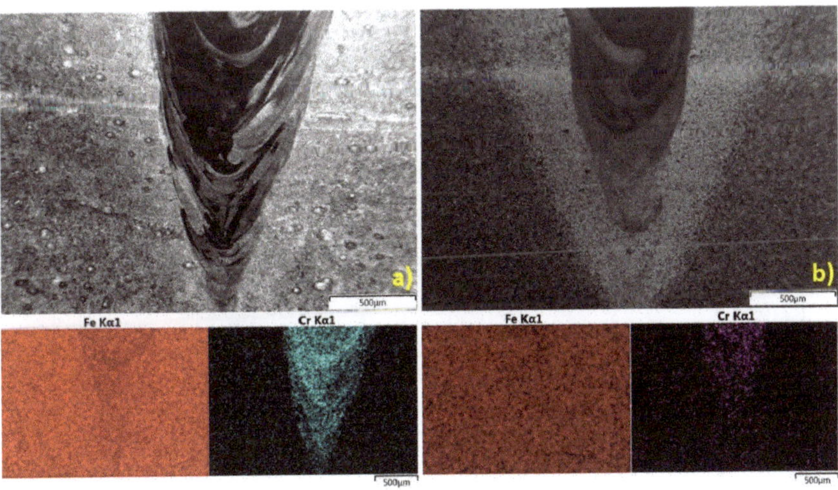

Figure 22. Bottom fusion zone iron and chromium map distribution (region S1): (**a**) 1st configuration (S355J2 steel on top), (**b**) 2nd configuration (316L steel on top).

The linear distribution of Fe, Cr, and Ni in the upper plate (region S3) showed greater differences for the 1st configuration (low-carbon steel as the topsheet), while in the 2nd configuration, the distribution was more uniform, however, an analysis showed largedifferences in the fusion line region. Moreover, according to the iron and chromium distributions, the joints made in the 1st configuration showed low iron content and a largeamount of chromium, while in the 2nd configuration, the iron distribution was almost uniform, with slightly increased chromium content.

4. Discussion

The welded materials have important differences in thermo-physical properties related to their chemical composition (Table 1). Both materials are characterized by good weldability, but due to the differencesdiscussed above, joining them by welding is problematic, especially when we consider a lap joint with partial penetration in the bottom plate, as described in this paper.

Global observation showed differences in the weld builds for both assumed joint configurations. The macroscopic analysis showed a potentially non-uniform structure in the fusion zones. Therefore, in order to investigate the quality level of the joints and to choose which configuration gives better results, numerical analysis and then metallographic studies were carried out.

Based on the developed numerical model, calibrated in the preliminary research, the welding parameters for the sealed lap joint were estimated. According to the established boundary conditions and the programmed simulation parameters, welding speed equal to 1 m/min with 6 kW of output power results in the assumed partial penetration. The performed simulations provided results where the differences in the width of the face of the weld were less than 14.5% compared to the trial joint for the 1st configuration and 1.62% for the 2nd configuration. The difference in the weld depth, calculated and obtained via experimental welding for the 1stconfiguration, was equal to 0.85% and for the 2nd configuration, to 2.1%. The width of the overlap region resulted in a 4.9% mismatch, comparing the simulation and the trial joint results for the 1st configuration, and 6.9% for the 2nd configuration. The numerical model obtained gave accurate results of the estimated weld bead geometry, however, the mismatch in the weld width for the 1st configuration, where low-carbon S355J2 steel was the topsheet, resulted in a 14.5% difference. The numerical model did not include the Marangoni effect and the surface tension, which has a great impact on the face of weld geometry. However, based on the small differences in other (Figures 7 and 8) weld geometries, the established model gave realistic results and was used for joint stress–strain analysis [68–70].

In order to define the properties of the obtained joints based on the simulation results, the distribution of the effective plastic strain and the maximum principal stress was studied according to the determined lines (Figure 9). The M lines define points distributed parallel to the weld, near the fusion and heat-affected zones lines, and D lines are related to the weld penetration depth and are distributed perpendicular to the weld axis. According to the M1 and M2 lines, the effective plastic strain analysis showed amaximum value equal to 0.035 and was similar for both joint configurations, however, in the 2nd case, the values increased more rapidly. For the 1st case, the increase was slower, however, after the end of the welding cycle (1.2 s), it was still rising in the cooling stage, to 2.6 s. For the M2 lines (located 3.6 mm from the weld axis), the effective plastic strains were much smaller and the maximum calculated value (which occurred in the 2nd configuration) did not exceed 0.015, while for the 1st configuration, it was equal to 0.03. The differences in the effective plastic strain can be related to the different values of thermal conductivity (curves from 0 to 1.2 s), which for low-carbon steel resulted in a higher accumulation of thermal energy in the fusion zone and phase transformation (after 1.2 s) in the cooling stage [71].

The maximum principal stress values across the Dlines showed that the stress characteristic was divided into two separate cycles. The first was related to heating and cooling during movement of the HS, however, this value did not exceed 150 MPa. The second was related to material cooling, when the heat source was turned off. The maximum principal stress increased on the surface of the topsheet and its value ranged from 60 to 290 MPa for the 1st configuration and 20 to 320 MPa for the 2nd (the curve characteristics were similar for all measurement points). In the center line (D2), a greater stress value occurred in joint configuration number 1 and was equal to 540 MPa, while for the 2nd configuration, the value did not exceed 400 MPa. For the D3 lines, the values were less than 240 MPa (1st configuration) and equal to 200 MPa (2nd configuration). The maximum principal stress had a greater value in the measurement points located closer to the weld axis, withthegreatest difference occurring across the D2 line, where at points 5 and 6 they were more than 5.5 times greater than in points 7 and 8. The calculated maximum principal stress had a greater value in the 1st configuration and was related to the thermal gradient and the phase transformation of low-carbon steel [72,73].

A metallographic analysis confirmed the typical structure of the welded materials in the BM region, austenitic in 316L steel and ferritic-pearlitic in S355J2 steel. Significant differences in the heat-affected zones were observed. In stainless steel, the HAZ was very narrow and no growth of austenite grains was observed, while the elongated grains of the ferrite formed a discontinuous network around the austenite grains. These regions occur in steels with the structure of a native material consisting of austenite with the participation of ferrite δ. At high temperatures near the fusion line, the transformation of $\gamma \rightarrow \delta$ occurs, which begins in the existing ferrite δ grains and progresses in areas with increased chromium concentration. After re-cooling, these areas do not achieve an equilibrium structure and the proportion of ferrite δ is increased. The HAZ of low-carbon steel is considerably wider and is made up of threeregions: an overheated zone, a normalization zone, and a partial recrystallization zone. These phenomena are related to phase transformation and structural changes during the solidification and cooling process. The obtained HAZs (including microstructure and geometry) of welded materials are typical for this steel grade welded using the LBW method [74].

In the global observation, some non-uniform microstructure of the fusion zones was detected, therefore, an extended weld microstructure analysis was carried out. The structure observed in the topsheet FZ had a typical uniform dendritic build, however, in the 1st configuration, coarse-graineddendritescontaining acicular ferrite were observed, while in the 2nd configuration, a typical pillar-dendritic build occurred. When the keyhole only partially penetrates the bottom workpiece, the influence of the flow field is clearly evident from the corresponding solidified structure. In the case of partial penetration, discrete growth bands occurred in the entire solidified weld bead, suggesting severe fluctuations in the flow field and the growth process [75]. On the contrary, in the topsheet, a full penetration keyhole was performed, where a columnar structure and an equiaxed zone along the centerline, similar to that shown in Figure 17b, can be observed. Discrete growth bands or striations occurred occasionally in this case, but the structural pattern was maintained within these bands. It should be noted that the laser output power in both cases was similar.

The fusion zone structure in the overlap region are similar to those observed in the FZ upper region, however, in the bottom plate, non-uniform structures were identified (Figure 18). There were differences in the fusion structure (Figure 19), where a pillar-dendritic build in region A and a cellular-dendritic structure in region B were observed. In the root of the weld, at the bottom of the fusion zone where weld penetration reaches, the structure was similar to the BM for stainless steel and the HAZ structure for low-carbon steel. In this region, the temperature exceeded the melting point, but only slightly, and according to the direction of heat diffusion in the XYZ axis, the heat flow to the BM was higher [76].

Homogenous weld builds are related to the uniform distribution of chemical elements. The welded materials have important differences in chemical composition, where S355J2 steel has a high content of iron and only trace amounts of chromium and nickel, while stainless steel contains more than 16% chromium, 11% nickel, and a balanced content of iron. Therefore, an analysis of defined regions in the cross-section of the welds was performed based on the linear and map distribution of these elements. The study showed bigger differences between the BM and the FZ in the distribution of Fe, Cr, and Ni in the 1st configuration. A clear boundary between the base material and the fusion zone was observed, however, the distribution curve had an almost vertical characteristic. For the 2nd configuration, the differences were smaller, with peak changes along the fusion line, which is related to chromium diffusion and concentration across the fusion line (phase transformation and increasing content of ferrite δ). An EDS analysis of the FZ bottom region showed further differences, particularly in chromium and iron distribution [77]. For the 1st configuration, small amounts of iron were detected and considerably greater differences in the detection of chromium were observed (Figure 22a). In the 2nd joint configuration, the iron distribution in the fusion zone and the base material was almost uniform, and some differences in chromium detection were observed (Figure 22b).

5. Conclusions

The study of laser lap joint welding, where low-carbon S355J2 and stainless 316L steel were joined together in two configurations (1st with low-carbon steel on the top, 2nd with stainless steel on the top), showed significant differences in the obtained results. According to the developed numerical model and the established boundary conditions, an accurate match of the simulated and experimental results of weld geometry was obtained. Assumingthat the obtained model gave accurate results, a numerical stress–strain analysis was performed. Higher values of effective plastic strain occurred in the 2nd configuration, however, the maximum principal stress values were greater in the 1st configuration. Greater differences in the measurement points occurredfor the configuration with low-carbon steel placed on the top. The microstructure analysis showed further differences, most of all in the fusion zone structure, where in the bottom plate, two separate structures were detected. A more uniform structure was observed in the 1st configuration, however, greater differences in Cr, Ni, and Fe distribution between the FZ and the BM were detected in this configuration. No porosity, cracks, or welding defects were identified, therefore, according to therequirement of PN-EN ISO 13919-1, a B quality level was assumed. According to the performed study, and based on the maximum principal stress distribution, differences in the weld structure, and the distribution of the alloying elements, the second configuration showed better properties. Therefore, the second configuration, where stainless steel is placed on the top, was chosen as a dominant joint. The microstructure study showeda non-uniform mixture of welded material in the root of the weld, however, no welding defects or precipitations in the inter-plate region were observed. This study confirmsthe impact of the welded material configuration on joint properties and shows which configuration ensures the lowest stress and strain concentration and a more uniform weld structure. Therefore, according to:

- calculated maximum principal stress value.
- lower differences in the chemical element distribution in theweld.
- more uniform structure and fusion rate in the weld.

When laser lap welding of S355J2 and 316L steels is considered, the configurationin which stainless steel is placed on top provides better joint properties and is recommended.The presented results showed that when ahigh-qualitysteel joint with good strength characteristics is required, such as pipeline components or crude oil storage tanks, the welding position should be related to the stainless steel side.

Further investigation of this joint type is required, therefore, the possibility of usingtwin spot laser welding optics is planned. Theresearch will also be extended to mechanical tests, including the tensile strength and hardness tests.

Author Contributions: Conceptualization, H.D., A.S., and R.N.; methodology, H.D., A.S., M.H., and P.D.; software, H.D., G.W., and S.T.; validation, H.D., A.S., M.H., P.D., and R.N.; formal analysis, H.D., A.S., R.N., S.T., and P.D.; investigation, H.D. and G.W.; resources, H.D.; writing—original draft preparation, H.D., A.S., M.H., P.D., and R.N.; writing—review and editing, H.D., A.S., M.H., and G.W.; visualization, H.D. and G.W.; ervision, A.S. and M.H.; project administration, H.D.; funding acquisition, H.D. All authors have read and agree to the published version of the manuscript.

Funding: This research was funded by NCBiR, grant number LIDER/31/0173/L-8/16/NCBR/2017: Technology of manufacturing sealed weld joints for gas installation by using concentrated energy source.

Conflicts of Interest: The authors declare no conflict of interest.

References

1. Mazumder, J. Laser Welding: State of the Art Review. *JOM* **1982**, *34*, 16–24. [CrossRef]
2. Mohanty, P.S.; Mazumder, J. Workbench for keyhole laser welding. *Sci. Technol. Weld. Join.* **1997**, *2*, 133–138. [CrossRef]
3. Matsunawa, A. Problems and solutions in deep penetration laser welding. *Sci. Technol. Weld. Join.* **2001**, *6*, 351–354. [CrossRef]
4. Yang, R.-T.; Chen, Z.-W. A study on fiber laser lap welding of thin stainless steel. *Int. J. Precis. Eng. Manuf.* **2013**, *14*, 207–214. [CrossRef]
5. Khan, M.; Romoli, L.; Fiaschi, M.; Sarri, F.; Dini, G. Experimental investigation on laser beam welding of martensitic stainless steels in a constrained overlap joint configuration. *J. Mater. Process. Technol.* **2010**, *210*, 1340–1353. [CrossRef]
6. Meco, S.; Pardal, G.; Ganguly, S.; Williams, S.W.; McPherson, N. Application of laser in seam welding of dissimilar steel to aluminium joints for thick structural components. *Opt. Lasers Eng.* **2015**, *67*, 22–30. [CrossRef]
7. Takano, K.; Koizumi, N.; Serizawa, H.; Tsubota, S.; Makino, Y. Development of laser welding technology for fully austenitic stainless steel. *Weld. Int.* **2017**, *31*, 827–836. [CrossRef]
8. Tan, W.; Shin, Y.C. Laser keyhole welding of stainless steel thin plate stack for applications in fuel cell manufacturing. *Sci. Technol. Weld. Join.* **2015**, *20*, 313–318. [CrossRef]
9. Kouraytem, N.; Li, X.; Cunningham, R.; Zhao, C.; Parab, N.; Sun, T.; Rollett, A.D.; Spear, A.D.; Tan, W. Effect of Laser-Matter Interaction on Molten Pool Flow and Keyhole Dynamics. *Phys. Rev. Appl.* **2019**, *11*, 064054. [CrossRef]
10. Lisiecki, A.; Wójciga, P.; Kurc-Lisiecka, A.; Barczyk, M.; Krawczyk, S. Laser welding of panel joints of stainless steel heat exchangers. *Weld. Technol. Rev.* **2019**, *91*, 7–19. [CrossRef]
11. Szczepaniak, A.; Fan, J.; Kostka, A.; Raabe, D. On the Correlation between Thermal Cycle and Formation of Intermetallic Phases at the Interface of Laser-Welded Aluminum-Steel Overlap Joints. *Adv. Eng. Mater.* **2012**, *14*, 464–472. [CrossRef]
12. Oussaid, K.; El Ouafi, A.; Chebak, A. Experimental Investigation of Laser Welding Process in Overlap Joint Configuration. *J. Mater. Sci. Chem. Eng.* **2019**, *7*, 16–31. [CrossRef]
13. Noga, P.; Węglowski, M.; Zimierska-Nowak, P.; Richert, M.; Dworak, J.; Rykała, J. Influence of welding techniques on microstructure and hardness of steel joints used in automotive air conditioners. *Met. Foundry Eng.* **2017**, *43*, 281. [CrossRef]
14. Guo, W.; Kar, A. Determination of weld pool shape and temperature distribution by solving three-dimensional phase change heat conduction problem. *Sci. Technol. Weld. Join.* **2000**, *5*, 317–323. [CrossRef]
15. Wang, R.; Lei, Y.; Shi, Y. Numerical simulation of transient temperature field during laser keyhole welding of 304 stainless steel sheet. *Opt. Laser Technol.* **2011**, *43*, 870–873. [CrossRef]
16. Dal, M.; Fabbro, R. An overview of the state of art in laser welding simulation. *Opt. Laser Technol.* **2016**, *78*, 2–14. [CrossRef]
17. Shanmugam, N.S.; Buvanashekaran, G.; Sankaranarayanasamy, K.; Manonmani, K. Some studies on temperature profiles in AISI 304 stainless steel sheet during laser beam welding using FE simulation. *Int. J. Adv. Manuf. Technol.* **2008**, *43*, 78–94. [CrossRef]
18. Schöler, C.; Haeusler, A.; Karyofylli, V.; Behr, M.; Schulz, W.; Gillner, A.; Niessen, M. Hybrid simulation of laser deep penetration welding. *Mater. Werkst.* **2017**, *48*, 1290–1297. [CrossRef]
19. Miyashita, Y.; Mutoh, Y.; Akahori, M.; Okumura, H.; Nakagawa, I.; Jin-Quan, X. Laser welding of dissimilar metals aided by unsteady thermal conduction boundary element method analysis. *Weld. Int.* **2005**, *19*, 687–696. [CrossRef]
20. Evdokimov, A.; Springer, K.; Doynov, N.; Ossenbrink, R.; Michailov, V. Heat source model for laser beam welding of steel-aluminum lap joints. *Int. J. Adv. Manuf. Technol.* **2017**, *93*, 709–716. [CrossRef]
21. Ozkat, E.C.; Franciosa, P.; Ceglarek, D. Development of decoupled multi-physics simulation for laser lap welding considering part-to-part gap. *J. Laser Appl.* **2017**, *29*, 22423. [CrossRef]
22. EN 10025-2: Hot rolled products of structural steels. In *Technical Delivery Conditions for Non-Alloy Structural Steels*; PKN: Warsaw, Poland, 2007.

23. *ASTM A240/A240M-18: Standard Specification for Chromium and Chromium-Nickel Stainless Steel Plate, Sheet, and Strip for Pressure Vessels and for General Applications*; ASTM: West Conshohocken, PA, USA, 2018.
24. Chen, C.; Lin, Y.-J.; Ou, H.; Wang, Y. Study of Heat Source Calibration and Modelling for Laser Welding Process. *Int. J. Precis. Eng. Manuf.* **2018**, *19*, 1239–1244. [CrossRef]
25. Derakhshan, E.D.; Yazdian, N.; Craft, B.; Smith, S.; Kovacevic, R. Numerical simulation and experimental validation of residual stress and welding distortion induced by laser-based welding processes of thin structural steel plates in butt joint configuration. *Opt. Laser Technol.* **2018**, *104*, 170–182. [CrossRef]
26. De, A.; Maiti, S.K.; Walsh, C.A.; Bhadeshia, H.K.D.H. Finite element simulation of laser spot welding. *Sci. Technol. Weld. Join.* **2003**, *8*, 377–384. [CrossRef]
27. Phanikumar, G.; Chattopadhyay, K.; Dutta, P. Joining of dissimilar metals: Issues and modelling techniques. *Sci. Technol. Weld. Join.* **2011**, *16*, 313–317. [CrossRef]
28. Mayboudi, L.; Birk, A.; Zak, G.; Bates, P. A Three-Dimensional Thermal Finite Element Model of Laser Transmission Welding for Lap-Joint. *Int. J. Model. Simul.* **2009**, *29*, 149–155. [CrossRef]
29. Shanmugam, N.S.; Buvanashekaran, G.; Sankaranarayanasamy, K.; Kumar, S.R. A Transient Finite Element Simulation of the Temperature Field and Bead Profile of T-Joint Laser Welds. *Int. J. Model. Simul.* **2010**, *30*, 108–122. [CrossRef]
30. Farrokhi, F.; Endelt, B.; Kristiansen, M. A numerical model for full and partial penetration hybrid laser welding of thick-section steels. *Opt. Laser Technol.* **2019**, *111*, 671–686. [CrossRef]
31. Mazumder, J.; Steen, W.M. Heat transfer model for cw laser material processing. *J. Appl. Phys.* **1980**, *51*, 941. [CrossRef]
32. Torkamany, M.J.; Sabbaghzadeh, J.; Hamedi, M. Effect of laser welding mode on the microstructure and mechanical performance of dissimilar laser spot welds between low carbon and austenitic stainless steels. *Mater. Des.* **2012**, *34*, 666–672. [CrossRef]
33. Sudnik, W.; Radaj, D.; Breitschwerdt, S.; Erofeew, W. Numerical simulation of weld pool geometry in laser beam welding. *J. Phys. D Appl. Phys.* **2000**, *33*, 662–671. [CrossRef]
34. Phaonaim, R.; Yamamoto, M.; Shinozaki, K.; Yamamoto, M.; Kadoi, K. Development of a Heat Source Model for Narrow-gap Hot-wire Laser Welding. *Q. J. Jpn. Weld. Soc.* **2013**, *31*, 82s–85s. [CrossRef]
35. Carmignani, C.; Mares, R.; Toselli, G. Transient finite element analysis of deep penetration laser welding process in a singlepass butt-welded thick steel plate. *Comput. Methods Appl. Mech. Eng.* **1999**, *179*, 197–214. [CrossRef]
36. Kano, S.; Oba, A.; Yang, H.-L.; Matsukawa, Y.; Satoh, Y.; Serizawa, H.; Sakasegawa, H.; Tanigawa, H.; Abe, H. Experimental assessment of temperature distribution in heat affected zone (HAZ) in dissimilar joint between 8Cr-2W steel and SUS316L fabricated by 4 kW fiber laser welding. *Mech. Eng. Lett.* **2016**, *2*, 15. [CrossRef]
37. Shanmugam, N.S.; Buvanashekaran, G.; Sankaranarayanasamy, K. Experimental investigation and finite element simulation of laser beam welding of AISI 304 stainless steel sheet. *Exp. Tech.* **2009**, *34*, 25–36. [CrossRef]
38. Danielewski, H.; Skrzypczyk, A. Steel Sheets Laser Lap Joint Welding—Process Analysis. *Materials* **2020**, *13*, 2258. [CrossRef]
39. Oh, R.; Kim, D.Y.; Ceglarek, D. The Effects of Laser Welding Direction on Joint Quality for Non-Uniform Part-to-Part Gaps. *Materials* **2016**, *6*, 184. [CrossRef]
40. *PN-EN ISO 15609-4: 2009: Specification and Qualification of Welding Procedures for Metallic Materials-Welding Procedure Specification-Part 4: Laser Beam Welding*; PKN: Warsaw, Poland, 2009.
41. *PN-EN ISO 13919-1: 2002: Welding-Electrons and Laser Beam Welded Joints-Guidance on Quality Levels for Imperfections-Part 1: Steel*; PKN: Warsaw, Poland, 2002.
42. *PN-EN ISO 17639:2013: Destructive Tests on Welds in Metallic Materials-Macroscopic and Microscopic Examination of Welds*; PKN: Warsaw, Poland, 2013.
43. Rong, Y.; Xu, J.; Lei, T.; Wang, W.; Sabbar, A.A.; Huang, Y.; Wang, C.; Chen, Z. Microstructure and alloy element distribution of dissimilar joint 316L and EH36 in laser welding. *Sci. Technol. Weld. Join.* **2018**, *23*, 454–461. [CrossRef]
44. Kuryntsev, S.V. Microstructure, mechanical and electrical properties of laser-welded overlap joint of CP Ti and AA2024. *Opt. Lasers Eng.* **2019**, *112*, 77–86. [CrossRef]
45. Tsoukantas, G.; Chryssolouris, G. Theoretical and experimental analysis of the remote welding process on thin, lap-joined AISI 304 sheets. *Int. J. Adv. Manuf. Technol.* **2008**, *35*, 880–894. [CrossRef]

46. Kubiak, M.; Piekarska, W.; Stano, S.; Saternus, Z. Numerical Modelling of Thermal And Structural Phenomena In Yb:YAG Laser Butt-Welded Steel Elements. *Arch. Met. Mater.* **2015**, *60*, 821–828. [CrossRef]
47. Ferro, P.; Bonollo, F.; Tiziani, A. Laser welding of copper–nickel alloys: A numerical and experimental analysis. *Sci. Technol. Weld. Join.* **2005**, *10*, 299–310. [CrossRef]
48. Huang, B.-S.; Yang, J.; Lu, D.-H.; Bin, W.-J. Study on the microstructure, mechanical properties and corrosion behaviour of S355JR/316L dissimilar welded joint prepared by gas tungsten arc welding multi-pass welding process. *Sci. Technol. Weld. Join.* **2016**, *21*, 381–388. [CrossRef]
49. Brand, M.; Siegele, D. Numerical Simulation of Distortion and Residual Stresses of Dual Phase Steels Weldments. *Weld. World* **2010**, *51*, 56–62. [CrossRef]
50. Mochizuki, M.; Katsuyama, J.; Higuchi, R.; Toyoda, M. Study of Residual Stress Distribution at Start-Finish Point of Circumferential Welding Studied by 3D-Fem Analysis. *Weld. World* **2005**, *49*, 40–49. [CrossRef]
51. Hempel, N.; Nitschke-Pagel, T.; Dilger, K. Residual stresses in multi-pass butt-welded ferritic-pearlitic steel pipes. *Weld. World* **2015**, *59*, 555–563. [CrossRef]
52. Benyounis, K.Y.; Olabi, A.G.; Hashmi, M. Residual Stresses Prediction for CO_2 Laser Butt-Welding of 304-Stainless Steel. *Appl. Mech. Mater.* **2006**, *3–4*, 125–130. [CrossRef]
53. Rong, Y.; Xu, J.; Huang, Y.; Zhang, G. Review on finite element analysis of welding deformation and residual stress. *Sci. Technol. Weld. Join.* **2018**, *23*, 198–208. [CrossRef]
54. Kong, F.; Ma, J.; Kovacevic, R. Numerical and experimental study of thermally induced residual stress in the hybrid laser–GMA welding process. *J. Mater. Process. Technol.* **2011**, *211*, 1102–1111. [CrossRef]
55. Buschenhenke, F.; Hofmann, M.; Seefeld, T.; Vollertsen, F. Distortion and residual stresses in laser beam weld shaft-hub joints. *Phys. Procedia* **2010**, *5*, 89–98. [CrossRef]
56. Nilsson, P.; Hedegård, J.; Al-Emrani, M.; Atashipour, S.R. The impact of production-dependent geometric properties on fatigue-relevant stresses in laser-welded corrugated core steel sandwich panels. *Weld. World* **2019**, *63*, 1801–1818. [CrossRef]
57. Wang, J.; Ren, L.; Xie, L.; Xie, H.; Ai, T. Maximum mean principal stress criterion for three-dimensional brittle fracture. *Int. J. Solids Struct.* **2016**, *102–103*, 142–154. [CrossRef]
58. Ben Salem, G.; Chapuliot, S.; Blouin, A.; Bompard, P.; Jacquemoud, C. Brittle fracture analysis of Dissimilar Metal Welds between low-alloy steel and stainless steel at low temperatures. *ProcediaStruct. Integr.* **2018**, *13*, 619–624. [CrossRef]
59. Casalino, G.; Angelastro, A.; Perulli, P.; Casavola, C.; Moramarco, V. Study on the fiber laser/TIG weldability of AISI 304 and AISI 410 dissimilar weld. *J. Manuf. Process.* **2018**, *35*, 216–225. [CrossRef]
60. Eisazadeh, H.; Achuthan, A.; Goldak, J.; Aidun, D. Effect of material properties and mechanical tensioning load on residual stress formation in GTA 304-A36 dissimilar weld. *J. Mater. Process. Technol.* **2015**, *222*, 344–355. [CrossRef]
61. Anawa, E.; Olabi, A.-G. Control of welding residual stress for dissimilar laser welded materials. *J. Mater. Process. Technol.* **2008**, *204*, 22–33. [CrossRef]
62. Yao, C.; Xu, B.; Zhang, X.; Huang, J.; Fu, J.; Wu, Y. Interface microstructure and mechanical properties of laser welding copper–steel dissimilar joint. *Opt. Lasers Eng.* **2009**, *47*, 807–814. [CrossRef]
63. Prabakaran, M.; Kannan, G.R. Optimization and metallurgical studies of CO_2 laser welding on austenitic stainless steel to carbon steel joint. *Ferroelectrics* **2017**, *519*, 223–235. [CrossRef]
64. Perricone, M.J.; Dupont, J.N.; Anderson, T.D.; Robino, C.V.; Michael, J.R. An Investigation of the Massive Transformation from Ferrite to Austenite in Laser-Welded Mo-Bearing Stainless Steels. *Met. Mater. Trans. A* **2010**, *42*, 700–716. [CrossRef]
65. Nikulina, A.A.; Bataev, I.; Smirnov, A.I.; Popelyukh, A.I.; Burov, V.; Veselov, S.V. Microstructure and fracture behaviour of flash butt welds between dissimilar steels. *Sci. Technol. Weld. Join.* **2014**, *20*, 138–144. [CrossRef]
66. Taban, E.; Deleu, E.; Dhooge, A.; Kaluc, E. Laser welding of modified 12% Cr stainless steel: Strength, fatigue, toughness, microstructure and corrosion properties. *Mater. Des.* **2009**, *30*, 1193–1200. [CrossRef]
67. Rossini, M.; Spena, P.R.; Cortese, L.; Matteis, P.; Firrao, D. Investigation on dissimilar laser welding of advanced high strength steel sheets for the automotive industry. *Mater. Sci. Eng. A* **2015**, *628*, 288–296. [CrossRef]
68. Ceglarek, D.; Colledani, M.; Váncza, J.; Kim, D.Y.; Marine, C.; Kogel-Hollacher, M.; Mistry, A.; Bolognese, L. Rapid deployment of remote laser welding processes in automotive assembly systems. *CIRP Ann.* **2015**, *64*, 389–394. [CrossRef]

69. Bakir, N.; Üstündağ, Ö.; Gumenyuk, A.; Rethmeier, M. Experimental and numerical study on the influence of the laser hybrid parameters in partial penetration welding on the solidification cracking in the weld root. *Weld. World* **2020**, *64*, 501–511. [CrossRef]
70. Kik, T.; Górka, J. Numerical Simulations of Laser and Hybrid S700MC T-Joint Welding. *Materials* **2019**, *12*, 516. [CrossRef] [PubMed]
71. Nagaraju, U.; Gowd, G.H.; Ahmad, R.M.I. Parametric analysis and evaluation of tensile strength for laser welding of dissimilar metal. *Mater. Today Proc.* **2018**, *5*, 7898–7907. [CrossRef]
72. Prabakaran, M.P.; Kannan, G.; Lingadurai, K. Microstructure and mechanical properties of laser-welded dissimilar joint of AISI316 stainless steel and AISI1018 low alloy steel. *Caribb. J. Sci.* **2019**, *53*, 978–998.
73. Ki, H.; Mazumder, J.; Mohanty, P.S. Modeling of laser keyhole welding: Part II. Simulation of keyhole evolution, velocity, temperature profile, and experimental verification. *Met. Mater. Trans. A* **2002**, *33*, 1831–1842. [CrossRef]
74. Esfahani, M.R.N.; Coupland, J.; Marimuthu, S. Microstructure and mechanical properties of a laser welded low carbon–stainless steel joint. *J. Mater. Process. Technol.* **2014**, *214*, 2941–2948. [CrossRef]
75. Pekkarinen, J.; Kujanpää, V. The effects of laser welding parameters on the microstructure of ferritic and duplex stainless steels welds. *Phys. Procedia* **2010**, *5*, 517–523. [CrossRef]
76. Vitek, J.M.; Dasgupta, A.; David, S.A. Microstructural modification of austenitic stainless steels by rapid solidification. *Met. Mater. Trans. A* **1983**, *14*, 1833–1841. [CrossRef]
77. Sinha, A.K.; Kim, D.Y.; Ceglarek, D. Correlation analysis of the variation of weld seam and tensile strength in laser welding of galvanized steel. *Opt. Lasers Eng.* **2013**, *51*, 1143–1152. [CrossRef]

Review

Manufacture and Performance of Welds in Creep Strength Enhanced Ferritic Steels

Jonathan Parker * and John Siefert

Electric Power Research Institute, Charlotte, NC 28262, USA
* Correspondence: jparker@epri.com; Tel.: +1-704-595-2791

Received: 3 June 2019; Accepted: 11 July 2019; Published: 13 July 2019

Abstract: Welding is a vital process required in the fabrication of 'fracture critical' components which operate under creep conditions. However, often the procedures used are based on 'least initial cost'. Thus, it is not surprising that in many high energy applications, welds are the weakest link, i.e., damage is first found at welds. In the worst case, weld cracks reported have had catastrophic consequences. Comprehensive Electric Power Research Institute (EPRI) research has identified and quantified the factors affecting the high temperature performance of advanced steels working under creep conditions. This knowledge has then been used to underpin recommendations for improved fabrication and control of creep strength enhanced ferritic steel components. This review paper reports background from this work. The main body of the review summarizes the evidence used to establish a 'well engineered' practice for the manufacture of welds in tempered martensitic steels. Many of these alternative methods can be applied in repair applications without the need for post-weld heat treatment. This seminal work thus offers major benefits to all stakeholders in the global energy sector.

Keywords: steel; weld; high temperature; creep; fracture; advanced methods

1. Introduction

EPRI is a not for profit organization that has been providing independent technical support to global stakeholders in the electricity supply industry for over 40 years. Within EPRI's generation sector, a key research imperative is knowledge creation and technology transfer linked to the reliable, safe, and economically flexible operation of power plants. Collaborative achievements have included contributions to the development of databases of key properties for high temperature alloys, publication of recommended guidelines for design and fabrication as well as compiling case studies of in-service problems and facilitating expert root cause assessment. Technology transfer has ensured that lessons learned can be used to establish best practices; these activities include annual workshops, the publication of summary documents and identification of additional research. In all cases, excellence in science and engineering is necessary to underpin technology which will help to meet challenges associated with the safe and reliable operation of a plant.

The present paper reviews key findings from over 10 years of collaborative research which was carried out to identify and quantify the factors affecting the high temperature behaviour of tempered martensitic steels. In particular, selected information is provided from a series of EPRI-coordinated, industry-sponsored projects on the manufacture and performance of welds made in 9%Cr creep strength-enhanced ferritic (CSEF) steels. The initial work in this area provided the basis for a meaningful asset management strategy for Grade 91 steel components. This seminal project involved more than 40 participants providing over $4 million of industry funding. The learnings and findings were realized with direct input and perspective from stakeholders representing the entire electricity supply chain. The knowledge created established that in addition to operating variables such as stress state and temperature, the high temperature behaviour of CSEF steels was dependent on a

Metallurgical Risk Factor. Meaningful assessment of component creep behaviour could thus only be achieved by integrating information from steel making and fabrication with microstructure and operating conditions, with particular emphasis on complex transient loading.

A key outcome from this research was that for metallurgically complex steels such as CSEFs, it is vital that research programmes should establish high temperature performance on steel sections which are carefully chosen and well characterized [1]. Thus, it is important that the full pedigree of the steel heat or cast selected for a research project is known or checked. For example, when research is assessing the creep behaviour of welds, the welds must be made in the same base substrate so that results can be directly compared. Similarly, it should not be assumed that the behaviour of welds made in plate will exhibit the same performance trends as welds made in tubes or pipes.

Achievements directly linked to factors controlling the high temperature performance of welds in CSEF steels are reviewed in this paper with detailed background provided in the original reports and papers [1–3]. The primary areas covered are summarized below:

1. Detailed research examining the microstructure of tempered martensitic ferritic steels has provided key information concerning the rate controlling damage mechanisms for different creep conditions. Initially sections from early in life component failures were examined to establish damage patterns and to highlight the critical evidence, see for example Figure 1. The root cause analyses performed [4] have established that the inherent cavity susceptibility of the Grade 91 base steel is a key factor which is directly linked to damage development in the heat affected zone (HAZ) of welds. This susceptibility also influences behaviour under cyclic creep conditions (often described as creep fatigue).
2. The microstructure in the heat-affected zone (HAZ) of welds made in 9% chromium martensitic steels is complex. Systematic investigations using advanced methods have defined the different regions of the microstructure across the HAZ as functions of the precipitate condition and the welding process [5,6]. These regions are categorized in terms of the influence of the thermal cycles on the sub-structural parameters controlling creep deformation and fracture. Results from cross weld feature tests then provided direct evidence to rationalize the apparently conflicting published evidence regarding nucleation, growth and linkup of creep cavities in the HAZ of welds.
3. Continuum damage mechanics (CDM) based methods have been available for some time. However, recent research has focused on technically relevant descriptions to evaluate both when a component will fail and also how it will fail [7]. For many operators and end users, there is a significant benefit to selecting component design and manufacturing methods which are inherently Damage Tolerant, i.e., there is a long period between crack initiation and fracture. Application of CDM has shown that the inherent advantages of reducing the risks of catastrophic fracture do not require significant greater financial investment [8,9]. These benefits can instead be realized by the application of sound engineering approaches to weld design and manufacture.
4. A clear technical route has established the improved design and fabrication for well-engineered welds in CSEF steels. Applying engineering best practice to welds for high energy applications is always of value. However, particular benefit has been found with establishing three alternative options for repair of Grade 91 steel components [10]. In particular, research, conducted over the last five years, has identified that repair welds in CSEF steels can be made without the need for post weld heat treatment. Indeed, detailed tracking of real components showed that badly controlled PWHT actually resulted in problems. Expert review of the EPRI weld repair methods has culminated in the acceptance within the National Board Inspection Code, USA, of Welding Method 6 for tubes and Welding Supplement 8 for thick section parts [11].

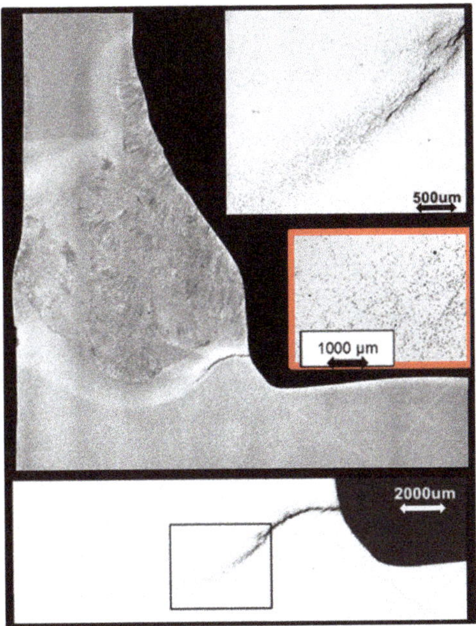

Figure 1. Micrographs showing detail of cracking at a stub tube to header weld. Cracking found at this location at an earlier inspection had been removed by local grinding. It is noteworthy that the re-cracking event did not take place at the base of the excavation. Instead, the creep damage is focused on the metallurgically susceptible structures in the weld HAZ.

It is apparent that the uncertainty regarding component performance is in part a consequence of inadequacies during the design and fabrication. Sometimes less than optimal approaches are justified on the basis of least cost. Failure to use best practice specification, design and manufacturing methods may also be the result of a lack of knowledge or lack of control or both. In all cases, excessive variability in the condition of a component or structure at the start of service life results in major downstream uncertainties which may include problems with fractures and forced outages. With particular reference to the fabrication and high temperature performance of CSEF steels, it is apparent that the EPRI knowledge base offers technical information and guidance to minimize variability in behaviour and thus greatly simplifies asset management [12]. The information described in this review paper should be used by designers, manufactures and end users to help to meet the challenges of economic, safe and reliable operation of high energy components.

2. Metallurgical Risk Factor

The high temperature performance of CSEF steels is dependent on the details of steel making, steel processing, composition, and heat treatment [1]. Moreover, it is not possible to simply 'recover' the performance of a poorly made steel batch by application of a subsequent heat treatment. Thus, the final quality or the steel to be used starts with high quality in composition control, steel making, poring etc. While a steel section with a poor microstructure cannot be easily fixed by heat treatment, it is now very well established that the poor control of heat treatment can lead to different problems. Thus, for example, when irregularities of heat treatment result in a ferrite microstructure (rather than the martensitic structure expected), the creep strength is significantly reduced compared to the creep strength of tempered martensite.

EPRI recommendations clearly advocate performing detailed metallographic characterization and documentation of the damage present after creep testing. This assessment is particularly important for tests of long duration since these results are generally considered most representative of behaviour of components in service. In CSEF steels these long-term tests generally fracture with very little reduction of area. This creep brittle behaviour is a consequence of the formation and growth of cavities and the subsequent formation of micro and macro cracks. Careful examination using modern techniques has shown that creep cavities are formed both on prior austenite grain boundaries and other features in the martensitic microstructure such as at lath boundaries and at inclusions. The diversity of the microstructural sites which develop cavities is illustrated in Figure 2. It appears that the formation of voids at these diverse sites makes the formation of micro cracks more difficult.

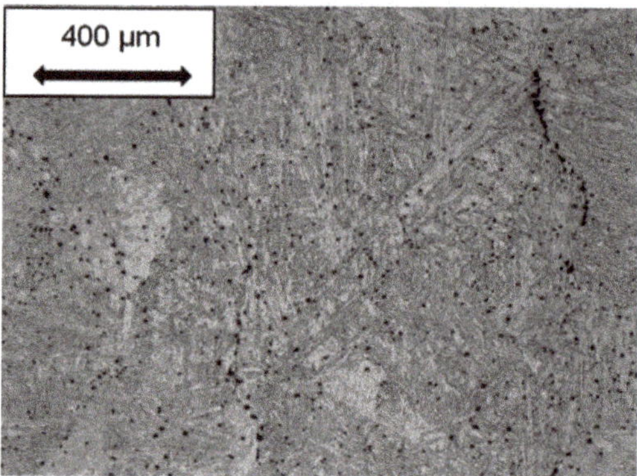

Figure 2. Micrograph showing detail of creep cavities formed in the base metal of a Grade 92 CSEF steel. These creep voids nucleate at a size below 1 μm and grow during component life. The sample shown here is close to final fracture, yet no macro cracks had formed.

The diversity of void nucleation sites creates a challenge to tracking in-service component damage using traditional inspection methods. While the details of the number of voids formed, and the tendency for reductions in strain to fracture, is different for the different CSEF steels, research to date [1] shows that void nucleation is related to the presence of trace elements and hard nonmetallic inclusions. In Figure 3 detailed characterization of the inclusions present has been performed. The steel identified as Barrel 2 has a relatively high inclusion number density and exhibits low creep ductility. In contrast, the steel identified as Tee piece 1 has a relatively low number density of inclusions and exhibits a high creep ductility.

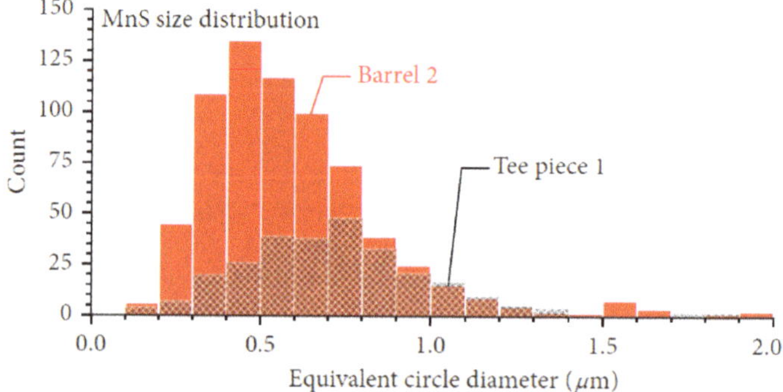

Figure 3. Number density of inclusions for two different but ASME code acceptable Grade 91 steels. A higher density of MnS was measured in the creep damage susceptible steel which contained 0.009 wt.% S (labelled barrel 2) as compared to the cavity resistant steel which contained 0.002 wt.% S (labelled tee piece 1).

An example of the link between the nucleation of a creep void, the presence of inclusions and segregation of trace elements is illustrated by the analysis results shown in Figure 4. The analysis reveals that high levels of copper are found associated with the creep void. Copper in this location can act to promote cavitation by reducing the local surface energy. A further key factor in determining whether inclusions aid the nucleation of voids is the particle size. Thus, only inclusions of a sufficient size (the critical inclusion size is directly linked to the creep stress) will be thermodynamically stable and thus able to act directly as nucleation sites.

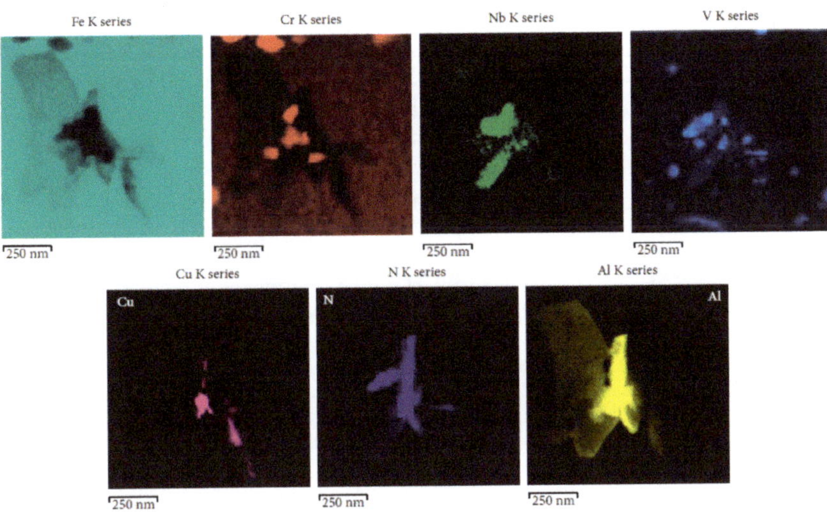

Figure 4. Detailed micrograph showing inclusions associated with an individual creep cavity as well as compositional maps showing the distribution of the elements Cr, Nb, V, Cu, N, and Al associated with this location.

It is clear from root cause analyses of Grade 91 steel components [1,4] that steel composition and processing variables are linked to low creep ductility in base metal and creep cracking in weld HAZs. The observed in-service component damage cannot be simply explained as a 'one off' anomaly. Examination of failures in several Grade 91 welded components [13,14] indicates a trend where an increasing number of failures of Grade 91 welds are occurring in times below that expected based on simple design rules. The trend in the reduced creep ductility in Grade 91 steel and the link to very low HAZ creep life may be a characteristic of a significant number of components which entered service with compositions which show a high susceptibility for cavity formation. It should be emphasized that the poor creep performance of weld HAZs is a key reason for the introduction of weld efficiency factors. These factors were aimed at reducing the risk of in-service damage by increasing the component thickness. Further, reductions in weld creep performance or indeed greater uncertainty over long term performance would be expected to lead to increases in recommended weld strength reduction factors (WSRF) or weld-efficiency factors. The evidence from EPRI research shows using Grade 91 steel with low densities of inclusions and controlled levels of deleterious trace elements should significantly reduce the risk of creep cracking associated with weldments. Selection of improved base material would provide significant benefit for high energy systems.

In order to properly assess metallurgical risk, it is necessary to obtain a full chemical composition for Grade 91 steel base material. As highlighted previously [1,15,16], there is concern that the influence of tramp elements such as As, S, Sn, Sb and Cu has been underappreciated and that these elements are playing a role in the reduction in creep ductility in martensitic CSEF steels. In general, the analysis of elements can be grouped into two sets: elements required by common specifications for Grade 91 steel

(14 total elements) [1] and elements for informational purposes (typically 10+ additional elements). The following approaches are typically used to determine the composition of each of the elements in EPRI research. Inductively coupled plasma optical emission spectrometry (ICP-OES) was utilized to determine the values for: Al, B, Ca, Co, Cr, Cu, La, Mn, Mo, Nb, Ni, P, Si, Ta, Ti, V, W, Zr. Inductively coupled plasma mass spectrometry (ICP-MS) was used to determine the amounts of As, Bi, Pb, Sb, Sn. Finally, combustion was necessary to determine the C, S levels while insert gas fusion (IGF) to assess the amount of O and N in the steel. It should be noted that, to ensure that sufficient information is provided for each of the requested elements, the number of required significant digits should also be specified in any specification to the laboratory performing the analysis.

Discussions of industry experience in general, and the EPRI recommendations in particular, have now resulted in agreement within ASME that for Grade 91 steels there should be a Type I and a Type II designation [16]. It has been recommended that the steel made with greater controls designated Type II, will have higher Allowable Design stress values [16]. This recommendation is reflected in the ASME Code Case 2864, Table 1 and associated documentation. Similar specifications have been approved by ASTM for the component specific requirements. Discussions continue about the influence of trace elements on creep performance of other CSEF steels. However, the position recommended by EPRI is clear. In all cases, purchase of CSEF steel components should recognize the complexity of the metallurgy, the need for care to achieve excellent creep performance without excessive variability. Thus, well controlled composition and fabrication should be mandated.

Table 1. Composition for Controlled Quality Grade 91 steel specified in ASME Code Case 2864 [16].

C	Mn	Cr	Ni	Mo	V	Nb	N	B	W
0.08–0.12	0.30–0.50	8.0–9.5	0.20 max	0.85–1.05	0.18–0.25	0.06–0.10	0.035–0.070	0.001 max	0.010 max
Si	**Al**	**Cu**	**S**	**P**	**As**	**Sn**	**Sb**	**Pb**	–
0.20–0.40	0.020 max	0.10 max	0.005 max	0.020 max	0.010 max	0.010 max	0.003 max	0.001 max	–

3. Weld Manufacture and HAZ Microstructural Characterization

Proper characterization of the microstructure in metallurgically complex steels is complicated by the fact that most of the features which define creep strength and ductility cannot be resolved or studied using optical metallography. Moreover, the diversity of thermal cycles experienced by multi-pass fusion welds further complicates meaningful characterization because without care there is a very significant spatial variation throughout the weld and HAZ.

The preferred approach to overcome the problems of relevant examination and recording meaningful information, it is usually necessary to balance the results from macro-, micro- and nano-evaluation with appropriate analysis. Thus, it is important to take an overall view to microstructure and then increase detail based on the information recorded. In view of the challenges associated with nano level examination this level of detail can only be performed selectively. It is thus vital to make sure that the methods used for this selection are rigorous.

This section summarizes information regarding EPRI recommended approaches [17] for this characterization for welds manufactured in CSEF steels. It should be emphasized that in all cases the procedures used have been validated and checked against calibrated sections.

3.1. Weld Manufacture

EPRI has published a series of documents which describe procedures for weld manufacture. The outline below is included here because it is critical to research projects that the base sections used is well pedigreed and the subsequent welding is performed in a controlled manner. Lack of control of fusion welding will lead to a very wide range of local temperature cycles and therefore microstructures in any weld.

In thick section components such as pipes EPRI has typically used a machined U-groove with a 15° bevel and using best practice guidance for the shielded metal arc welding (SMAW) process as detailed previously [10]. For Grade 91 steels the process included a minimum preheat temperature of 150 °C (300 °F), a maximum interpass temperature of 315 °C (600 °F), stringer beads only, and removal of slag after each weld layer through light grinding. The filler material used to make the weldments was consistent with an American Welding Society (AWS) type E9015-B9 filler material. Stipulating stringer beads without weaving and using only 3.2 mm (0.125 inches) diameter electrodes limited the variability in the heat input. The completed weldment including a macro sample, documented fill sequence, and the recorded data for amperage, voltage, travel speed, and interpass temperature are provided elsewhere [10].

Following welding, the weldment was allowed to cool to room temperature. In some cases, EPRI research into weld repair investigated the performance of welds without post weld heat treatment (PWHT). In other cases, and for the weld shown in Figure 5, a relatively low temperature of PWHT, namely at 675 °C (1250 °F) held for 2 h, was used. Full details of the development and testing of welds made using alternative weld and PWHT procedures have been reported [10].

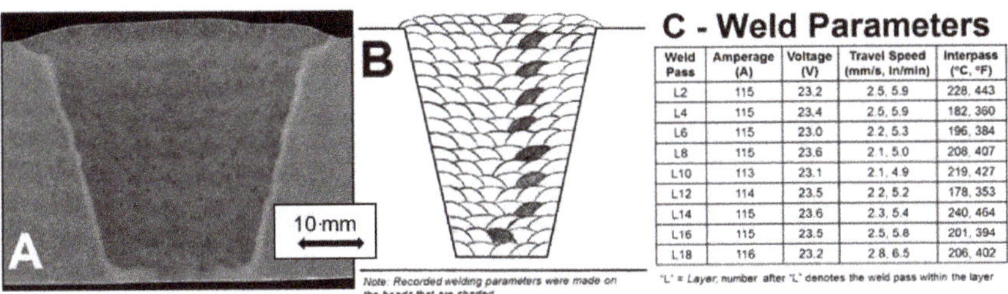

Weld Pass	Amperage (A)	Voltage (V)	Travel Speed (mm/s, In/min)	Interpass (°C, °F)
L2	115	23.2	2.5, 5.9	228, 443
L4	115	23.4	2.5, 5.9	182, 360
L6	115	23.0	2.2, 5.3	196, 384
L8	115	23.6	2.1, 5.0	208, 407
L10	113	23.1	2.1, 4.9	219, 427
L12	114	23.5	2.2, 5.2	178, 353
L14	115	23.6	2.3, 5.4	240, 464
L16	115	23.5	2.5, 5.8	201, 394
L18	116	23.2	2.8, 6.5	206, 402

"L" = Layer, number after "L" denotes the weld pass within the layer

Figure 5. (**A**) Macro Sample of the As-fabricated Weldment in the Post Weld Heat Treated Condition. (675 °C, 1250 °F for 2 h) as shown the pipe thickness is 50 mm; (**B**) Fill Sequence used to complete the Weldment. Note that the darkened fill passes constitute a fill pass that was monitored for voltage, amperage, travel speed and interpass; (**C**) Details for the Monitored Fill Passes.

The weld macro section shown in Figure 5A demonstrates that the appearance of the weld was consistent with the expected bead sequence. This is shown in Figure 5B. Thus, it was clear that the requirements of the procedure had been followed. No obvious regions of inhomogeneity in the weld macrostructure were identified. Because of the limitations of optical metallography regarding defining microstructural differences, to provide visual images of the macroscopic bead size and shape and the overall pattern of structure EPRI has developed procedures for macro-analysis which includes hardness mapping [17]. The equipment utilized for the hardness mapping characterization was a LECO Automatic Hardness Tester, Model AMH-43. Hardness mapping was conducted so that the requirements in both ASTM E384-11 (ASTM 2011) and ISO 6507 (ISO 2005) were met. One of the key requirements in these two standards is that for a given hardness load (e.g., for this study 0.5 kgf), the indents should be at least 2.5 d apart (where d = mean diagonal distance of the measured Vickers indent in the material being examined).

To ensure that sufficient resolution in the data was obtained, i.e., to achieve enough indents in the heat affected zone, a very large area around the weld fusion boundary was analyzed. Thus, in contrast to simple hardness line scans carried out in some traditional studies, in the present research representative portions of the base metal, heat affected zone (HAZ) and deposited filler metal were captured in the hardness map. This approach resulted in a final hardness map size that was 25 mm × 25 mm and included a total of 10,000 indents. The location examined is shown as the highlighted box in Figure 6A with the hardness map produced shown in Figure 6B.

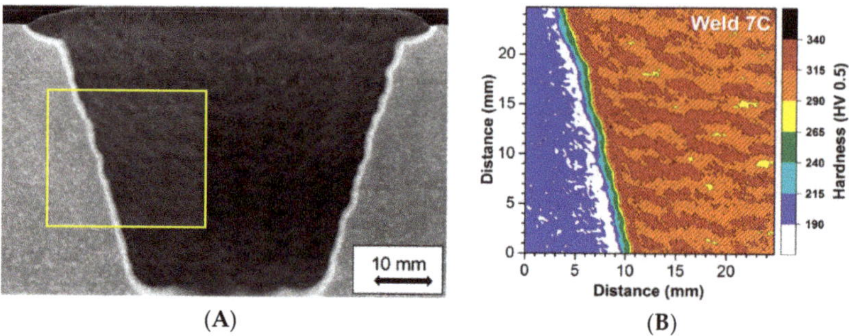

Figure 6. (**A**) Macro Sample of the As-fabricated Weldment in the Post Weld Heat Treated Condition (675 °C, 1250 °F for 2 h) with the yellow box showing the region examined by hardness mapping; (**B**) The hardness map produced with a scale showing hardness ranges for each colour.

It should be noted that the regions of highest hardness in the weld were recorded at, or close to, the centre of the individual beads. The pattern exhibited by these regions shows the uniformity of the beads deposited. The highest hardness recorded were values that are <350 HV 0.5. This observed maximum value is higher than for typical weldment manufacturing where the hardness values are reduced to ~<300 HV 0.5. The difference is attributed to the specified PWHT namely (675 °C for 2 h), which is consistent with the recognized minimum in the National Board Inspection Code Part 3 Repairs and Alterations Supplement 8 (NBIC 2017). This Code provides post construction repair requirements for power generation components and materials including Grade 91 steel and is commonly utilized in North America and in some southern European countries.

It should also be noted that recently ASME B&PV Code Section I reduced the specified minimum PWHT requirements for Grade 91 steel type welds in new construction. The stipulated minimum conditions are 705 °C (1300 °F) for weld thickness > 13 mm (0.50 inches) and 675 °C (1250 °F) for weld thickness ≤ 13 mm (0.50 inches) [18].

Heat Affected Zone Microstructure

Historically, the microstructures in Low Alloy Steel welds have been classified based on the prior austenite grain size and whether the original precipitates present had tempered. This type of characterization is not technically justified for tempered martensitic steels since properties are controlled by substructure as well as the type and size of precipitates. Recent research has therefore been performed to properly characterize the HAZ regions in 9 wt. % Cr CSEF steels. This research [5,6] involved systematically investigating the microstructural distribution in the HAZ of single pass and multipass welds and performing microstructural simulations under known conditions. The multipass welds capture the accumulated influences from multiple weld thermal cycles.

The advanced characterization techniques used included a Nova 600 Nanolab dual beam (Thermo Fisher Scientific, Hillsboro, OR, USA) focused ion beam field emission gun scanning electron microscope was used to collect the electron backscatter diffraction (EBSD) data from the matrix. Ion-beam-induced secondary electron imaging was used to evaluate the distribution of precipitate particles. EBSD maps were collected using an EDAX Hikari camera (EDAX Inc., Mahwah, NJ, USA), at an accelerating voltage of 20 kV and a nominal beam current of 24 nA.

The output of the characterization made using these advanced techniques is illustrated with reference to the images shown in Figure 7. The observations and subsequent analysis of the results obtained [5,6] demonstrated that a new description for the HAZ regions in martensitic CSEF steels based on the transformation behaviour was required. This description which emphasizes the effect of the welding thermal cycles on the precipitates present is summarized as follows:

- Completely transformed zone (CTZ)—This region occurs closest to the weld fusion line. Thus, the temperatures range from close to the melting point of the base steel down to about 1020 °C. The combination of time/temperature experienced is sufficient that the original base metal is fully re-austenitized. The application of advanced electron optic techniques reveals that under typical conditions all of the precipitates formed during initial fabrication are dissolved.
- Partially transformed zone (PTZ)—This region occurs where the peak temperatures range from about 1020 °C to about 950 °C. The combination of time/temperature experienced is such that the original parent metal is only partially re-austenitised. In addition, the application of advanced electron optic techniques reveals that under typical conditions there is insufficient thermal energy to dissolve the pre-existing precipitates.
- Over-tempered zone (OTZ)—This region occurs where the peak temperatures experienced are below about 950 °C. The combination of time/temperature experienced is such that the original parent metal is not modified when viewed in the optical microscope. However, the application of advanced electron optic techniques reveals the coarsening of secondary precipitates

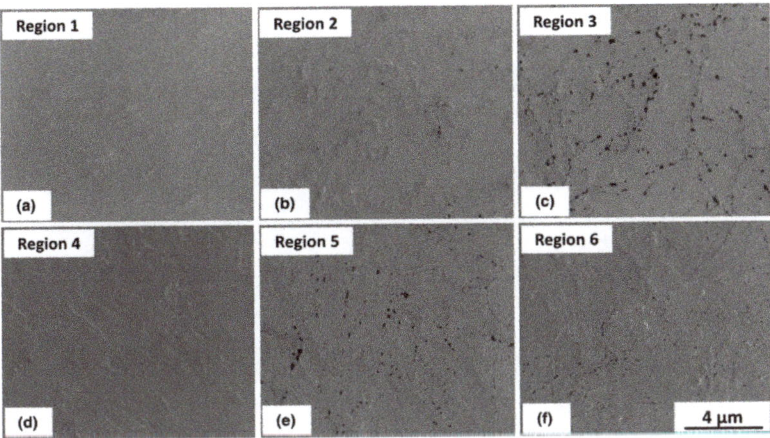

Figure 7. Ion-beam-induced secondary electron micrographs (**a**–**f**) showing the secondary precipitate particles in the HAZ.

These descriptions more closely match the original regions described in the documentation from ORNL [19] and are justified by the extensive nature of the characterization techniques used [5,6].

3.2. Heat Affected Zone Damage

The heterogeneous microstructures in weld HAZ's are such that during creep, multiaxial stresses can develop in large cross weld specimens under uniaxial loading. These multiaxial stresses are established because the large samples constrain the deformation until cracking occurs. Metallographic examination of feature test samples has shown that creep damage in the HAZ is a function of the metallurgical risk inherent in the base steel and the local stress state. In the HAZ the welding thermal cycles modify the parent microstructure. The local stress state established during creep is influenced by the weld geometry, orientation and the properties of the specific microstructural zones.

After the creep failure or termination of the creep test, a macro sample was removed from the post-test feature creep test using fine wire, electrostatic discharge machining (EDM). All specimens were removed from the approximate center (the line in the top image in Figure 8) to analyze the most representative distribution of damage well-controlled, feature-type cross-weld creep tests.

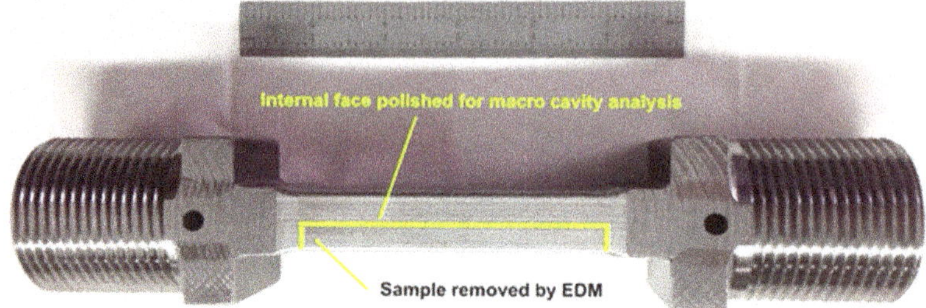

Figure 8. Example of Feature Creep Tests Used in the Evaluation of Damage in the Heat Affected Zone of Grade 91 Steel. Note: the feature test sample contains the entirely of the weld, the HAZ on both sides of the weld and sufficient base metal to promote stress states in the weldment.

When these feature specimens are creep tested to the end of life for low stresses, i.e., long times the fracture path ran through the weld HAZ, Figure 9. Fractures of this nature take place with very little ductility or tearing, and the cross sections shows very little or no reduction in area. Fractures of this type are typical of the damage observed in CSEF welds in service.

Figure 9. Macro Image from a feature sample of Grade 91 Weldment which was creep tested to failure.

Details of the creep damage developed in the specimens have been establish using specialist metallographic techniques. Information regarding the sample preparation and characterization methods have been published [17]. The present paper therefore only summarizes key information used for the laser metallography. However, even a basic appreciation of these factors emphasizes the level of effort invested to obtaining relevant and accurate results. It is apparent that the usefulness of the data is dependent upon careful preparation, observation, recording of data and then analysis. Other researchers are encouraged to comment on specific techniques used in studies documenting the microstructure and creep damage in tempered martensitic steels. The following are the main stages involved with the sample preparation and examination in EPRI research:

- Following the fine wire electro discharge machining the samples were prepared using metallographic preparation procedures which included initial grinding on 240 to 1200 grit SiC and then subsequent polishing on standard cloths with diamond suspension down to 1 μm. A final extended chemo-mechanical polishing procedure was carried out. This was performed using 0.06 μm colloidal silica suspension. This final polish ensured that all evidence of surface deformation that was introduced in the abrasive stage of preparation was eliminated.

- A Keyence VK-X105 Confocal Laser Microscope was used for capturing images from the surface of the prepared specimen. The specimen was mounted on a precision stage which facilitated controlled movement and location of a specific region. In the present case the stage used had a maximum travel distance of 100 mm in both the X and Y orientations.
- For most assessments two pieces of software associated with the Keyence microscope were used for image analysis. These packages were the VK Image Stitching Software and the VK Image Analyzer Software. The VK Image Stitching Software was used to merge the individual images collected into a single compiled image. In general, the images are overlapped by about 12%–15% to ensure that the detailed of the compiled image was accurate. The VK Analyzer Software provided the ability to adjust details of the overall image. Specific examples of the features which could be adjusted include the brightness, contrast, laser intensity and the high dynamic range.

An example of the examination approach is shown in the highlighted region of Figure 10. As with all metallographic techniques there is frequently the need to balance being able to resolve specific items in the specimen and obtaining sufficient images to ensure meaningful observations. Very high magnification can aid in the identification of fine detail, but the small fields involved makes capture of sufficient numbers of features difficult. In the present research a 20× objective was established as appropriate for the analysis of creep cavitation in the HAZ of Grade 91 steel welds. This objective provided a magnification of ~400× on a 15-inch monitor [20].

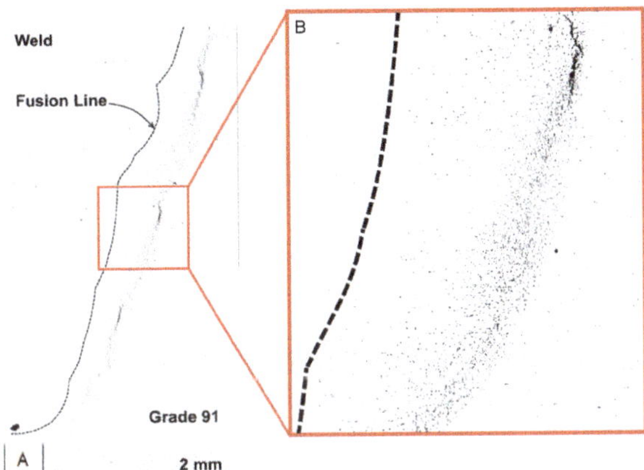

Figure 10. (**A**) Macro Image from the Examined Grade 91 Weldment and Highlighted Region of Interest along the fusion line and HAZ of the Weld; (**B**) Magnified View of Creep Damage in the HAZ of the Weldment.

The magnification is not the only critical variable in the assessment of creep cavitation since the number of pixels in the obtained image can also be altered. The default size for each image collected is 1024 × 768 pixels. The complied macro image can be saved using the full resolution. The number of pixels used for analysis of creep cavitation was 6346 × 3303 pixels. The image size (i.e., the number of pixels) is an important factor in determining the threshold for counting cavities in any material.

The damage developed in cross weld feature testing of a Grade 91 steel which is inherently damage susceptible is illustrated in Figure 11. The creep voids present are shown as dark contrast in Figure 11a. For this feature test, it is apparent that even though high number densities of voids have developed in the weld HAZ there are no cracks present. Thus, for structures which have a uniform susceptibility to void formation, and a similar stress across the section, only very limited periods of stable creep crack growth can take place. The specific region within the HAZ where the highest number of voids was developed is shown in Figure 11b. Based on the detailed analysis performed, it appears that the highest damage levels are found in the region which experienced a thermal cycle in the range 1050 to 920 °C during welding. Whereas, traditional descriptions have termed this as 'fine grained' or 'intercritical', recent EPRI research has established that this region should more correctly be defined as partially transformed [5,7].

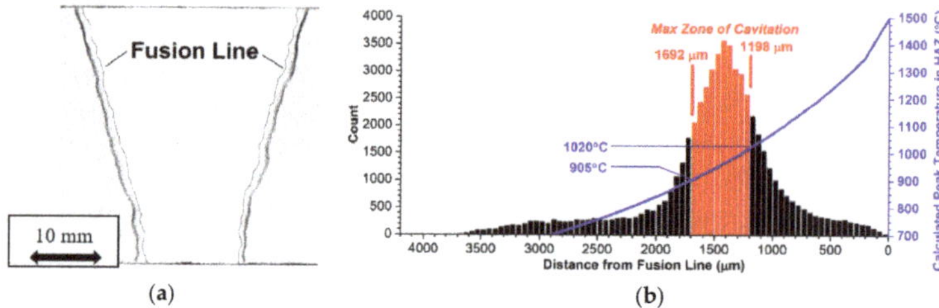

Figure 11. Macrograph showing a Grade 92 steel cross weld feature test sample after long term testing, note using the EPRI classification system the steel parent is considered as 'susceptible' to void formation (**a**) and, histogram showing the number of voids present as a function of the distance from the weld fusion line (**b**). The basic EPRI procedures used for sample preparation, data capture and analysis have been detailed previously [17].

4. Continuum Damage Mechanics

A creep continuum damage mechanics (CDM) constitutive model for Grade 91 steel developed as part of a multi-year collaborative research project has been used to assess the creep performance of structures and components. As described in previous papers [9,20], EPRI research in the area of CDM has been guided by a systematic, step wise methodology. The following list summarizes the important steps followed during the development and testing of this model, and its ability to describe high temperature behaviour:

1. The foundational step involved identification of relevant creep data for 9% Cr tempered martensitic steels. In addition to identifying results from reputable world-wide organizations, EPRI undertook high temperature testing on well pedigreed Grade 91 and Grade 92 steels. The overall results were used to create a database of creep behaviour that included long term results. The availability of the long-term data with failure lives in excess of 10,000 h limited the need for extrapolation. This increased confidence in using the results to describe in-service behaviour.
2. In addition to initial characterization linking details of composition and fabrication to microstructure, all of the EPRI creep tested specimens were evaluated after testing. Evaluation of these data permitted the rate controlling mechanism to be established. The ability to link stress, stress, temperature and metallurgical variables with creep strain and rupture is critical to ensuring that the formulation of the model is based on good science. Thus, knowledge established regarding these factors was used to underpin the formulation of the equations used for the CDM.
3. The rigorous approach to model development advocated by EPRI ensures not only that the expressions used are justified on a mechanistic basis but also that the approaches are not unnecessarily complicated. Initially, the proposed expressions were tested against overall trends in the results rather than immediately seeking to describe the creep behaviour for a given set of conditions. As described in Section 2, there is definitive evidence that individual heats of steel exhibited specific behaviour. Thus, the formulations and the associated constants were initially tested against the creep behaviour of selected heats of Grade 91 steel tested using simple specimens under constant load conditions.
4. To increase the component relevance of the validation creep behaviour under complicated stress states was assessed. EPRI had undertaken controlled testing using creep specimens with notches. The geometry of the specimens and the notches as well as the applied load control the stresses developed. In the present research the predictions from the CDM were compared against data for more complex stress states.

5. As described in Section 3 of this paper, EPRI research has involved of feature sized specimens. The constraint in these specimens is introduced because the cross-sectional area in the gauge is up 80 times that of a small cylindrical sample. In cross weld tests this constraint introduces complex stress because the microstructural heterogeneity acts as a metallurgical notch.
6. As described in Section 5 of this paper, EPRI research also full-size vessel tests undertaken on components typical of those found in plant. These tests require expert design, fabrication and execution but since all aspects of the tests are well controlled the results provide a unique opportunity for evaluation of predictions.
7. The knowledge developed from root cause analysis of in-service failures offers perhaps the most important opportunity for model validation. These examples of course involve some uncertainty as in an operating power plant specific details of the stress and temperature are not known exactly.

The broad scope of these tasks requires an integrated and sustained approach implemented and reviewed over several years. The basis for this research has been described in an EPRI published report [9]. The overall plan required that there would be key points over the course of the model development and validation where the details of the work were reviewed and further evaluated. At each of these review points, important knowledge gaps or needs, such as the requirement for additional metallurgical data, were identified and the necessary testing and analysis undertaken.

Using the logical approach outlined above the expression for the CDM was formulated recognizing that the creep deformation and fracture behavior was different at high stresses and at low stresses. The relationship used to describe the stress dependence of the minimum creep rate involved summing two power laws. This overall relationship is shown in Equation (1). This considers the two different high temperature damage mechanisms, namely strain softening and cavitation.

$$\dot{\varepsilon}_{min} = A_{HT}\sigma^{n_H} + A_{MT}\sigma^{n_m} \tag{1}$$

where

$$A_{HT} = A_H \mathbf{exp}(-Q_H/RT) \tag{2}$$

and

$$A_{MT} = A_M \mathbf{exp}(-Q_M/RT) \tag{3}$$

In these equations, $\dot{\varepsilon}_{min}$ is the minimum creep rate, R is the ideal gas constant, Q represents the Activation Energy and T is the temperature in Kelvin.

The damage and strain softening state variable are incorporated into the strain rate equation as shown in Equation (4). Damage as a result of strain softening was defined by the state variable G, which is 0 in the initial condition and when ductile failure occurs. The softening rate is proportional to the strain rate, see Equation (5). Cavitation damage was described associated with the formation and growth of creep cavities. Following the work of Kachanov [21], cavitation was represented by the state variable, ω, and this also varies from 0 to 1. The classical damage rate equation with $(1-\omega)$ is used to describe the relationship between the rate of cavitation and the value of the maximum principal stress, see Equation (6).

$$\dot{\varepsilon}_{ij}^{c} = \frac{3s_{ij}}{2\sigma_e}\left(A_{HT}{\sigma_e}^{n_H} + A_{MT}\left(\frac{\sigma_e}{(1-\omega)(1-G)}\right)^{n_M}\right) \tag{4}$$

$$\dot{G} = k\dot{\varepsilon}_e^{c} \tag{5}$$

$$\dot{\omega} = \frac{A_{DT}\sigma_I^{x}}{(1-\omega)^{\varphi}} \tag{6}$$

In these equations $\dot{\omega}$ is the rate of damage due to cavitation,σ_I is the maximum principal stress, $\dot{G}$ is the strain softening rate, $\dot{\varepsilon}_e^{c}$ is the equivalent creep strain rate, $\dot{\varepsilon}_{ij}^{c}$ is the creep strain tensor, s_{ij} is the deviatoric stress tensor, and σ_e is the von Mises equivalent stress.

Extensive analysis of the Grade 91 heat affected zone microstructure by EPRI has identified distinct microstructural regions in the HAZ. The first, termed the completely transformed zone (CTZ), is located adjacent to the weld fusion line. This zone has historically been referred to as the coarse-grained heat-affected zone is exposed to the highest temperatures during the welding process and the microstructure which develops in the CTZ is similar to the microstructure observed in the base metal. For this reason, in CDM analyses of cross-weld behavior, material properties in this zone have been assumed to be the same as those of the base metal. The second zone is termed the partially transformed zone (PTZ). The temperatures reached in this region are less than those achieved in the completely transformed zone and a desired dislocation and precipitate substructure is not present. The PTZ is associated with lower strength and greater susceptibility to creep cavitation than in Grade 91 base metal or the CTZ.

The structure of the constitutive model for the partially transformed zone is the same as that of the Grade 91 base metal model. In order to represent the difference in strength between the partially transformed zone and the base metal, a strength factor F was introduced, where $0 < F \leq 1$. When $F = 1$, base metal strength is achieved; when $F < 1$, the creep strength is less than that of base metal:

$$\dot{\varepsilon} = A_{HT}\left(\frac{\sigma}{F}\right)^{n_H} + A_{MT}\left(\frac{\sigma}{F(1-G)(1-\omega)}\right)^{n_M} \tag{7}$$

The strength difference observed between the partially transformed zone and the base metal is described using the factor, F. Since F is a function of temperature an Arrhenius temperature dependence was incorporated as follows:

$$F = 1 - F_0 \exp(-Q_F/RT) \tag{8}$$

The stress redistribution leads to enhancement of the maximum principal stress in the PTZ. This was considered when determining material parameters for the damage mechanism.

Determining material property coefficients for the PTZ model requires data from a minimum of three test types (each performed over a range of temperatures and stresses): uniaxial smooth bar tests of the base metal to define base metal creep strength; uniaxial smooth bar tests of simulated PTZ material to define the creep strength of the PTZ; and a test imposing triaxial constraint on the PTZ or simulated PTZ material to determine the cavitation damage response. At present, cross-weld tests have been used to define the damage response of the PTZ. To ensure the coherence and relevance of the developed data, all the proceeding tests should be performed on material from the same heat.

Testing according to the above methodology was performed by Hongo et al. [22], representing a coherent and relevant dataset that could be used to determine material parameters for the partially transformed zone (note that the partially transformed zone is referred to as the fine-grained HAZ in [22]). A comparison of the measured and predicted time to rupture for uniaxial creep rupture tests of Grade 91 steel base metal, simulated HAZ (PTZ), and cross-welds is shown in Figure 12. A key item to note is that the model predicts failure by the development of damage in the HAZ for cross-welds tested at low stresses, but failure by creep rupture of the base metal at high stresses. The behavior is thus directly in accordance with the trends noted in tests reported elsewhere [22].

As an example of the application of CDM to the behaviour of a component, the physically-based creep continuum damage constitutive model was applied to the serviceability assessment of a 90° large bore, welded branch connection [9]. This connection had developed a steam leak which occurred as the result of in-service cracking. The connection was removed from the component and selected samples were metallographically prepared and examined in detail, Figure 13. The model based predicted time to crack initiation (when), location of crack initiation (where), and direction of subsequent crack growth (how) were shown to agree with the observed trends in the reported failure. The agreement between observation and analysis underpins confidence that cracking initiated at the surface. This indicates that nondestructive examination techniques such as surface metallurgical replication and magnetic particle testing can be used to detect damage. However, some caution is required since the operating time between initiation and through-wall cracking can be short. For the exemplar component a leak-break-before break assessment was performed to provide an estimate of damage tolerance, i.e., an estimate of how long the crack would take from initiation to through wall growth. This further analysis demonstrated that the crack growth rate was relatively slow and there was a very low risk of the connection to fracture catastrophically. The expectation for a leak type failure was based on the fact that crack extension of greater than 90° around the branch pipe would be necessary for catastrophic rupture of the connection to occur. The prediction of leak not break is consistent with experience of cracks in similar components. Damage for the typical component geometries is found by steam leaks, not catastrophic rupture.

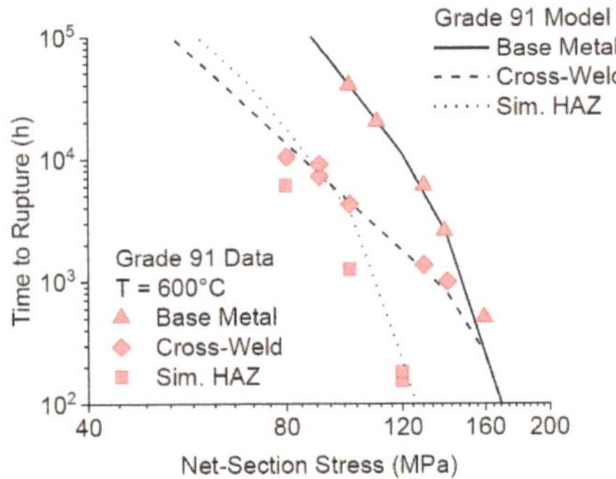

Figure 12. Comparison of measured and model predicted time to rupture for Grade 91 base metal, simulated HAZ, and cross-welds all tested at 600 °C and under uniaxial loading. The data shown in this Figure were published previously [22], the lines shown are as estimated from the CDM analysis.

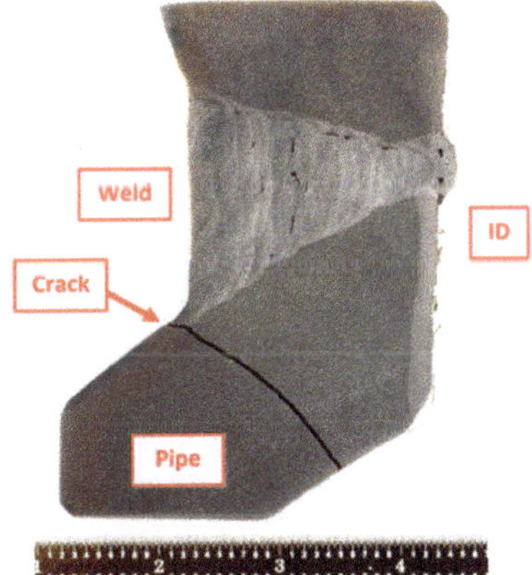

Figure 13. Metallographic section showing in-service cracking at a welded tee piece connection. The scale shown in graduated in inches.

5. Advanced Weld Repair Technologies

EPRI has a long-established reputation in the development and validation of weld repair methodologies. Research supporting fossil-fired asset management has developed viable weld repair techniques for mainstay power generation CrMo steels which have been widely used for coal-fired boiler or combined cycle heat recovery steam generator (HRSG) components. A series of reports provide research results and document weld repair procedures specific to CrMo steel Grades 11, 12 and 22 [23]. It is clear the knowledge base published provides an important strand for cost-effective life management, i.e., informed run/repair/replace decision-making. The research performed through the EPRI studies, in part, led to the acceptance of Welding Method 4 and Welding Method 5 in the NBIC Part 3 Repairs and Alterations. These weld repair methods are often cited as 'temperbead methods'. Although the primary consideration for CrMo and CrMoV repair is the avoidance of reheat cracking through the concept of 'grain refinement' rather than simply tempering the microstructure.

EPRI-coordinated, industry-sponsored research projects in CSEF steels began in 2007 with a major effort to establish sound life management approaches for Grade 91 steel components. The learnings and findings, realized with direct input and perspective from over 40 stakeholders representing the entire electricity supply chain, underpin component specific repair methodologies for the CSEF steel Grade 91 in use today. Key evidence for these 'alternative' repair methods are summarized below.

5.1. Development of Alternative Repair Methodologies

Historically, it has often been the case that weld repairs performed after periods of in-service operation were based on the procedures mandated for new construction. Clearly since the equipment is not new it is important to consider whether this approach is reasonable. A major EPRI initiative investigating and quantifying the factors affecting weld repair of CSEF steels started in 2011. After this initial thrust, a series of research programs has been undertaken to develop well-engineered weld repair procedures, initially for Grade 91 steel components [10]. The variables studied in these studies are summarized in Table 2. It is important to emphasize that this research resulted in acceptance of weld repair procedures which could be applied without the need for PWHT. The development of weld repair methodologies has been largely proactive, so that accepted procedures were established before widespread failures were reported. Today, the research continues to evolve the repair methods for dissimilar metal welds to Grade 91 steel and for the next generation of 9 to 12%Cr CSEF steels such as Grade 92.

Table 2. Variables Studied in the Examination of Well-Engineered Repair Methods in Grade 91 Steels.

Excavation	Base Metal	Weld	Weld Metal	Process
Minor (50% depth)	Tee Piece 2 to Barrel 3 (Groove in Barrel 3)	1C	E9015-B91	675 °C PWHT
		2C	E8015-B8	Controlled Fill
		3C	EPRI P87	Controlled Fill
Partial (50% depth)	Barrel 1 to Tee Piece 1	4C	E9015-B91	675 °C PWHT
		5C	E8015-B8	Controlled Fill
		6C	EPRI P87	Controlled Fill
Full Repair (~95% depth)	Tee Piece 1	7C	E9015-B91	675 °C PWHT
		8C	E8015-B8	Controlled Fill
		9C	EPRI P87	Controlled Fill
Full Repair (100% depth)	Barrel 1 to Barrel 2	10C	E9015-B91	675 °C PWHT
		11C	E8015-B8	Controlled Fill
		12C	EPRI P87	Controlled Fill
Refill+ Repair (50% depth)	Barrel 1 to Barrel 2 (Groove in Barrel 2)	13C	E8015-B8	Controlled Fill
		14C	EPRI P87	Controlled Fill
Partial (25% depth)	Tee Piece 2 to Barrel 3	15C	E8015-B8	Controlled Fill
		16C	EPRI P87	Controlled Fill

Note: Welds made using Controlled Fill were NOT heat treated prior to testing.

The power generation industry is now recording operating times in boilers and piping systems using Grade 91 steel which exceed 100,000 h. The likelihood for damage increases with increasing operating time, so that as time in service increases so does the need for more regular and widespread repairs. More than two dozen EPRI reports underpin the recommended repair practices now recognized by the NBIC Part 3 Repairs and Alterations as Welding Method 6 (first published in 2015 edition) and Welding Supplement 8 (first published in 2017 edition). These methods include options without a mandatory post weld heat treatment (PWHT). Examples of publicly available information can be found EPRI documents for example [10,24,25].

The design and implementation of well-engineered weld repairs in Grade 91 steel components required a number of considerations be taken into account to provide a best practice approach. It was apparent that there is no one-size-fits-all procedure for all alternative weld repairs. The critical decisions and recommendations that were reviewed during the code approval process for Welding Method 6 and Welding Supplement 8 have been explained in a key document. Some of the important questions raised and answered during the code approval process have been documented [26]. These summaries concerned the following issues on performance:

- Responsibilities of authorized inspectors and jurisdictions;
- Welding qualifications and monitoring repair quality;
- Documenting repair information and weld repair performance;
- Key definitions such as "well-engineered repair" and "impractical";
- Filler metal selection;
- Defect removal and excavation practices and partial or localized excavations;
- Applicability to creep strength enhanced ferritic steels other than Grade 91;
- Susceptibility to stress corrosion cracking;
- Limited access repairs;
- Damage and root cause analyses;
- Using a minimum post weld heat treatment of 1250 °F (675 °C).

The knowledge base supporting these repair methods will be updated periodically as needed to address questions specific to new or emerging issues associated with alternative weld repairs in Grade 91 steel.

5.2. *Validation of Alternative Repair Methodologies*

Although significant information was produced to validate the alternative repair methods during the first phases of research on-going studies include refinement of performance expectations for the weld repairs. These further approaches for quantification of behavior include the use of state-of-the-art testing and evaluation methods to provide a comprehensive basis for informed decision making. These independent approaches include:

- feature-type cross-weld creep testing including assessment of the creep behaviour of dissimilar metal welds (Figure 14);
- full-size vessel tests undertaken on components typical of those found in plant (Figure 15), and
- continuum damage mechanics and constitutive modeling of the performance of repairs in the structural components (Figure 16).

A typical feature test cross weld creep specimen in shown in Figure 14A. This specific example shows a weld which simulates the manufacture of repair in a thick section joint between Grade 91 and Grade 22 steels. The width of the repair is purposely designed to extend beyond the HAZ of the original weld profile. This extended fill creates a step in the repair profile, and this has been shown to increase the damage tolerance of the joint. As shown in Figure 14B, creep damage which develops in the HAZ of the repair cannot easily propagate because the base metal exhibits greater resistance to creep damage than the HAZ. This is a very important benefit because in the majority of cases weld repair is required after removal of damage, i.e., the location needing to be repaired is known to be susceptible to forming damage. In the current case, the weld HAZ has developed stable micro and even macro defects which should be readily detected using recommended nondestructive testing methods.

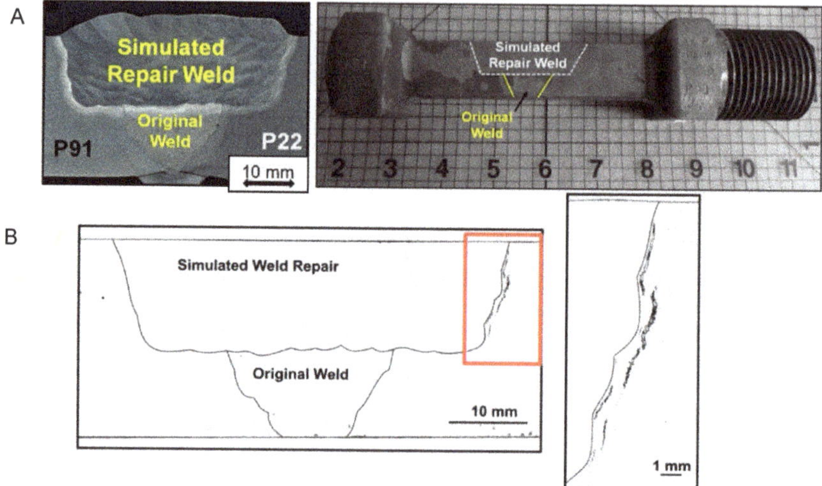

Figure 14. Example of a feature test cross weld specimens used by EPRI, (**A**) containing a simulated weld repair and (**B**) showing HAZ creep damage in a dissimilar metal repair weld between Grade 91 and Grade 22 Steels.

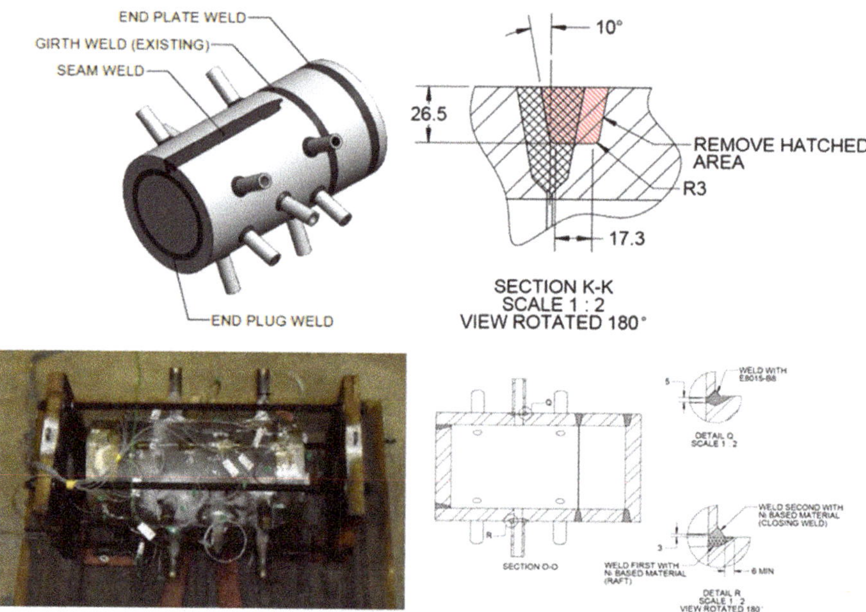

Figure 15. Example of a full-size component vessel test containing innovative weld repairs in Grade 91 Steel performed without post weld heat treatment.

An example of one of the EPRI designed vessel creep tests is given in Figure 15. This vessel was manufactured from a cylindrical section of Grade 91 steel taken from a superheater outlet header. The vessel contains several features representative of the geometric complications found in power and petro-chemical plant. The features included in the vessel were a flat end cap and an end plug, a longitudinal seam weld and stub tube to header welds. In the original pressure boundary circumferential weld, a partial weld repair was made using one the newly approved repair approaches.

The repaired region is shown in the schematic section of Figure 15. As previously the excavation and subsequent fill extend beyond the HAZ of the original weld. This vessel has been on test at 625 °C and the results will be reported in detail in due course.

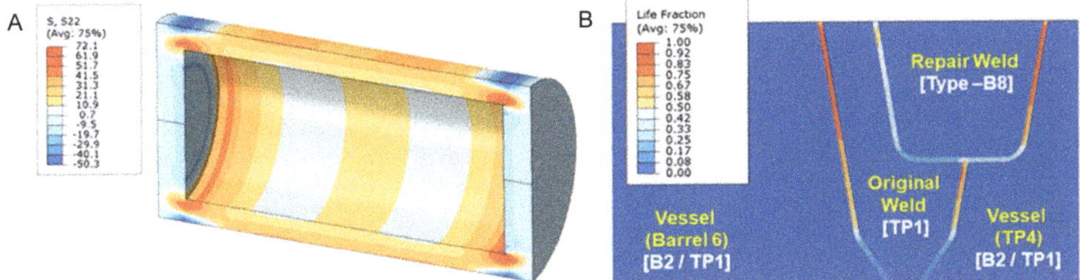

Figure 16. Evaluation of vessel test behavior using finite element analysis coupled with physically-informed constitutive models for creep damage initiation and crack growth, (**A**) showing local stress concentrations at the end caps and (**B**) showing damage estimates for the weldments.

The vessel type creep tests offer very valuable direct evidence of high temperature performance both in terms of the time to cracking, the locations of first cracking and cracks growth leading to failure. As these tests are undertaken in purpose-built facilities it is possible to run tests to fracture as all risks are minimized. It is, however, not possible to perform these specialist experiments on all combinations of pressure, temperature and vessel design. Thus, in addition to extracting maximum value from the direct experimental observations further benefits are derived if the actual results from the vessels are assessed using the CDM analysis methods described earlier. The EPRI recommended approach has been to seek to benchmark this type of analytical assessment using results from in-service Case studies and vessel data. Figure 16A shows the results of stresses developed in the vessel as a consequence of internal pressure. Details of damage analysis for the HAZ of pressure boundary welds are presented in Figure 16B.

A further important method for evaluation of performance involves collating and analysis of information regarding the behaviour of actual repairs in installed components. Development of this database permits analysis of the performance of the overall performance of repairs and more detailed assessment by component. In the case of Grade 91 steel, EPRI has documented thousands of repairs over the last five years in multiple counties which were fabricated in accordance with the NBIC and EPRI reports. Tracking these cases is vital for the industry to provide best practices should issues arise, and such that the continued success of specific approaches is supported by statistical relevant information.

Future work is necessitated to evaluate repair options for other CSEF steels allowed by Design Codes. These include Grade 92, and a new generation of emerging CSEF steels such as VM12SHC (ASME Code Case 2781), Grade 93 (ASME Code Case 2839) and THOR 115 (ASME Code Case 2890). Furthermore, common place failures in traditional and A-ASS materials in tube to attachment welds is necessitate the development of novel test techniques to screen material performance and provide insightful guidance for remediation using weld repair.

6. Discussion

6.1. Creep and Fracture of Welds

Assessment of the creep performance of welds is frequently studied using cross weld testing. Critical issues in setting up these programmes include selecting test conditions and specimen geometries that result in damage mechanisms which are relevant to long term service. It is now widely accepted that the results of creep tests at relatively high stress and temperature are not typical of long-term damage in component welds. Thus, information from tests under non-representative conditions should not be used as a guide to in-service behaviour. EPRI has established a testing protocol which involves performing cross weld creep tests on feature test type specimens. These sample examined in this study has a cross sectional area in the gauge of ~965 mm^2 (1.5 $inch^2$). This is about 30× greater than typical cross-sectional areas for standard round bar creep tests machined to a diameter of 6.35 mm (0.252 inches).

The characterization of damage in these cross weld tested samples typically involves [17,27] visual examination followed by optical metallography. The microscopy was carried out on sections which were prepared using careful polish/etch techniques to reveal damage present on the central plane of each specimen. This approach has been shown to be necessary since the tendency to form service relevant HAZ creep damage is related to hydrostatic stresses and is thus minimized at the specimen surfaces.

Experience has shown that as a rule the welds present in the pressure boundary of CSEF steel components are susceptible to creep damage and fracture before base metal locations. Observations of this type have therefore focused research to evaluate the factors affect weldment performance. It is historically the case that research of this type involves undertaking laboratory creep tests of specimens cut from the selected welds. In most cases decisions regarding the geometry of the samples are made considering the size of the welded section available and the load capacity of the available creep test machines. In most cases then the specimen geometry chosen is that of a cylindrical section with a diameter in the range 4 to 10 mm. EPRI has established a testing protocol which involves performing cross weld creep tests on feature test type specimens. The types of specimen involved in EPRI research assessing the creep behaviour of welds are shown in Figure 17. These samples typically have a cross sectional area in the gauge of ~965 mm^2 (1.5 $inch^2$). This is about 30× greater than typical cross-sectional areas for standard round bar creep tests machined to a diameter of 6.35 mm (0.252 inches). Figure 17 compares the feature type specimens with a small cylindrical sample of the type used in other testing.

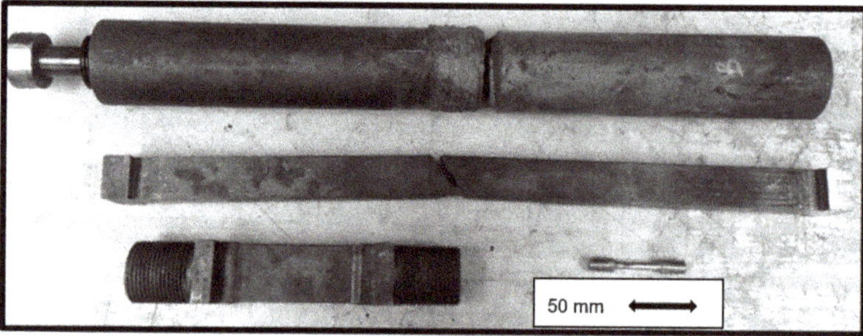

Figure 17. Examples of the feature test cross weld specimens used by EPRI to evaluate the factors affecting the creep behaviour of pressure boundary joints in tempered martensitic steels.

In addition to selection of a meaningful specimen geometry it is also important to consider the test conditions since all of these factors will influence whether damage mechanisms in the laboratory tests are relevant to long term service. Experience has shown that the results of cross weld creep tests which are performed at relatively high stress and temperature are not typical of long-term damage in component welds. Thus, information from tests under non-representative conditions should not be used as a guide to in-service behaviour. It is critical then that the details of the fracture location and relationship of the damage present with the constituent microstructure of cross weld tests is established using accurate posttest metallographic preparation and examination. This type of characterization provides important knowledge as to how to use cross weld creep test results.

Details of the post-test characterization methods recommended by EPRI regarding the sample preparation and characterization methods have been published previously [17]. It is clear that to accurately document the creep damage developed in the specimens requires the use of specialist metallographic techniques. The information summarized in the present paper provides an important appreciation of the level of effort invested to obtaining relevant and accurate results using laser metallography. It should be emphasized that the usefulness of the data is dependent upon careful preparation, observation, recording of results and then analysis. This process must start by ensuring that the sectioning is performed to reveal damage present on the central plane of each specimen. This is necessary since the tendency to form service relevant HAZ creep damage is related to hydrostatic stresses and is these hydrostatic stresses are minimized at the specimen surfaces. In addition to the need for care during sectioning and specimen preparation, it is important when documenting creep cavity size and shape that no etching of the specimen is undertaken. It is clear that etching will modify both the size and character of cavities. In general, etching will increase the size of cavities present and will tend to 'round' the cavity shape. While the use of laser microscopy provides a reasonable record of the number density of creep voids it is clear that the most accurate observations of creep void shape are obtained using ion beam techniques [17,27]. In this advanced research, the cavities are shown to be angular in shape and usually associated with the inclusions or other hard particles present.

In addition to care in preparation, examination and recording of the creep cavities present effort should also be invested in careful selection of both magnification and image resolution. Earlier research [17] has shown that for images recorded from the same specimen using the same magnification, the cavitation densities shown were markedly different as a consequence of image resolution. Thus, for the lower resolution used, the theoretical minimum cavity diameter that was counted was 2.04 μm. In contrast, using an improved image resolution, improved the minimum cavity diameter that was counted to 0.70 μm. Because the digital processing of images using light microscopes is becoming increasingly common, it is critical that that appropriate settings and procedures are selected. In the case of the equipment used in EPRI studies, there are options for utilizing "super fine" image capturing that would further increase the resolution of the selected objective to 0.34 μm. The difference in cavity size recorded between the low resolution and "super fine" resolution represents a range of almost an order of magnitude.

The parent metal condition has a direct effect on the cross-weld creep performance in Grade 91 steel. This is most often put into the context of a deformation-related mechanism (i.e., strength), but as presented in this review paper there is a clear influence of the damage resistance (i.e., ductility) on cross-weld creep performance. Thus, the results from EPRI research demand that designers and alloy developers place an equal emphasis on parent metal creep strength and ductility. Inherent base metal performance is a key aspect linked to the susceptibility to damage in the HAZ. Detailed testing supported by comprehensive metallurgical characterization has established the factors which influence cross-weld creep performance. For a set of Grade 91 materials which exhibited the same strength, an increase in the ductility from 15% to 83% ROA resulted in increased the cross-weld creep life by 5× at a test condition of 625 °C (1157 °F) and 60 MPa (8.7 ksi). Ongoing assessment is working to more definitively link risk factors in the microstructure which are responsible for damage (i.e., inclusions and/or other cavitation-susceptible features in the material). However, it is apparent that although weld performance maybe life limiting for in-service components a significant contribution effecting the life and manner of creep fracture can be directly attributed to the base steel. As shown in Figure 18, the creep performance of cross weld tests and of simulated HAZ microstructures are very similar under low stress long–time conditions.

The effect of filler metal strength on weldment performance has also been included in EPRI research. Cross weld creep tests on samples made using a weld filler metal with an over-matching strength to the base metal did not show any trend on life. For example, there was no significant influence on life for creep tests performed at an applied stress of 60 MPa (8.7 ksi). In the tests at a higher applied stress of 80 MPa (11.6 ksi), there was some evidence of reduced creep life as filler metal strength increased but the results of all tests were still above the lower bound of acceptable performance, i.e., the test duration was greater than the mean-20% bound. In contrast, cross-weld testing of samples made with under-matching filler material consistently failed at the mean-20% bound, Figure 18. It appears that the fact that welds with lower creep strength filler show creep failure in the HAZ can be explained by the fact that the B8 type filler material used was close to the strength of the PTZ region of the HAZ. It was also of potential practical benefit to note that the cross-weld tests for the under-matching filler exhibited the greatest damage tolerance of all the weld metals studied. Thus, in these tests there appeared to be evidence that macro-cracking started at an earlier life fraction than in tests with the higher strength filler metals. It is clear that a relatively slow rate of crack propagation increases the opportunity for defect detection by non-destructive testing during service. It is apparent therefore that designers should consider the geometry of the weld and the specific properties of the filler metal on component creep performance. Ensuring damage tolerance is a key benefit to component design for long-term operation under high temperature creep conditions.

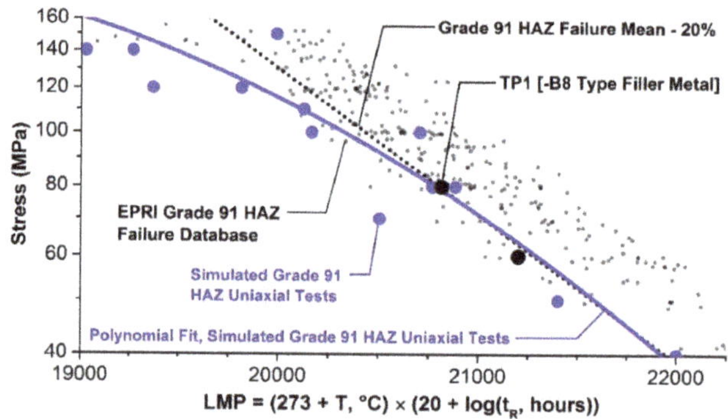

Figure 18. Relationship showing the creep performance of the HAZ defines the properties of cross weld specimens in long times.

6.2. Management of Assets Made from CSEF Steels

The life management approach at the fleet-, plant-, system- or component-level is conventionally divided into a staged approach [24]. The Level 0, I, II or III methodology incorporates important concepts such as risk-ranking or risk-based inspection (RBI) and fitness for service (FFS). The level approach is summarized in the following information:

- Level 0. Macro screening and risk-ranking of utility fleet and/or of the systems in the plant.
- Level I. Conservative and refined risk-ranking of susceptible plants and/or components in the system.
- Level II. Assessment of plant-specific information to provide information on 'where' and 'when' to look for damage in the plant, system or location in the system.
- Level III. Detailed assessment of plant-specific information and consideration of unique material properties including the potential disposition of defects in the system.

Fundamentally, this level of detail helps to refine prior recommendations and assess 'how' the component will fail, e.g., leak or break. The end user must be involved in performing a meaningful risk assessment, particularly to ensure that the consequences of fracture are properly established and factored into the risk calculation.

As a minimum EPRI recommends that defining the consequence of the event should consider the following factors in relation to component failure:

- Safety. Proximity of the location to a high traffic area or control room, etc.;
- Collateral damage. Such as the proximity of the location to a critical component or support equipment that may be damaged in a leak or failure;
- Financial impact. This can include lost generation, replacement generation or other considerations that are difficult, and in many cases impossible, for a vendor or independent assessor to define.

Risk-ranking can be performed to a progressively more detailed level of probabilistic and consequential assessment which typically requires more certain engineering factors (e.g., stress/temperature/metallurgy/dimensional measurements). Basic concepts which should be considered in the risk-ranking of 9%Cr steel components are covered in more detail in [10,12,28,29]. The general approach to risk-ranking is defined in ASME Post Construction Code 3 (PCC-3) with a more thorough approach defined in EN 16991:2018 [29]. Initially, effective asset management should target components which have a high expectancy to fail early-in-life. This decision-making is predicated on identification of the factors which contribute to reduced performance (e.g., design, fabrication, operation and metallurgy). In many cases, a detailed understanding of the historical issues identified through available service experience reports, such as in [13,14], are of immense value to the engineer to provide needed perspective and begin to target the highest risk locations

7. Concluding Comments

Components in power generating plants must operate safely under complex conditions, including a high temperature, high pressure and severe environments. New or post construction codes provide a basic set of rules to ensure that a minimum expected performance will be achieved. Fabrication of power generation components necessitates joining sections to establish a suitable pressure boundary through conventional fusion welding processes. Throughout the course of a component's life, damage at the most susceptible, high-risk locations will necessitate run/repair/replace decision-making as part of an integrated life management philosophy.

The present review emphasizes that the high temperature performance of tempered martensitic steels is complex. However, by appropriate planning and introducing good practice procedures, significant benefit to 'through life' costs can be achieved. These savings can be realized without having to compromise equipment safety and reliability.

Author Contributions: J.P. and J.S. contributed equally to Conceptualization of the work; establishing appropriate Methodologies for the research; Validation, of methods and Preparation of the manuscript. Submission of the manuscript and final Editing of the proofs was performed by J.P.

Funding: The original research described in this review paper was funded through the support to EPRI from Stakeholders in the Electricity Supply Industry.

Conflicts of Interest: The authors declare no conflict of interest.

References

1. EPRI. *The Influence of Steel Making and Processing Variables on the Microstructure and Properties of Creep Strength Enhanced Ferritic (CSEF) Steel Grade 91*; 3002004370; EPRI: Palo Alto, CA, USA, 2014; 20p.
2. Parker, J.D.; Siefert, J.A. Metallurgical and Stress State Factors Which Affect the Creep and Fracture Behavior of 9% Cr Steels. *Adv. Mater. Sci. Eng.* **2018**, *2018*, 6789563. [CrossRef]
3. EPRI. *Life Management of 9%Cr Steels: Evaluation of Metallurgical Risk Factors in Grade 91 Steel Parent Material*; 3002009678; EPRI: Palo Alto, CA, USA, 2018; 198p.
4. Parker, J.D.; Brett, S.J. Creep performance of a Grade 91 header. *Int J. Press. Vessel. Pip.* **2013**, *111*, 82–88. [CrossRef]
5. Xu, X.; West, G.D.; Siefert, J.A.; Parker, J.D.; Thomson, R.C. The Influence of Thermal Cycles on the Microstructure of Grade 92 Steel. *Metall. Mater. Trans. A* **2017**, *48*, 5396–5414. [CrossRef]
6. Xu, X.; West, G.D.; Siefert, J.A.; Parker, J.D.; Thomson, R.C. Microstructural characterization of the heat-affected zones in grade 92 steel welds: Double-pass and multipass welds. *Metall. Mater. Trans. A* **2018**, *49*, 1211–1230. [CrossRef]
7. EPRI. *Development of a Creep Continuum Damage Mechanics Model for Creep Strength Enhanced Ferritic Steels*; 3002009232; EPRI: Palo Alto, CA, USA, 2016.
8. EPRI. *Life Management of 9%Cr Steels—Continuum Damage Mechanics assessment of Novel Step Weld Geometry for Girth Welds in Thick-Section Components*; 3002011053; EPRI: Palo Alto, CA, USA, 2017.

9. Pritchard, P.G.; Perrin, I.J.; Parker, J.D.; Siefert, J.A. Application of a Physically-Based Creep Continuum Damage Mechanics Constitutive Model to the Serviceability Assessment of a Large Bore Branch Connection. In Proceedings of the 2018 ASME Symposium on Elevated Temperature Applications of Materials for Fossil, Nuclear, and Petrochemical Industries, Seattle, WA, USA, 3–5 April 2018. ETAM2018-6719.
10. EPRI. *Integrated Life Management of Grade 91 Steel Components: A Summary of Research Supporting the Electric Power Research Institute's Well-Engineered Approach*; 3002012262; EPRI: Palo Alto, CA, USA, 2018; 36p.
11. The National Board of Boiler and Pressure Vessel Inspectors. *NBBI NB23-2017, National Board Inspection Code—NBIC*; 2017 Edition (Four Volumes); The National Board of Boiler and Pressure Vessel Inspectors: Columbus, OH, USA, 2017.
12. EPRI. *Life Management of 9Cr Steels: Basic Approach to Risk Ranking Systems of Components*; 3002009231; EPRI: Palo Alto, CA, USA, 2016; 60p.
13. Parker, J.D. In-service behaviour of creep strength enhanced ferritic steels Grade 91 and Grade 92—Part 2 weld issues. *Int. J. Press. Vessel. Pip.* **2014**, *114*, 76–87. [CrossRef]
14. EPRI. *Service Experience with Creep Strength Enhanced Ferritic Steels in Power Plants in the Asia-Pacific Region*; 3002005089; EPRI: Palo Alto, CA, USA, 2015; 142p.
15. EPRI. *Life Management of 9%Cr Steels–Guidelines for the Assessment of Composition Using Shavings, Scoop, or Bulk Samples*; 3002009682; EPRI: Palo Alto, CA, USA, 2018; 124p.
16. ASME. *Code Case 2864 in ASME Boiler and Pressure Vessel Code–Code Cases: Boilers and Pressure Vessels*; ASME: New York, NY, USA, 2017.
17. Siefert, J.A.; Parker, J.D. Evaluation of the creep cavitation behavior in Grade 91 steels. *Int. J. Press. Vessel. Pip.* **2016**, *138*, 31–44. [CrossRef]
18. ASME. *Boiler and Pressure Vessel Code–Code Cases: Boilers and Pressure Vessels*; ASME: New York, NY, USA, 2017.
19. DiStefano, J.R.; Sikka, V.K.; Blass, J.J.; Brinkman, C.R.; Corum, J.M.; Horsk, J.A.; Huddleston, R.L.; King, J.F.; McClung, R.W.; Sartory, W.K. *Summary of Modified 9Cr-1Mo Steel Development Program: 1975–1985, in ORNL-6303*; Oak Ridge National Laboratory: Oak Ridge, TN, USA, 1986.
20. EPRI. *Alternative Well-Engineered Weld Repair Options for Grade 91 Steel—Continuum Damage Mechanics Assessment Supporting Alternative Weld Repairs in Tube-to-Header Connections*; 3002009688; EPRI: Palo Alto, CA, USA, 2017.
21. Kachanov, L.M. Time of the Rupture Process under Creep Conditions. *Izv. Akad. Nauk SSR Otd Tekh. Naul* **1958**, *8*, 26–31.
22. Hongo, H.; Tabuchi, M.; Watanabe, T. Type IV creep damage behavior in Gr.91 steel welds. *Metall. Mater. Trans. A* **2012**, *43*, 1163–1173. [CrossRef]
23. EPRI. *Temperbead Welding of P-Nos. 4 and 5 Materials*; Product ID: TR-111757; EPRI: Palo Alto, CA, USA, 1998; 29p.
24. EPRI. *Alternative Well-Engineered Weld Repair Options for Grade 91 Steel: An Executive Summary of Results from 2010 to 2014*; 3002006403; EPRI: Palo Alto, CA, USA, 2015; 16p.
25. EPRI. *Four Utilities Successfully Performed Alternative Weld Repair Methods in Grade 91 Steel Components*; 3002009810; EPRI: Palo Alto, CA, USA, 2017; 2p.
26. EPRI. *Alternative Well-Engineered Weld Repair Options for Grade 91 Steel—Responses to Common Questions and Concerns*; 3002012182; EPRI: Palo Alto, CA, USA, 2018.
27. Siefert, J.A.; Parker, J.D.; Thomson, R.C. Linking Performance of Parent Grade 91 Steel to the Cross-weld Creep Performance Using Feature Type Tests. In *Advances in Materials Technology for Fossil Power Plants, Proceedings of the Eighth International Conference, Albufeira, Portugal, 11–14 October 2016*; Parker, J.D., Shingledecker, J., Siefert, J.A., Eds.; EPRI: Palo Alto, CA, USA, 2016.
28. EPRI. *AEP and LG&E and KU Reduce O&M Costs through the EPRI Recommended Integrated Approach to Life Management of CSEF Steels*; 3002015066; EPRI: Palo Alto, CA, USA, 2019; 2p.
29. CEN. *CEN Standard–EN 16991:2018 on Risk-Based Inspection Framework*; CEN: Brussels, Belgium, 2018.

Review

Welding Joints in High Entropy Alloys: A Short-Review on Recent Trends

Fabio C. Garcia Filho * and Sergio N. Monteiro

Department of Materials Science, Military Institute of Engineering—IME, Praça General Tibúrcio 80, 22290-270 Rio de Janeiro/RJ, Brazil; snevesmonteiro@gmail.com
* Correspondence: fabiogarciafilho@gmail.com

Received: 13 February 2020; Accepted: 16 March 2020; Published: 20 March 2020

Abstract: High entropy alloys (HEAs) emerged in the beginning of XXI century as novel materials to "keep-an-eye-on". In fact, nowadays, 16 years after they were first mentioned, a lot of research has been done regarding the properties, microstructure, and production techniques for the HEAs. Moreover, outstanding properties and possibilities have been reported for such alloys. However, a way of jointing these materials should be considered in order to make such materials suitable for engineering applications. Welding is one of the most common ways of jointing materials for engineering applications. Nevertheless, few studies concerns on efforts of welding HEAs. Therefore, it is mandatory to increase the investigation regarding the weldability of HEAs. This work aims to present a short review about what have been done in recent years, and what are the most common welding techniques that are used for HEAs. It also explores what are the measured properties of welded HEAs as well as what are the main challenges that researchers have been facing. Finally, it gives a future perspective for this research field.

Keywords: high entropy alloys; welding techniques; welding zone microstructure; welding joint properties

1. Introduction

High entropy alloys (HEAs) emerged in the beginning of this century as promising materials for engineering applications. Their unique compositions, microstructure, and properties are the most attractive characteristics and an incredible number of papers have been published regarding this subject. According to the Scopus database, publications include more than 7,000 research papers, 300 review articles, and eight books over the 16 years, since these alloys were firstly reported by Cantor's [1] and Yeh's research groups [2–6]. One should be wondering what make these alloys so special. In fact, the multi-main element in contrast to the traditional just one or two main elements that mankind have been using in the development of metallic alloys over the centuries leads to interesting discoveries. Multi-main elements would be responsible for increasing the configurational entropy of the alloy and, therefore, favoring a single phase material instead of a microstructure with several phases [6]. This incredible microstructure led to the remarkable mechanical and functional properties reported for these alloys [7–13]. Gludovatz et al. [7] studied the fracture resistance of HEAs. The authors reported that these HEAs were able to combine high fracture toughness, above 10^2 MPa.m$^{1/2}$, with high yield strength, above 1.1 GPa. Moreover, these HEAs exhibit the highest relation between strength and ductility among all materials when compared to other class of materials as stainless steels, low alloy steel, nickel-based super alloys, metallic glasses, polymers, and ceramics, even at cryogenic temperature. This is a notable characteristic of these alloys that encourages their use in engineering applications. Nevertheless, it is critical to understand the behavior of these alloys during the processing and fabrication in order to achieve practical use.

Welding is a fabrication technology that is used in a variety of industries. Moreover, it is certainly the most used jointing techniques for metallic materials. Welding techniques could be divided into solid and liquid state processes. Liquid-state welding relies on the fusion of the metal to make the weld. Techniques that are based on gas oxyacetylene, shield metal arc (SMAW), gas-tungsten arc (GTAW), gas-metal arc (GMAW), electroslag (ESW), and others, as well as high-energy beam such as electron beam (EBW) and laser beam (LBW) welding are the most commonly used liquid-state ones [14]. Solid-state welding is defined as a joining process without any liquid/vapor phase formation and with the use of pressure. Friction (FW) and friction stir (FSW) welding are the some notable techniques in such class [15–18]. It is obvious that each welding technique presents significant differences in terms of joint preparation and sample thickness, as well as speed and energy input into the welding joint. For instance, the heat density of a GMAW technique is around 10^5 W/cm^2, while, for the EBW, this density can reach 10^8 W/cm^2 [19]. Such differences will directly impact the quality and properties of the welded joint [20,21].

Few works actually combined both subjects despite the aforementioned importance of the HEAs and the commonly used welding techniques for jointing metallic materials. When comparing the number of publications regarding HEAs properties or characterization and those that somehow discussed the welding in these alloys, less than 0.5% indeed concerned it. The behavior of a novel alloy during welding is considered to be a key technological issue. HEAs are no exception, and the behavior of the alloy when exposed to a welding thermal cycle needs to be explored and understood. The investigation of the alloy can be welded or joined without degradation, whether that is detrimental to the weldment microstructure or properties (during/after welding) and for the duration of intended service, is mandatory for its potential use as an engineering material for structural applications [22]. In this context, this short-review objective is to cover what has been reported on HEAs welded so far, including the influence of the welding techniques and the parameters in the material's properties and microstructure, as well as to critically assess the achievements reported and outlook what of is yet to come in future years.

2. Welding Techniques in HEAs

Gas tungsten arc welding (GTAW) is a well-known technique among the arc welding techniques for jointing metallic materials, especially high alloy materials [14]. In this process, a non-consumable tungsten electrode is used to produce the welded joint by melting the base metal. In addition, the weld area is protected from atmospheric contamination by an inert shielding gas, such as argon or helium. In one of the first papers regarding the welding of HEAs, Sokkalingam et al. [23] used the GTAW technique to weld an $Al_{0.5}CoCrFeNi$ alloy. The parameters of 40 A for current, 12 V for voltage, and 80 mm min^{-1} for weld velocity were chosen to joint 2.5 mm HEA plates. The main achievement of this work was a remarkable refinement of the grains from 60 μm in the base metal to a range of 8–12 μm grain size in the fusion zone. Nevertheless, a reduction of approximately 6.4 and 16.5 % were reported for strength and ductility, respectively in comparison with the base metal. Figure 1 displays the microstructure of the welded zones as well as the measured tensile properties of that work. Similarly, Wu et al [24] welded a CoCrFeMnNi alloy while using GTAW. Sheets with 1.6 mm of thickness were butt-welded using a voltage of 8.4 V, current of 75 A, and velocity of 25.4 mm min^{-1} as the welding parameters. No cracks or significant microsegregation were produced.

Recently, high-energy beam welding were reported in several publications regarding high entropy alloys weld [25–28]. The high concentrated heat from laser or electron beam create a narrow heat affected zone in the material and allow for high welding rates to be achieved in comparison to arc welding techniques. Chen et al. [25] were able to successfully use high-power solid-state laser to weld CoCrFeMnNi high entropy alloy plates. The laser power was 2 kW, defocusing amount of +2 mm, laser spot moving rate of 1 m min^{-1}, and shielding argon gas flow rate of 30–40 L $min.^{-1}$ were the parameters used in YLS10000 fiber laser (American IPG Company, Oxford, MA, USA to produce the welded joint. When considering the same composition HEA Cantor alloy processed in two different conditions (cast and rolled), Nam et al. [26] used laser beam welding to study the impact of different welding velocities in their microstructures. The butt welding specimens dimensions were $100 \times 20 \times 1.5$ mm^3 and Nd:YAG laser power of 3.5 kW, beam diameter of 300 µm, and focal legth of 304 mm, with no shielding gas was used for welding. The welding velocity was in the range of 6–10 m min^{-1}. Figure 2 exhibits the result of such investigation. The authors reported the good weldability of the CoCrFeMnNi alloy with no macro-defects, such as internal pores and cracks under all welding conditions. Furthermore, shrinkage voids were observed in the interdendritic region near the center line of the weld metal, and the volume fraction of these voids decreased as the welding velocity increased.

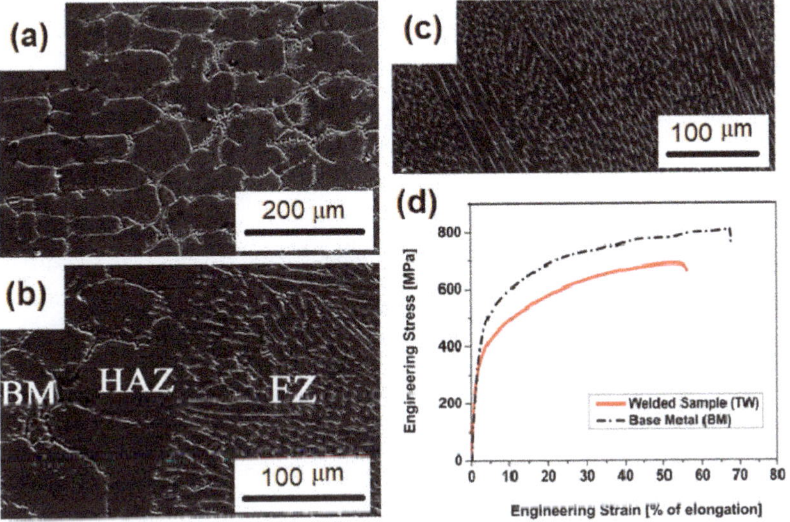

Figure 1. Gas-tungsten arc (GTAW) technique to weld an $Al_{0.5}$CoCrFeNi alloy (**a**) base metal zone (BM) microstructure, (**b**) BM+HAZ+FZ region, (**c**) FZ microstructure, and (**d**) comparison of the mechanical strength of the BM and welded sample. Adapted from [23].

Sokkalingam et al. [27] welded an $Al_{0.5}$CoCrFeNi HEA with plate thickness of 2.5 mm while using an optic fiber laser source with a beam power and transverse speed of 1.5 kW and 600 mm min^{-1}, respectively. Figure 3 shows that the laser weld resulted in a microstructure of the welding zone (WZ) with a lower degree of Al-Ni segregation in comparison with the base metal zone (BM). This microstructural difference was associated with the lack of diffusion in Al-Ni phase formation at rapid solidification.

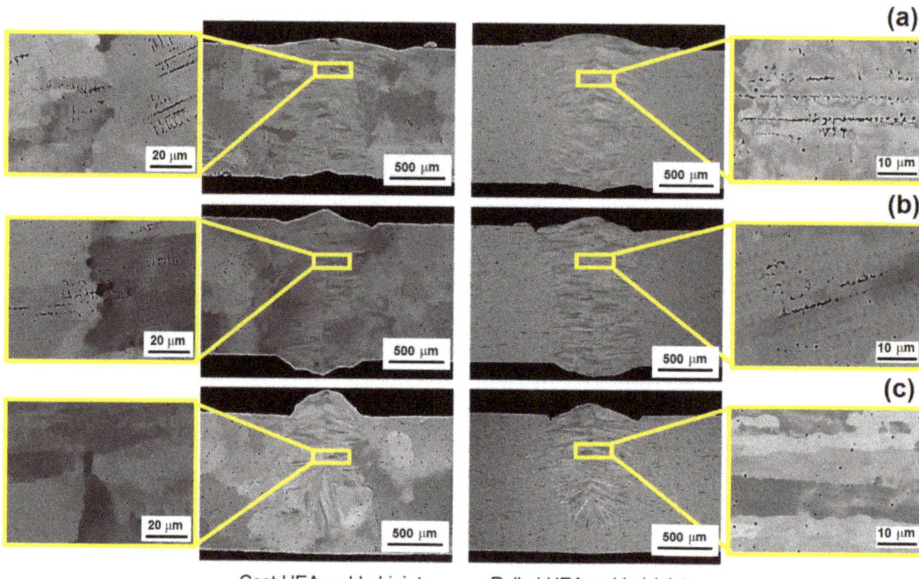

Figure 2. Comparison of the welding microstructure of the CoCrFeMnNi alloy as cast (left) and rolled (right). Samples welded by LBW in different velocities (**a**) 6 m min^{-1}, (**b**) 8 m min^{-1}, and (**c**) 10 m min^{-1}. Adapted from [26].

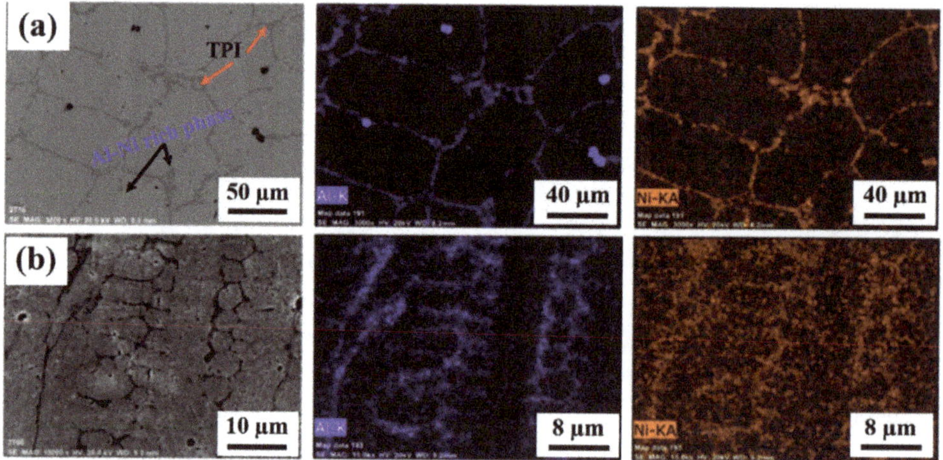

Figure 3. Comparison of the microstructure of the (**a**) BM and (**b**) WZ of $Al_{0.5}$CoCrFeNi HEA. Inset of EDX mapping of Al and Ni of both regions. Adapted from [27].

Regarding solid-state welding techniques for HEAs, Zhu et al. [29,30] discussed the possibility of using friction-stir welding (FSW) as welding technique to obtain sound joints in HEAs. Indeed, the authors were able to weld a $Co_{16}Cr_{28}Fe_{28}Ni_{28}$ and $CoCrFeNiAl_{0.3}$ high entropy alloys while using this technique. Figure 4a shows how this process runs, the parameters, and the microstructural zones generated through this technique. The welding process was conducted using a load-controlled FSW machine with the WC-Co based welding tool. The welding speeds were 30 mm min^{-1} and 50 mm min^{-1}, while the rotation rate and load force were kept constant at 400 rpm and 1500 kg, respectively. The diameter of the shoulder and pin are 12 mm and 4 mm, respectively, with a pin length of 1.8 mm. The authors reported that W-rich particles were detected within the stir zone, which was associated with the friction between the WC-Co based rotating tool and the material, although, in both cases, the weld was considered to be successful. Li et al [31] studied a variant of the Friction-Stir welding by a rotary friction welder (HSMZ-20, Harbin Welding Institute, Harbin, China) to weld an $AlCoCrFeNi_{2.1}$ alloy. Such a welding technique is a solid-state welding process with some interesting advantages, such as: high welding productivity, low heat input, excellent welding quality, and, unlike FSW, no contamination was reported. Figure 4b illustrates the Rotary Friction Welding (RFW) process, as well as the samples that were produced by Li et al in that work. As for the parameters, the rotation speed was kept constant at 1500 rpm, while the friction pressure varies from 80 to 200 MPa. The authors showed that four different zones could be observed in the microstructure of these materials: base metal (BM), heat-affected zone (HAZ), thermal-mechanically affected zone (TMAZ), and dynamic recrystallization zone (DRZ). In addition, it was disclosed that an increase of the pressure up to 200 MPa notably enhanced the quality of the weld in a way that the tensile test for that condition resulted in the fracture being propagated in the BM zone, unlike the other conditions in which the specimens fractured in the welding zone (WZ).

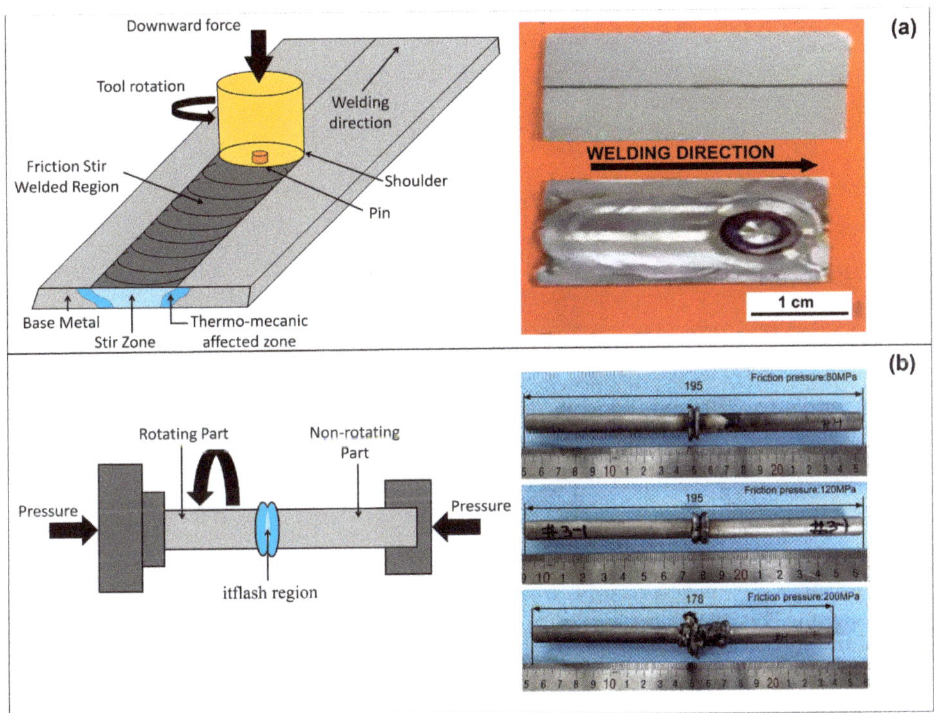

Figure 4. Schematic illustration and produced specimens of (**a**) friction stir (FSW) and (**b**) Rotary Friction Welding (RFW). Adapted from [29–31].

Another interesting welding technique that was recently reported for HEAs is the diffusion bonding [32,33]. As one should expect, in this technique the metallurgical bonding of the alloys depend on the diffusion of the elements from one metallic block to the other. Therefore, two metallic blocks are placed in a vacuum chamber, under a certain pressure, temperature (between 0.6 and 0.8 of the melting temperature), and during a determinate time. The applied parameters depend on the materials that have been welded in this diffusion-base welding method. Lei et al. [32] investigated the dissimilar joint of a single phase face center cubic $Al_{0.85}$CoCrFeNi alloy and a TiAl intermetallic while using direct diffusion bonding under vacuum. For this work, temperatures in the range from 750–1050 °C, holding time of 30–120 min. and constant pressure of 30 MPa, were used to evaluate the weldability of this dissimilar joint. Figure 5a shows the typical interfacial microstructure that was observed for the TiAl/$Al_{0.85}$CoCrFeNi. One should notice that the emergence of such graded microstructure is directly related to the diffusion velocity of each element of the HEA, as well as the TiAl. The authors suggest four stages for the diffusional bonding of the HEA/TiAl joint. In the first stage, there was physical contact of the base materials with a low degree of atomic diffusion and no reaction layer formed. In the second stage, a large number of Ni and Co atoms diffused into the TiAl. Consequently, it is observed that the formation of α_2-phase and the solid solution strengthened γ-TiAl. In the third stage, it is noticed the formation of Ti(Ni, Co)$_2$Al and Cr(Fe, Ni)ss layer. At this stage, the diffusional layers are formed and a reliable metallurgical bond can be observed. Finally in the fourth stage the growth of diffusion layers takes place. This late grow could be associated with the sluggish diffusion character of FCC-structured AlCoCrFeNi HEA. Figure 5b presents the compositional elemental distribution over the graded microstructure marked with yellow points in Figure 5a. The Figure 5b helps to understand the microstructural evolution of the joint and with the appearance of intermetallic phases. Such intermetallics as Cr- rich, Ti_3Al, and FeNi phases were held responsible for increasing the hardness in each region, as shown in Figure 5c.

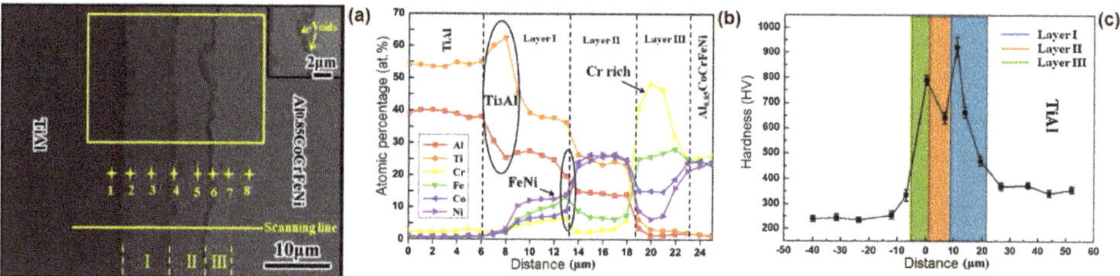

Figure 5. Typical insterfacial microstructure of a dissimilar diffusion bonding (**a**), compositional elemental distribution through the graded microstructura (**b**), and (**c**) measured Vickers hardness in the different layers that formed the welded joint. Adapted from [32].

3. Properties of Welded HEAs

The welding process of HEAs presents direct impact in several properties of such materials. In fact, many of these properties are measured to assess whether the quality of the welding was, indeed, satisfactory. Microstructure might present grain size modification, secondary phases precipitation, segregation, and lattice distortions. Therefore, corrosion, fatigue, and creep service conditions are significantly modified. In addition, mechanical properties, such as tensile strength, ductility, and hardness, are also important parameters to appraise the welding.

$Al_{0.5}$CoCrFeNi alloy was reported to present better corrosion resistance than 304 stainless steel, but with an increased strength [34]. Sokkalingam et al. [27] studied the corrosion resistance of that welded alloy to verify whether the welding process could deteriorate it. It was observed that the WZ exhibited higher corrosion current density (2.83×10^{-5} mA/cm^2) than the BM (8.63×10^{-6} mA/cm^2), which showed a higher corrosion rate. The welded joint BM+WZ resulted in the WZ acting like cathode and BM acting as anode, which could result in the secondary phases and particles in the BM been corroded first, as the weldment is exposed to a corrosive environment. Furthermore, deep pit corrosion was observed in the interface between the WZ and the BM zone. This phenomenon was associated with the dissolution of Al-rich particles at the BM zone near the interface. Figure 6a shows the corrosion behavior of such welded joint, while, in Figure 6b, one might observe the deep pit corrosion that occurs near the interface WZ+BM.

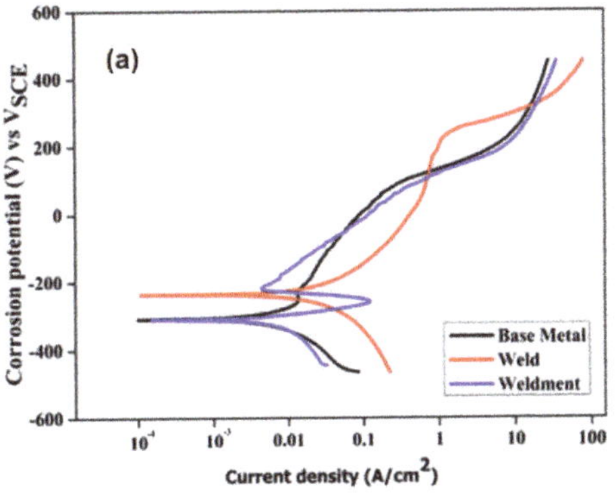

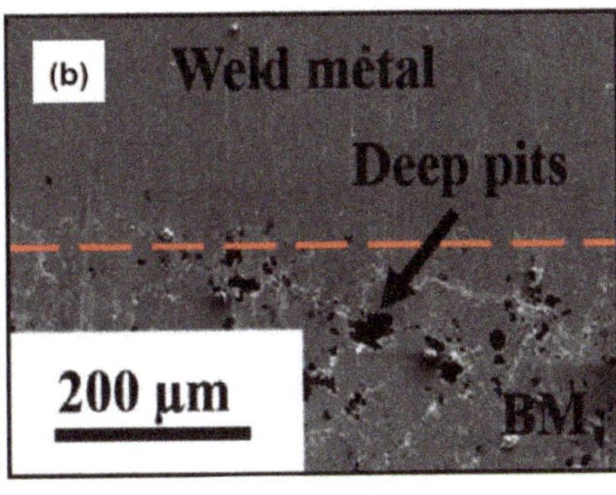

Figure 6. (**a**) Corrosion behavior and (**b**) deep pits near the interface between WZ and BM of $Al_{0.5}CoCrFeNi$ HEA. Adapted from [27].

In another study regarding the properties of HEAs welded joints, Wu et al. [24] compared the microstructure and mechanical properties of the CoCrFeMnNi alloy welded by GTAW and electron beam welding (EBW). In spite of the low process velocity for the EBW process, 38 mm min^{-1}, which was compared to the GTAW process, 25.4 mm min^{-1}, a remarkable difference in the microstructure could be observed in comparison to each case. The GTAW welding zone is at least 2.5 times wider than the EBW welding zone, 3.3 and 1.3 mm respectively which is, obviously, associated with the heat input in each technique. Moreover, both of the welds exhibit yield columnar grains that grows towards the maximum temperature gradient following the solid-liquid interface direction. The mechanical properties of tensile strength and ductility were analyzed, Figure 7a–c. For both GTAW and EBW conditions, the yield strain was increased when compared to the base metal, but only the EBW weld was able to keep similar ultimate tensile strength properties. For both cases, the ductility was decreased, 15% for GTAW, 27% for EBW against 38% for the BM. Furthermore, a compositional mapping of the welding produced revealed a depletion of Mn in the weld zone for the EBW technique, where the atomic percentages in the range from 13–18 at% of Mn were reported. This depletion was associated with the evaporation of Mn due to the high power density of the EBW welding process. Indeed, this should be expected, since Mn is the element with the lowest melting temperature and highest evaporation pressure in the Cantor alloy. On the other hand, the energy input for the GTAW technique did not impact the local composition of the Cantor alloy and the atomic percentage of all elements was kept around 20 at%.

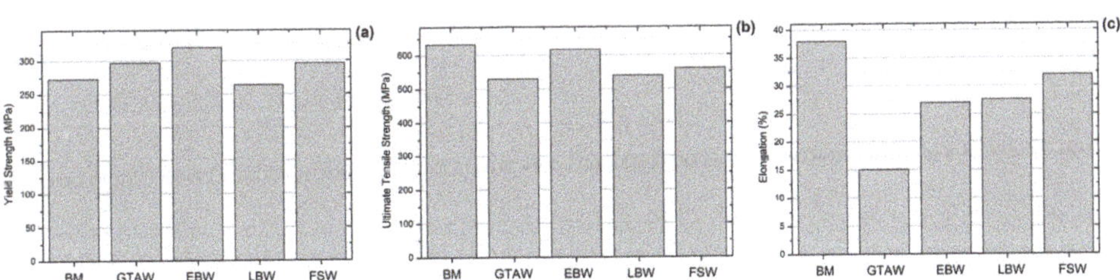

Figure 7. Comparison of mechanical properties of CoCrFeMnNi HEAs welded by different techniques (**a**) yield strength, (**b**) ultimate tensile strength, and (**c**) elongation.

Jo et al. [28] also undertook a comparative study on the properties of CrMnFeCoNi HEA welded by FSW and LBW. It was not observed macroscopic defects for both FSW and LBW HEA. The strength and ductility of the welded specimens were comparable with that of the BM. The FSW specimen had relatively higher yield strength (296 MPa) when compared with that of the BM (272 MPa). The loss in ductility in the FSW specimen (9%) was less than that in the LBW specimen (16%) when compared with the BM, was associated with the grain refinement due to dynamic recrystallization by the FSW process. Similarly to what Wu et al. [24] reported for EBW, the LBW technique also addressed a high energy input, which leads to a fluctuation in the composition of the FZ with the emergence of Mn-rich and Fe-rich phases. The tensile fracture tended to occur in the BM away from the weld center in the FSW specimen, while, in the LBW specimen, fracture occurred in the FZ. The fracture surface in both FSW and LBW specimens showed the features of dimpled rupture, which is typical of ductile fracture. Figure 7a–c compare the mechanical properties that were reported for the Cantor alloy joint by these different welding techniques.

Kashaev et al. [35] used LBW with a laser power of 2kW, 300μm of core diameter, 300 mm of focal length, focus position of 0.0 mm, and welding velocity in range of 3–6 m min^{-1} to butt joints CoCrFeMnNi alloys. It was observed that the welding process resulted in the precipitation of M_7C_3 carbides along the fcc matrix. The precipitation of this secondary phase enhanced the hardness from 150 to 205 HV, in the BM and WZ, respectively. The authors also evaluated the influence of such a welding technique in the fatigue behavior of the alloy. No significant difference was observed between the investigated conditions. Moreover an endurance limit of 200 MPa was determined for both conditions. Due to the higher strength of the weld, the failure occurs in the BM, and any possible stress concentrators in the weld as well at the WZ/HAZ or HAZ/BM boundaries do not play any significant role. Figure 8a–d present the results of that investigation. Figure 8a shows the hardness distribution of along the material. Figure 8b the behavior of the material is showed under fatigue cycles. It is important to notice that, for cycles above 10^7, both of the conditions reached to a plateau which was associated with the endurance limit of the material. Finally, Figure 8c,d show the microstructure of the materials as-sintered (BM) and the LBW material after 10^7 cycles, respectively. One should observe a high density of dislocation in both case, but, in Figure 8d, it is arrowed the presence of the M_7C_3 carbides that seem to lock the dislocations around it.

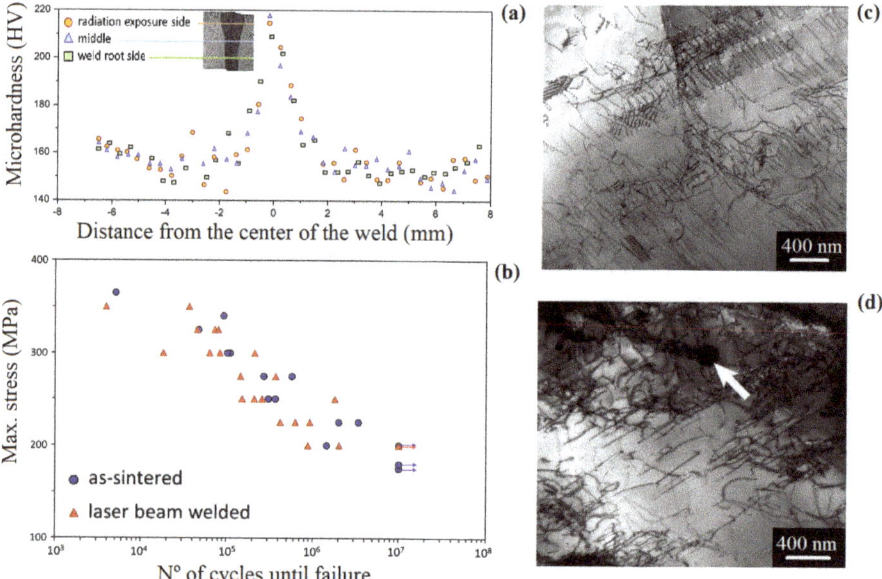

Figure 8. (**a**) Hardness distribution along different regions of CoCrFeMnNi HEA, (**b**) fatigue behavior of the this alloy as-sintered and LBW, TEM microstructure of the alloy (**c**) as-sintered, and (**d**) laser beam welding (LBW) after 10^7 cycles. Adapted from [35].

In some cases, even an enhancement of the properties was obtained in WZ in comparison with the BM. Shaysultanov et al. [36] used FSW to butt joint 2 mm thickness of a modified CoCrFeMnNi alloy. Along with the main elements, 0.9 at% of C was added to the alloy. This resulted in a microstructure of the HEA alloy that consisted of face centered cubic matrix and fine Cr- rich $M_{23}C_6$ carbides. The use of FSW for a butt-jointing of carbon-doped CoCrFeNiMn HEA specimens permitted the formation of a sound weld without any cracks or pores. Moreover, a moderate microstructure refinement was observed as the grain size in the BMl was measured to be 9.2 µm, while being 4.6 µm for the WZ. The grain refinement was claimed to be one of the advantages of the FSW and similar results were reported for the $Co_{16}Cr_{28}Fe_{28}Ni_{28}$ [29], $CoCrFeNiAl_{0.3}$ [30] and CoCrFeNiMn [28]. Furthermore, this microstructural change leads to a notable enhancement of mechanical properties, such as yield strength and ultimate tensile strength. By contrast, the ductility was impaired. However, it is also important to notice that the failure occurred in the BM and not in the WZ. Figure 9 shows the grain size in three different regions along the welding joint and also display the tensile strength of both the BM and WZ.

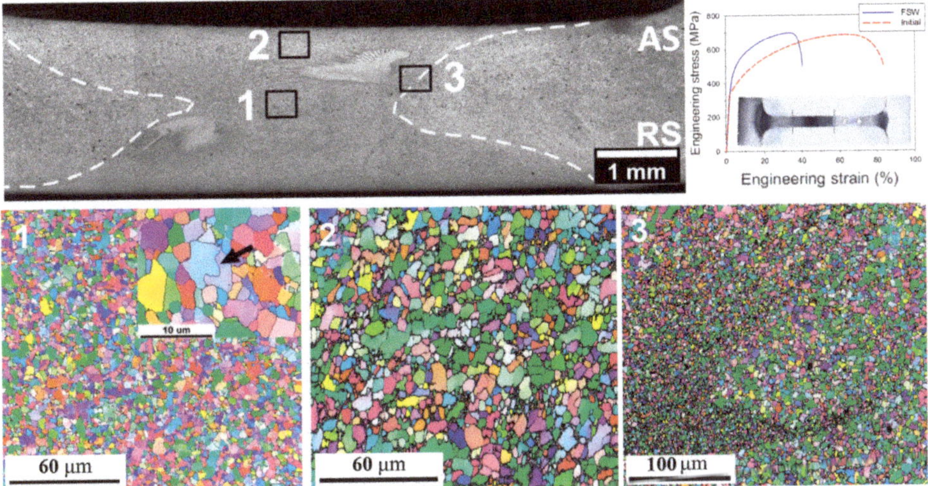

Figure 9. Grain size in three different regions of CoCrFeMnNi alloy welded by FSW and mechanical properties. Adapted from [36].

Table 1 summarizes the main parameters, the range, and references of the alloys that were welded by each technique discussed in this short-review.

Table 1. Welding parameters and techniques reported for high entropy alloys (HEAs).

	Welding Technique	Welding Parameters							References
Liquid-state Welding	GTAW	Voltage (V)	Current (A)	Velocity (mm min^{-1})	Gas Shield (L min^{-1})				[23,24,37,38]
		8.4–16	40–75	25.4 - 80	0 - 16				
	LBW	Power (kW)	Focus (mm)	Velocity (mm min^{-1})	Gas Shield (L min^{-1})				[25–28,35,39]
		1.5–3.5	0–2	600–10000	0–40				
	EBW	Voltage (kV)	Current (mA)	Velocity (mm min^{-1})					[22,24]
		125	2.2–5.0	38–570					
Solid-state Welding	FSW	Rotation (rpm)	Shoulder Diameter (mm)	Pin Length (mm)	Pin Diameter (mm)	Force (kgf)	Tilt Angle (°)	Velocity (mm min^{-1})	[28–30,36]
		400 -1000	12–12.5	1.5–1.85	4–5.76	1130–1500	2–3	30–150	
	RSW	Friction Pressure (MPa)	Friction Time (s)	Forging Pressure (MPa)	Forging Time (s)	Rotation Speed (rpm)			[31]
		80–200	3	120–400	15	1500			
	Diffusion Bonding	Temperature (°C)	Time (min)	Pressure (MPa)	Dissimilar Welding				[32,33]
		750–1050	30–120	15–30	TiAl alloy				

4. Challenges and Future Perspective

In an innovative work, Hao et al. [39] suggested the possibility of using a $(CoCrFeNi)_{100-x}Cu_x$ HEA as a filler metal in a hybrid structure between 304 stainless steel and TC4 titanium alloy, as in Figure 10a. Figure 10b shows that the weld reinforcement presented some undercut defects, pores, and slag inclusion. No obvious interface could be seen between the WZ and 304 stainless steel, which suggested that reliable metallurgical bonding was achieved for these materials. However, a thin transition layer was formed between TC4 titanium alloy and the WZ. This transition layer could be divided into two different regions, a Ti-depleted and a Ti-rich layer, as shown in Figure 10c. All of the joints failed through the Ti/Cu transition zone, exhibiting a brittle nature with typical cleavage fracture characteristics. In spite of the unsuccessful attempt of joint this hybrid structure, such an approach with careful evaluation of the microstructural evolution of the BM/WZ interface as well as compositional matching between the materials to be welded and the filler metal could represent an important strategy for hybrid structure welding.

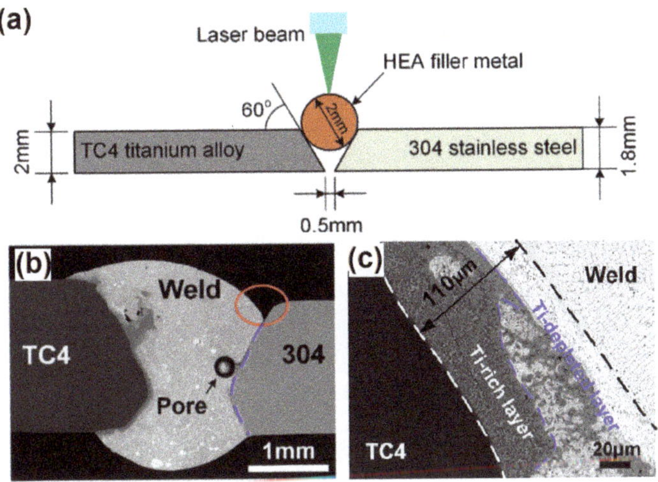

Figure 10. (a) Schematic illustration of dissimilar welding between TC4 Ti alloy and 304 stainless steel using a HEA as filler metal, (b) welding macroscopic aspect, and (c) layer formed between the WZ and the TC4 titanium. Adapted from [39].

The use of brazing welding has also been reported along with HEAs [40,41]. Lin et al. [40] investigated the dissimilar infrared brazing of CoCrFeMnNi equiatomic high entropy alloy and 316 stainless steel. In that study, two nickel-based fillers (BNi-2 and MBF601) were investigated as candidates for producing the CoCrFeMnNi / 316 SS joint. As expected, microstructural evolution through the joint tends to occurs and P-rich compounds were observed precipitated in the grain boundaries of the CoCrFeMnNi base metal as well as 316 SS substrate. The highest shear strength was obtained for CoCrFeMnNi / BNi-2 / 316 SS joint, 374 MPa when brazed at 1020 °C for 600 s, against 324 MPa that was obtained for the CoCrFeMnNi/MBF601/316 SS brazed at 1080 °C for 600 s. Using a different approach, Bridges et al. [41] studied the parameters and effects of using a NiMnFeCoCu HEA as a filler metal for laser brazing Inconel®718 nickel superalloy. The authors were able to achieve a reliable metallurgical bond, Figure 11a. Moreover, it was observed that, if the brazing temperature is way above the Liquidus temperature, the shear strength of the brazing decreased, as shown in Figure 11b.

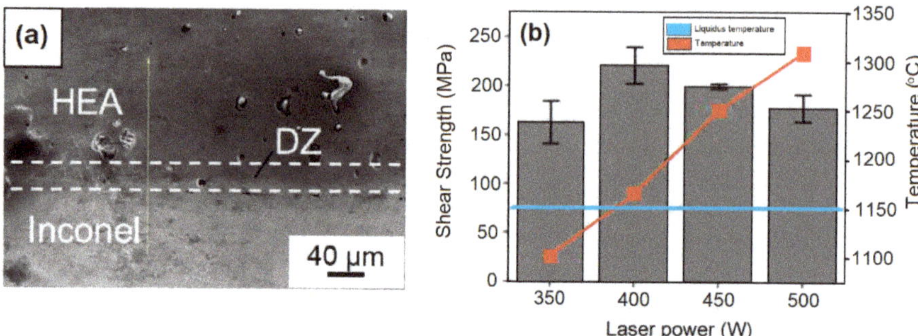

Figure 11. Brazing dissimilar welding between HEA and Inconel 718. (**a**) Metallurgical bond and (**b**) relationship between shear strength and brazing temperature. Adapted from [41].

Abed et al. [37] used the GTAW process to add a hardfacing HEA layer into a carbon steel substrate. This could be understood as another interesting strategy for the welding of HEAs into different substrate, as the HEA filler rod was added layer after layer, as in Figure 12a. In this investigation a $Fe_{49}Cr_{18}Mo_7B_{16}C_4Nb_6$ was the chosen HEA filler material. The produced microstructure consisted of α-Fe matrix with Mo_2FeB_2 and NbC precipitated particles that substantially increased the hardness and wear resistance of the material. It was observed high-quality multi-layer deposits free of cracking with an excellent metallurgical bonding to the carbon steel base metal, as in Figure 12b.

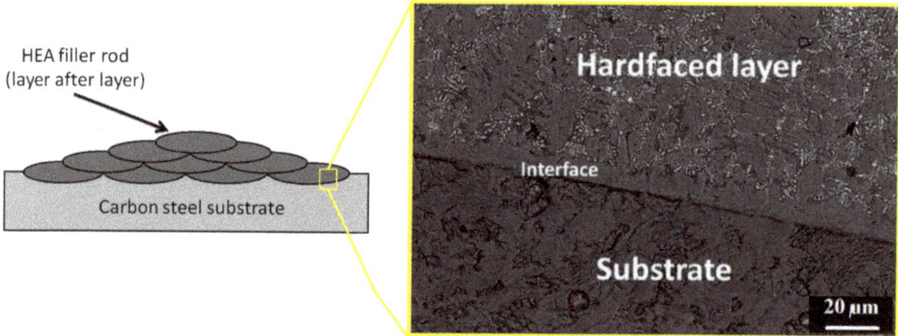

Figure 12. Schematic illustration of the use of HEA welded layer after layer in a carbon steel substrate and inset of the formed interface. Adapted from [37].

Many interesting techniques, properties, and possibilities have been discussed regarding the welding in HEAs in the present work. Yet, one should be aware that most of the reported papers are based on the application of the Cantor alloy, its variants, or its modification. The CoCrFeMnNi HEA is, by far, the most investigated among all possible alloys due to both its properties at room and cryogenic temperature, compromise between strength and toughness, as well as the synergy of their main elements [42]. Nevertheless, this is just one possibility in a limitless compositional hyper-space. Miracle and Senkov [43] classified the HEAs in seven possible families and suggested that over 500,000 HEAs are possible to be produced if only equiatomic configurations with five main elements be considered. Hence, it is clear that, so far, only a minimum number of the potential application of HEAs is reported and discussed. This is particularly true regarding the welding of such materials, where only the "easiest" cases have been investigated so far.

The welding of HEAs, in which the main elements present significant differences in the melting temperature could be considered as a big challenge. Chen et al. [44] suggested that the evaporation of elements with low melting points lead to a difficult proper control of the chemical composition of the produced alloy. The aforementioned work from Wu et al. [24] proves that hypothesis. In fact, in their work was revealed the depletion of Mn in the WZ due to the high energy input through the EBW welding and the lowest melting temperature of Mn when compared to the other main elements in the Cantor alloy. Moreover, Stepanov et al. [45] was able to produce an AlNbTiV high entropy refractory alloy with over 1 GPa of compression strength and density of approximately 5.6 g/cm^3. Despite these remarkable properties, one should be wondering how to weld such material, once the melting temperature for Nb is 247 °C and the vaporization temperature for Al is 2470 °C. Adapting solid-state welding techniques, such as FSW or RFW, could come up with possible solutions for this case. New processing techniques, such as metal additive manufacturing that are emerging rapidly could be helpful in the manufacturing of HEAs with controllable microstructure and enhanced properties, as well as high complex geometry components and high freedom of design [46]. In this layer-wise fabrication process, the metallic materials are bonded together by sintering or melting using high energy source, such as high power laser, electron beam, or plasma arc. Joseph et al. [47] faced some difficult obtain desired microstructure of an $Al_{0.6}CoCrFeNi$ that is produced by direct laser deposition (laser power of 800 W, beam focus diameter of 4mm, and velocity of 800 mm min^{-1}). The reason was associated with the higher cooling rate and much larger thermal gradient than traditional methods, such as arc melting. It is interesting to notice that the parameters are in the same magnitude of LBW, as showed in Table 1. Therefore, this process could be seen as a localized, layer-by-layer welding technique. In fact, most of challenges faced on the welding of HEAs are also observed for additive manufacturing of these alloys. Ocelik et al. [48] focused their study on the effects of laser processing parameters in the manufacturing of an AlCoCrFeNi HEA. Deviation from the original chemical composition, as the concentration of an element would be higher the lower its melting point, and porosity were some of the commonly reported defects. Thus, understanding and optimizing the parameters could be beneficial for both additive manufacturing and LBW in HEAs.

Finally, other important points to be looked into include the geometry of the welded specimens, as well as heat treatment and protection against possible contamination during the welding. In fact, most of the papers about HEAs welding discussed in this short-review limited their investigation to the butt-joint configuration, which is the simplest possible one. Complex configurations, closer to service conditions, such as corner, edge, and T- joints, and how heat input and amount of energy would impact in the quality of the welded joint should be further studied. The investigation on different heat treatment conditions, with or without pre- or post-weld heating associated with the welding technique, should be carried out in order to verify the effects on the mechanical properties and microstructure for each condition of HEAs welding.

5. Conclusions

The present short-review discussed the recent development in the welding of high entropy alloys (HEAs) and how the different possible welding techniques impact in the microstructure, mechanical properties, and service conditions, such as fatigue and corrosion behavior in this novel class of materials.

- Several techniques were reported to produce reliable metallurgical bonding in HEAs. Liquid-state welding, such as GTAW, LBW, and EBW, as well as solid-state welding, such as FSW, RSW, and diffusion bonding were discussed.
- Grain refinement, dynamic recrystallization, hardness enhancement, secondary phase precipitation, ductility decrease, and strengthening of the material were some of the detail-discussed phenomenon associated with the welding of HEAs. Moreover, the CoCrFeMnNi, which is one of most study HEAs, was also compared in terms of different welding techniques and parameter as well as mechanical properties and microstructural evolution.
- For welding techniques with high energy input, such as LBW and EBW, the loss of elements with low melting and evaporation points, such as Mn and Al, was presented as a challenge for controlling the chemical composition of the desired alloy.

- Different approaches were presented for the use of HEAs as filler materials for hybrid structure, as brazing welding dissimilar joints and layer-by-layer welding in comparison to additive manufacturing.

Finally, this specific field of welding still walks in baby steps, as the HEAs presents limitless possibilities of properties and perspectives for near future applications.

Author Contributions: Conceptualization, F.C.G.F., S.N.M.; Methodology, F.C.G.F.; Investigation, F.C.G.F., Writing—Original draft preparation, F.C.G.F.; Writing—Review and editing, S.N.M.; Supervision, S.N.M.; Project administration, S.N.M.; Funding acquisition, F.C.G.F., S.N.M. All authors have read and agreed to the published version of the manuscript.

Funding: This study was financed in part by the Coordenação de Aperfeiçoamento de Pessoal de Nível Superior-Brasil (Capes)-Finance Code 001.

Acknowledgments: The authors thank the support to this investigation by the Brazilian agencies: CNPq, CAPES and FAPERJ.

Conflicts of Interest: The authors declare no conflict of interest.

References

1. Cantor, B.; Chang, I.T.H.; Knight, P.; Vincent, A.J.B. Microstructural development in equiatomic multicomponent alloys. *Mater. Sci. Eng. A.* **2004**, *375–377*, 213–218. [CrossRef]
2. Chen, T.K.; Shun, T.T.; Yeh, J.W.; Wong, M.S. Nanostructured nitride films of multi-element high-entropy alloys by reactive DC sputtering. *Surf. Coat. Technol.* **2004**, *188–189*, 193–200. [CrossRef]
3. Hsu, C.Y.; Yeh, J.W.; Chen, S.K.; Shun, T.T. Wear resistance and high temperature compression strength of FCC $CuCoNiCrAl_{0.5}Fe$ alloy with boron addition. *Metall. Mater. Trans. A* **2004**, *35A*, 1465–1469. [CrossRef]
4. Huang, P.K.; Yeh, J.W.; Shun, T.T.; Chen, S.K. Multi-principal-element alloys with improved oxidation and wear resistance for thermal spray coating. *Adv. Eng. Mater.* **2004**, *6*, 74–78. [CrossRef]
5. Yeh, J.W.; Chen, S.K.; Gan, J.W.; Lin, S.J.; Chin, T.S.; Shun, T.T.; Tsau, C.H.; Chang, S.Y. Formation of simple crystal structures in Cu-Co-Ni-Cr-Al-Fe-Ti-V alloys with multiprincipal metallic elements. *Metall. Mater. Trans. A* **2004**, *35A*, 2533–2536. [CrossRef]
6. Yeh, J.W.; Chen, S.K.; Lin, S.J.; Gan, J.Y.; Chin, T.S.; Shun, T.T.; Tsau, C.H.; Chang, S.Y. Nanostructured high-entropy alloys with multiple principal elements: Novel alloy design concepts and outcomes. *Adv. Eng. Mater.* **2004**, *6*, 299–303. [CrossRef]
7. Gludovatz, B.; Hohenwarter, A.; Catoor, D.; Chang, E.H.; George, E.P.; Ritchie, R.O. A fracture-resistant high-entropy alloy for cryogenic applications. *Science* **2014**, *345*, 1153–1158. [CrossRef]
8. Zhang, Y.; Zuo, T.T.; Tang, Z.; Gao, M.C.; Dahmen, K.A.; Liaw, P.K.; Lu, Z.P. Microstructures and properties of high entropy alloys. *Prog. Mater. Sci.* **2014**, *61*, 1–93. [CrossRef]
9. Ye, Y.F.; Wang, Q.; Lu, J.; Liu, C.T.; Yang, Y. High-entropy alloy: Challenges and prospects. *Mater. Today* **2016**, *19*, 349–362. [CrossRef]
10. Miracle, D.B.; Miller, J.D.; Senkov, O.N.; Woodward, C.; Uchic, M.D.; Tiley, J. Exploration and development of high entropy alloys for structural applications. *Entropy* **2014**, *16*, 494–525. [CrossRef]
11. Lu, Z.P.; Wang, H.; Chen, M.W.; Baker, I.; Yeh, J.W.; Liu, C.T.; Nieh, T.G. An assessment on the future development of high-entropy alloys: Summary from a recent workshop. *Intermetallics* **2015**, *66*, 67–76. [CrossRef]
12. Tsai, M.H.; Yeh, J.W. High-entropy alloys: A critical review. *Mater. Res. Lett.* **2014**, *2*, 107–123. [CrossRef]
13. George, E.P.; Curtin, W.A.; Tasan, C.C. High entropy alloys: A focused review of mechanical properties and deformation mechanisms. *Acta Materialia* **2020**, *188*, 435–474. [CrossRef]
14. Kou, S. *Welding Metallurgy*, 2nd ed.; John Wiley & Sons, Inc.: Hoboken, NJ, USA, 2003; 0-471-43491-4.
15. Panaskar, N.; Terkar, R. A review on recent advances in friction stir lap welding of aluminum and copper. *Mater.Today Proc.* **2017**, *4*, 8387–8393. [CrossRef]
16. Singh, K.; Singh, G.; Singh, H. Review on friction stir welding of magnesium alloys. *J. Magnes. Alloy.* **2018**, *6*, 399–416. [CrossRef]
17. Shinde, G.; Gajghate, S.; Dabeer, P.S.; Seemikeri, C.Y. Low cost friction stir welding: A review. *Mater. Today Proc.* **2017**, *4*, 8901–8910. [CrossRef]

18. Meisnar, M.; Baker, S.; Bennett, J.M.; Bernard, A.; Mostafa, A.; Resch, S.; Fernandes, N.; Norman, A. Microstructural characterization of rotary friction welded AA6082 and Ti-6Al-4V dissimilar joints. *Mater. Des.* **2017**, *132*, 188–197. [CrossRef]
19. Vendan, S.; Gao, L.; Garg, A.; Kavitha, P.; Dhivyasri, G.; SG, R. *Interdisciplinary Treatment to Arc Welding Power Sources*; Springer: Singapore, 2018.
20. Monteiro, S.N.; Nascimento, L.F.C.; Lima, E.P., Jr.; Luz, F.S.; Lima, E.S.; Braga, F.O. Strengthening of stainless steel weldment by high temperature precipitation. *J. Mater. Res. Tech.* **2017**, *6*, 385–389. [CrossRef]
21. Jorge, L.J.; Candido, V.S.; Silva, A.C.R.; Garcia Filho, F.C.; Pereira, A.C.; Luz, F.S.; Monteiro, S.N. Mechanical properties and microstructure of SMAW welded and thermically treated HSLA-80 steel. *J. Mater. Res. Tech.* **2018**, *7*, 598–605. [CrossRef]
22. Wu, Z.; David, S.A.; Feng, Z.; Bei, H. Weldability of a high entropy CrMnFeCoNi alloy. *Scripta Materialia* **2016**, *124*, 81–85. [CrossRef]
23. Sokkalingam, R.; Mishra, S.; Cheethirala, S.R.; Muthupandi, V.; Sivaprasad, K. Enhanced relative slip distance in Gas-Tungsten-Arc-Welded $Al_{0.5}$CoCrFeNi High-Entropy Alloy. *Metall. Mater. Trans. A* **2017**, *48A*, 3630–3634. [CrossRef]
24. Wu, Z.; David, S.A.; Leonard, D.N.; Feng, Z.; Bei, H. Microstructures and mechanical properties of a welded CoCrFeMnNi high-entropy alloy. *Sci. Tech. Weld. Join.* **2018**, *23*, 585–595. [CrossRef]
25. Chen, Z.; Wang, B.; Duan, B.; Zhang, X. Mechanical properties and microstructure of laser welded FeCoNiCrMn high-entropy alloy. *Mater. Lett.* **2019**, *262*, 127060. [CrossRef]
26. Nam, H.; Park, C.; Moon, J.; Na, Y.; Kim, H.; Kang, N. Laser weldability of cast and rolled high-entropy alloys for cryogenic applications. *Mater. Sci. Eng. A* **2019**, *742*, 224–230. [CrossRef]
27. Sokkalingam, R.; Sivaprasad, K.; Duraiselvam, M.; Muthupandi, V.; Prashanth, K.G. Novel welding of $Al_{0.5}$CoCrFeNi high-entropy alloy: Corrosion behavior. *J. Alloy. Compd.* **2020**, *817*, 153163. [CrossRef]
28. Jo, M.G.; Kim, H.J.; Kang, M.; Madakashira, P.P.; Park, E.S.; Suh, J.Y.; Kim, D.I.; Hong, S.T.; Han, H.N. Microstructure and mechanical properties of friction stir welded and laser welded high entropy alloy CrMnFeCoNi. *Met. Mater. Int.* **2018**, *24*, 73–83. [CrossRef]
29. Zhu, Z.G.; Sun, Y.F.; Ng, F.L.; Goh, M.H.; Liaw, P.K.; Fujii, H.; Nguyen, Q.B.; Xu, Y.; Shek, C.H.; Nai, S.M.L.; et al. Friction-stir welding of a ductile high entropy alloy: Microstructural evolution and weld strength. *Mater. Sci. Eng. A* **2018**, *711*, 524–532. [CrossRef]
30. Zhu, Z.G.; Sun, Y.F.; Goh, M.H.; Ng, F.L.; Nguyen, Q.B.; Fujii, H.; Nai, S.M.L.; Wei, J.; Shek, C.H. Friction stir welding of a CoCrFeNi$Al_{0.3}$ high entropy alloy. *Mater. Lett.* **2017**, *205*, 142–144. [CrossRef]
31. Li, P.; Sun, H.; Wang, S.; Hao, X.; Dong, H. Rotary friction welding of AlCoCrFe$Ni_{2.1}$ eutectic high entropy alloy. *J. Alloy. Compd.* **2020**, *814*, 152322. [CrossRef]
32. Lei, Y.; Hu, S.P.; Yang, T.L.; Song, X.G.; Luo, Y.; Wang, G.D. Vacuum diffusion bonding of high-entropy Al0.85CoCrFeNi alloy to TiAl intermetallic. *J. Mater. Process. Tech.* **2020**, *278*, 116455. [CrossRef]
33. Li, P.; Wang, S.; Xia, Y.; Hao, X.; Dong, H. Diffusion bonding of AlCoCrFeNi2.1 eutectic high entropy alloy to TiAl alloy. *J. Mater. Sci. Tech.* **2020**, *45*, 59–69. [CrossRef]
34. Lin, C.M.; Tsai, H.L. Evolution of microstructure, hardness and corrosion properties of high entropy $Al_{0.5}$CoCrFeNi alloy. *Intermetallics* **2011**, *19*, 288–294. [CrossRef]
35. Kashaev, N.; Ventzke, V.; Petrov, N.; Hortmann, M.; Zherebtsov, S.; Shaysultanov, D.; Sanin, V.; Stepanov, N. Fatigue behavior of a laser beam welded CoCrFeNiMn-type high entropy alloy. *Mater. Sci. Eng. A* **2019**, *766*, 138358. [CrossRef]
36. Shaysultanov, D.; Stepanov, N.; Malopheyev, S.; Vysotskiy, I.; Sanin, V.; Mironov, S.; Kaibyshev, R.; Salishchev, G.; Zherebtsov, S. Friction stir welding of a carbon-doped CoCrFeNiMn high-entropy alloy. *Mater. Charact.* **2018**, *145*, 353–361. [CrossRef]
37. Abed, H.; Ghaini, F.M.; Shahverdi, H.R. Characterization of FE49Cr18Mo7B16C4Nb6 high-entropy hardfacing layers produced by gas tungsten arc welding (GTAW) process. *Surf. Coat. Tech.* **2018**, *352*, 360–369. [CrossRef]
38. Oliveira, J.P.; Curado, T.M.; Zeng, Z.; Lopes, J.G.; Rossinyol, E.; Park, J.M.; Schell, N.; Braz Fernandes, F.M.; Kim, H.S. Gas tungsten arc welding of as-rolled CrMnFeCoNi high entropy alloy. *Mater. Des.* **2020**, *189*, 108505. [CrossRef]
39. Hao, X.; Dong, H.; Xia, Y.; Li, P. Microstructure and mechanical properties of laser welded TC4 titanium alloy/304 stainless steel joint with $(CoCrFeNi)_{100-x}Cu_x$ high-entropy alloy interlayer. *J. Alloy. Compd.* **2019**, *803*, 649–657. [CrossRef]

40. Lin, C.; Shiue, R.K.; Wu, S.K.; Lin, Y.S. Dissimilar infrared brazing of CoCrFe(Mn)Ni equiatomic high entropy alloys and 316 stainless steel. *Crystals* **2019**, *9*, 518. [CrossRef]
41. Bridges, D.; Zhang, S.; Lang, S.; Gao, M.; Yu, Z.; Feng, Z.; Hu, A. Laser brazing of a nickel-based superalloy using a Ni-Mn-Fe-Co-Cu high entropy alloy filler metal. *Mater. Lett.* **2018**, *215*, 11–14. [CrossRef]
42. Li, Z.; Zhao, S.; Ritchie, R.O.; Meyers, M.A. Mechanical properties of high-entropy alloys with emphasis on face-centered cubic alloys. *Prog. Mater. Sci.* **2019**, *102*, 296–345. [CrossRef]
43. Miracle, D.B.; Senkov, O.N. A critical review of high entropy alloys and related concepts. *Acta Materialia* **2017**, *122*, 448–511. [CrossRef]
44. Chen, Y.Y.; Duval, T.; Hung, U.D.; Yeh, J.W.; Shih, H.C. Microstructure and electrochemical properties of high entropy alloys—A comparison with type-304 stainless steel. *Corros. Sci.* **2005**, *47*, 2257–2279. [CrossRef]
45. Stepanov, N.D.; Shaysultanov, D.G.; Salishchev, G.A.; Tikhonovsky, M.A. Structure and mechanical properties of a light-weight AlNbTiV high entropy alloy. *Mater. Lett.* **2015**, *142*, 153–155. [CrossRef]
46. Li, X. Additive manufacturing of advanced multi-component alloys: Bulk metallic glasses and high entropy alloys. *Adv. Eng. Mater.* **2018**, *20*, 1700874. [CrossRef]
47. Joseph, J.; Jarvis, T.; Wu, X.; Stanford, N.; Hodgson, P.; Fabijanic, D.M. Comparative study of the microstructures and mechanical properties of direct laser fabricated and arc-melted AlxCoCrFeNi high entropy alloy. *Mater. Sci. Eng. A* **2015**, *633*, 184–193. [CrossRef]
48. Ocelik, V.; Janssen, N.; Smith, S.N.; De Hosson, J.T.M. Additive manufacturing of high-entropy alloys by laser processing. *JOM* **2016**, *68*, 1810–1818. [CrossRef]

MDPI
St. Alban-Anlage 66
4052 Basel
Switzerland
Tel. +41 61 683 77 34
Fax +41 61 302 89 18
www.mdpi.com

Materials Editorial Office
E-mail: materials@mdpi.com
www.mdpi.com/journal/materials